Corrosion Control and Surface Finishing

Hideyuki Kanematsu • Dana M. Barry
Editors

Corrosion Control and Surface Finishing

Environmentally Friendly Approaches

Springer

Editors
Dr. Hideyuki Kanematsu, Ph.D.
Deputy President
Professor, Department of Materials Science & Engineering
National Institute of Technology, Suzuka College
Suzuka, Mie
Japan

Fellow
Institute of Materials Finishing
U.K.

Dr. Dana M. Barry, Ph.D.
Research Professor, Departments of Chemical & Biomolecular Engineering and Electrical & Computer Engineering
Clarkson University
Potsdam, NY
USA

Senior Technical Writer & Editor
Center for Advanced Materials Processing (CAMP)
Clarkson University
Potsdam, NY
USA

ISBN 978-4-431-56737-0 ISBN 978-4-431-55957-3 (eBook)
DOI 10.1007/978-4-431-55957-3

Softcover reprint of the hardcover 1st edition 2016

Printed on acid-free paper

This Springer imprint is published by Springer Nature
The registered company is Springer Japan KK

Acknowledgment

We the editors (who also serve as authors) would like to thank the National Institute of Technology, Suzuka College and its President Yasutsugu Nitta, Professor Toshihiro Tanaka and Professor Eiji Arai along with Osaka University, Clarkson University, CAMP (Clarkson's Center for Advanced Materials Processing), Clarkson's Department of Chemical & Biomolecular Engineering and Department of Electrical & Computer Engineering, the American Chemical Society, Professor Dr. Roger Haw, and Ansted University, for their greatly appreciated support. We also thank ISIJ (Iron and Steel Institute of Japan) and its Division of Process Evaluation and Material Characterization for their encouragement and financial support. Much of the knowledge and information in this book is derived from their forum activities in the Iron and Steel Society of Japan. Their funding and planning made it possible for us to discuss the book together in person. Also we are grateful to the Iketani Science and Technology Foundation. We thank the great surface finishing societies in the United States (National Association for Surface Finishing, NASF), the United Kingdom (Institute of Materials Finishing, IMF) and Japan (Surface Finishing Society of Japan, SFSJ) for their encouragement and cooperation.

Some information contained in this book was obtained from work carried out for national funding projects in Japan. Also we are grateful to all of our chapter authors, and to our colleagues and collaborators, especially those in Japan and the US.

Our sincerest appreciation is expressed to Springer Japan and to the Editor for Life Sciences, Dr. Mei Hann Lee.

In addition, we would like to thank our families for their continued interest and support: Dr. Kanematsu's parents (Shoji and Michiko Kanematsu), his wife (Reiko), and children (Hitomi and Hiroyuki); Dr. Barry's parents (Daniel and Celia Malloy), her husband (James), and children (James, Brian, Daniel, and Eric).

Hideyuki Kanematsu
Dana M. Barry

Contents

About the Editors

Dr. Hideyuki Kanematsu, Ph.D. FIMF (a Fellow of IMF) is a full professor and the Deputy President of the National Institute of Technology, Suzuka College, Japan. There he is a researcher in Environment Materials Engineering. He is interested particularly in the interfacial phenomena between the metallic surface and organisms from an environmental science viewpoint. He holds a B. Eng. (1981), a M. Eng. (1983) and a Ph.D. in Materials Science and Engineering (1989), all from Nagoya University. He is active as a member, board member, and editorial member of ASM International, the National Association of Surface Finishing, USA (NASF), the institute of Materials Finishing, UK (IMF), as well as the Minerals, Metals & Materials Society, USA (TMS), the American Chemical Society (ACS), the Japan Institute of Metals (JIM) and the Iron & Steel Institute of Japan (ISIJ). He is a NASF Scientific Achievement Award Winner and now literally the leading researcher for Surface Engineering in the world.

Dr. Dana M. Barry, Ph.D. is a Research Professor in Clarkson University's Departments of Chemical & Biomolecular Engineering and Electrical and Computer Engineering and Senior Technical Writer & Editor in the University's Center for Advanced Materials Processing. She is a Professor and Scientific Board President for Ansted University and has served as a Visiting Professor in the Department of Materials Science & Engineering at the National Institute of Technology, Suzuka College, Japan, numerous times (2002, 2005, 2008 where she was a Keynote Speaker, 2010, 2012, 2013, and 2014). Also Dana was a Visiting Professor in England (2003), Malaysia (2001, 2004, and 2007), and China (2011), where she served as a Keynote Speaker. She holds a B.A. degree (1971) and a M.S. degree (1972) in Science Education from SUNY Potsdam, a M.S. degree in Chemistry (1974) from Clarkson University, a Ph.D. in Education (1985) from Columbia Pacific, an honorary Ph.D. in Science Education (2001) from Ansted University, and Ph.D. in Engineering (2011) from Osaka University, Japan. She has over 200 professional publications and many honors including 20 Consecutive Apex Awards for Publication Excellence from Communications Concepts in Springfield, Virginia, and numerous listings in Marquis Who's Who publications. In addition, she is a Chemistry Ambassador for the American Chemical Society and serves as an officer for its Northern New York Section.

Introduction

Hideyuki Kanematsu and Dana M. Barry

Abstract In this chapter, we provide background information as to why we decided to launch on this book project. Also we present a general sketch about the organization of the book. Surface finishing technology is actually composed of many processes such as electroplating, electroless plating, hot dip galvanizing, painting, physical vapor deposition (PVD), chemical vapor deposition (CVD), carburizing, nitriding, and so on. Each process has already been described in other books. They have sought for various functionalities so far. However, a new important axis has been added to the field. It is called Environmental Friendliness. Therefore, we have to establish the new notion and idea very soon so that surface engineers and researchers, in the near future and beyond, will be able to use them, especially through our effective guidebook. In this chapter we discuss background information about environmental issues and current problems. We also introduce the organization of the book's contents, so that readers can easily find desired information.

1 Corrosion of Metallic Materials and Surface Finishing

Corrosion of metals seems to be inevitable. Long ago, our Earth did not have so much oxygen. In the beginning iron existed as itself in a primitive sea. Several hundred million years after the birth of our planet Earth, cyanobacteria appeared in the primitive sea. The bacteria produced oxygen by photonic synthesis. Thus the amount of oxygen increased. Thermodynamics tells us that iron oxide is more

H. Kanematsu (✉)
National Institute of Technology, Suzuka College, Suzuka, Japan
e-mail: kanemats@mse.suzuka-ct.ac.jp

D.M. Barry
Clarkson's Departments of Chemical & Biomolecular Engineering and Electrical & Computer Engineering, Clarkson University, Potsdam, NY, USA

Center for Advanced Materials Processing (CAMP), Clarkson University, Potsdam, NY, USA
e-mail: dmbarry@clarkson.edu

H. Kanematsu, D.M. Barry (eds.), *Corrosion Control and Surface Finishing*,
DOI 10.1007/978-4-431-55957-3_1

stable than iron itself (Eqs. 1 and 2). Therefore, elemental iron and the divalent iron ion could not have a stable existence in the sea any more.

$$2Fe + \frac{3}{2}O_2 \rightarrow Fe_2O_3 \tag{1}$$

$$\Delta G = -740\,\text{kJ} \tag{2}$$

Iron oxide and other iron compounds were deposited on the sea bed. After many years, crustal changes and land deformations occurred repeatedly. As a result iron ore appeared on land.

Corrosion of iron and steel is an oxidation reaction. When we come to think about the Earth's history, corrosion is an inevitable reaction. Iron and steel are the main materials in our civilization. We are still in the Iron Age, even though many new materials have been found and utilized. Therefore, corrosion problems occur on a daily basis. However, the problem is not only for iron and steel, but also for many other metallic materials. Usually, corrosion is not favorable for artificial metallic materials, when we use them for some purposes. For the materials (mainly metallic ones) are degraded as a result. Therefore, we have to protect materials from corrosion. As an effective countermeasure, one can mention surface finishing. The technology is classified into many detailed techniques and processes. Actually, it covers a broad range as shown in Fig. 1. A few main purposes are mentioned for practical applications in the surface finishing industry.

One of them is color tone. When we look at the materials of components, equipment, facilities, etc., our impressions of them are very important. The color

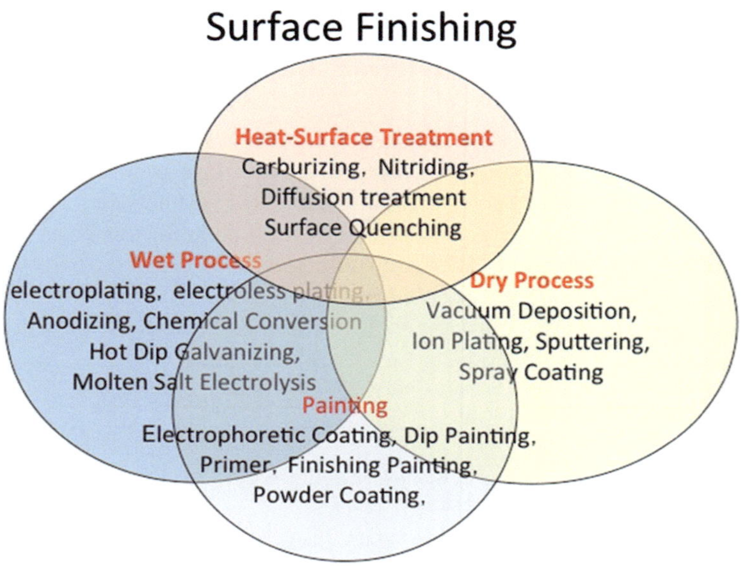

Fig. 1 Various surface finishing processes

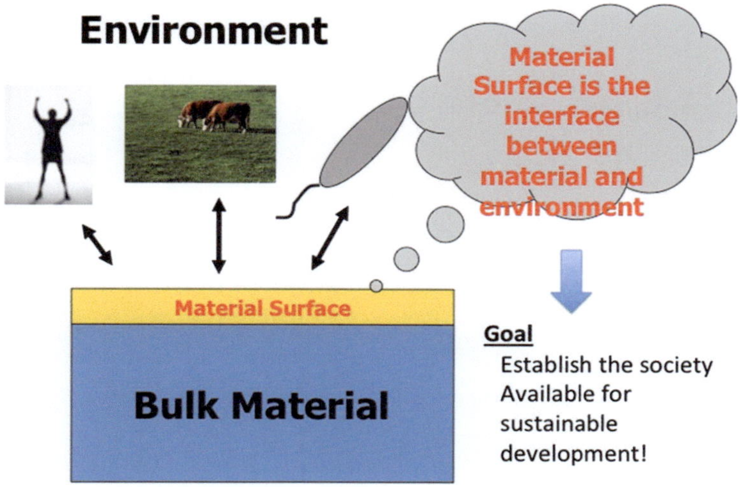

Fig. 2 Materials and environments

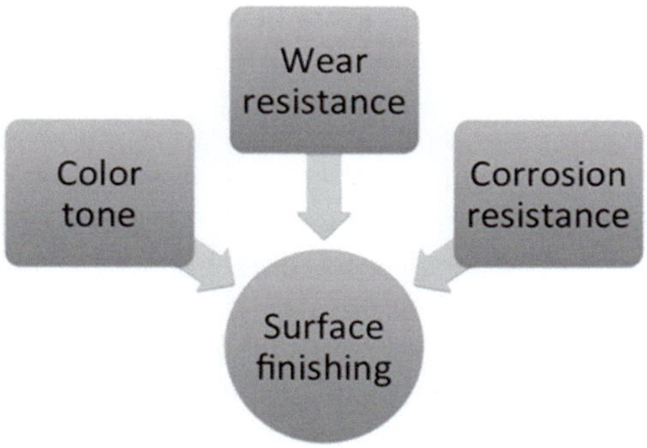

Fig. 3 Main performances of surface finishing

tone of a material's surface usually plays a very important role. Actually it can be said that a material's surface is the interface between human beings and the surrounding environment. Also for color tones, a material's surface is the most important. Therefore, surface finishing is important because it gives us impressions and feelings (Fig. 2).

Wear resistance is also a main purpose for surface finishing.

Hardness and other mechanical properties are also main players to determine the characteristics of a specific item. These properties ensure that the surface finished works and parts can be applied to the automobile and aviation industries, and to others requiring hard surfaces and tribological purposes.

Another important purpose for surface finishing is corrosion resistance. Actually, there are many types of surface finishing to control corrosion. For example in

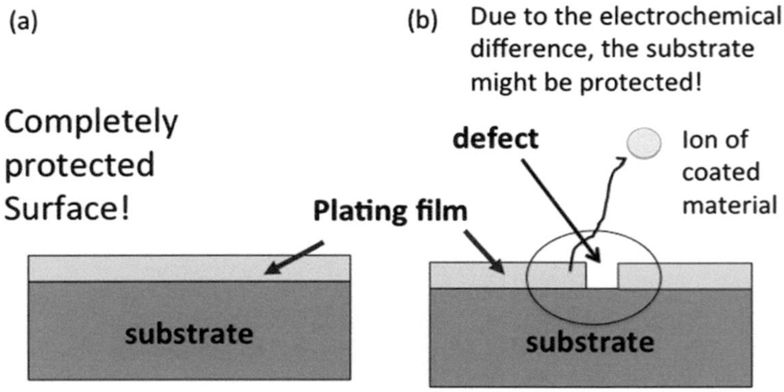

Fig. 4 (**a**) Defect-free plating and (**b**) defective plating from the viewpoint of corrosion resistance

plating technology, many kinds of plating have been used for corrosion protection. Figure 3 shows the importance of three purposes for practical plating and surface finishing. Generally speaking, zinc plating has a relatively high corrosion resistance. It is usually made by electroplating or hot dip galvanizing. Chromium plating and nickel plating can also display relatively good corrosion resistance. Usually, plating films cover the substrate as shown in Fig. 4a. If the film does not have any flaws and defects, then the substrate metal never has any contact with the environment. However, complete plating (100 %) does not exist in real life. Substrates contain lots of micro-defects or macro-defects inevitably. In such a case, the substrate metal dissolves and deteriorates (Fig. 4b).

For zinc plating, the substrate iron does not corrode, since zinc is less noble than iron. When the zinc film contains defects as shown in Fig. 4b, the environmental solution can have contact with the zinc film and iron substrate simultaneously. Then the local cell forms, where zinc becomes the anode and iron is a cathode, as shown in Fig. 5a. As a result, iron as the cathode never corrodes, until the zinc is completely gone. This is the principle for cathodic protection. On the other hand, when a more noble metal (such as tin) is coated onto the substrate, then the iron substrate corrodes. See Fig. 5b. However, the important thing from the viewpoint of environmental friendliness is that zinc corrodes and dissolves in the surrounding environment as ions. From the viewpoint of iron protection against corrosion, this phenomenon would be completely favorable. However, we need to determine if the dissolved zinc ions would be harmful to human beings, animals, and the environment. If the dissolved zinc ions would be harmful, the phenomenon should be investigated further and other coating processes might be proposed. This is a hypothetical example that will be discussed in some of the following chapters. However, we can see that the environmental factor has been added at this point. This new discipline is needed to solve problems. Therefore, we decided to write this special book (*Corrosion Control and Surface Finishing*: *Environmentally Friendly Approaches*) with our colleagues.

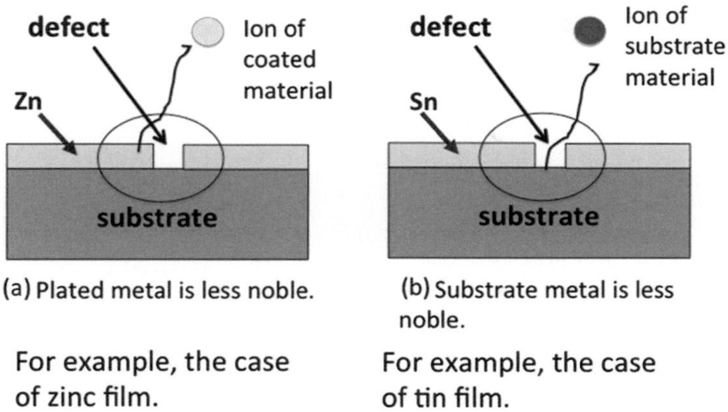

Fig. 5 A schematic illustration of sacrificial corrosion protection (**a**) and the opposite case (**b**). In (**a**), the *small arrow* on the left refers to the zinc layer. In (**b**), the *small arrow* on the left refers to the layer of tin

2 Environmentally Friendly Approach for Corrosion and Surface Finishing

Since this approach is novel, unique, and interdisciplinary, we need a new discipline to make the unique approach possible. Figure 6 shows the types of disciplines needed for the new approach. On one side, we definitely need biology. To clarify the interactions between environmental factors and materials, we have to use a biological approach, since organisms are always one of the main players at any time. For the best example, we have to investigate and discuss what kinds of metallic ions would be harmful for various kinds of organisms in this approach. How and why? We believe biology is the best discipline to solve the problems.

Another good example may be some liberal arts topics as shown in Fig. 6. As written above, the surface is the interface between human beings in the environment and materials. Our impressions come from a material's surface in most cases. To analyze human impressions, some psychological analyses would be needed. Actually, we utilized the method in the past to analyze human impressions for the color tone of plating films.

Since environmental friendliness affects business and managing industries tremendously, economics is also needed to evaluate the new technology and processes. As a concrete example, Life Cycle Assessment (LCA) can be mentioned. You can read about it in chapter "Life Cycle Assessment (LCA) of Surface Treatment Products". LCA is one of the interdisciplinary methods among economics, sociology, psychology, etc. It can be applied and combined to surface engineering for the environmental approach. In this way, the new approach contains various new factors and evaluation methods. This book describes many of them from the viewpoint of our new approach.

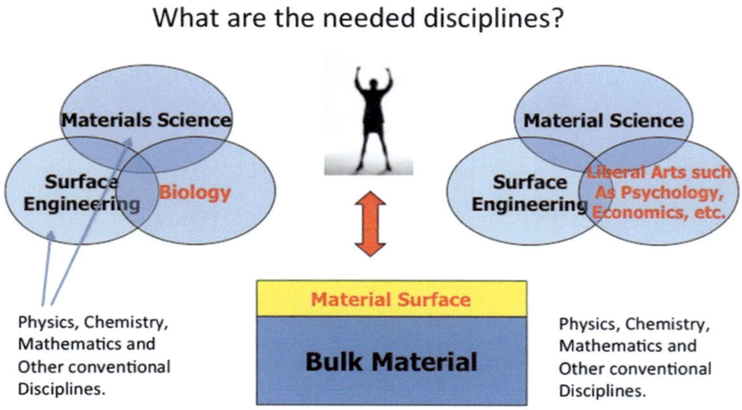

Fig. 6 Disciplines needed to understand environmental surface engineering

3 Organization of This Book

As described in the section above, the reader can understand why and how this book came about. Therefore, the appearance of this book is not the same as those of conventional handbooks for surface finishing and corrosion.

Our book contains five parts. The first part describes the fundamental aspects and knowledge for surface finishing and corrosion protection. In chapter "The Fundamentals of Corrosion Science and Engineering: Equilibrium Theory and Its Meaning", we show the corrosion phenomenon from the viewpoint of electrochemistry. Chapter "Measurement and Evaluation for Corrosion" describes the measurement methods for corrosion that mainly relate to electrochemistry. Polarization curves are included. In chapter "Industrial Surface Treatments", various surface finishing processes are introduced. Surface finishing is classified into a wet process and a dry process. These processes are explained in a summarized way. However, the details focus more on the wet process in this chapter. In chapter "Surface Analysis", various surface analyses are described along with the instruments and microscopes used. Chapter "Corrosion and Surface Finishing" mainly focuses on surface finishing for corrosion protection.

Part II is dedicated to the environmental regulations in the world. Chapter "Background of Environmental Regulations" presents the background and historical development of environmental protection and its meaning for surface finishing. The importance of the idea, "sustainable development" is stressed. In chapter "World Health Organization's Standards from the Viewpoint of Health Risks", the World Health Organization and its Standards are explained. In chapter "European Union (EU) Directives and Regulations", the European Union's Directives and regulations are described. You will see that the European Union's (EU) laws are becoming the global standards and leading the world in terms of environmental protection. In chapter "Regulations by the Environmental Protection

Agency in the US", the Environmental Protection Agency's (EPA) regulations in the US are described. The United States is a very advanced country and a major player in the world. Since the US stance in regards to industry and the environment differs from that of the EU, both are compared to provide an understanding of the political aspects for environmental regulations. Chapter "Japanese Environmental Regulations" shows how Japan, the other big player, has tackled with environmental problems to establish laws.

Part III provides several conventional, environmental evaluation processes for surface finishing. In chapter "The Necessity and Meaning", we describe the necessity and meaning of environmental evaluation for surface finishing. Also we discuss evaluation processes that have already been used. In chapter "Frequently Used Evaluations for Effluents", the Inductively Coupled Plasma Atomic Emission Spectrometer (ICP-AES) and the Inductively Coupled Plasma Mass Spectrometer (ICP-MS) are mainly described and discussed as conventional, but powerful analytical tools. In chapter "Frequently Used Evaluations for Aerial and Solid Pollution", the conventional analyses for aerial and solid pollution are presented. These types of pollution have caused very serious problems. Therefore, various countermeasures have been devised for them. And even nowadays, new problems such as air pollutants called particulate matter are emerging. These air pollutants include solid particles and liquid droplets that come in various sizes. The small particles that are 2.5 μm or less are called PM2.5. In this chapter, we also focus on gas chromatography/ion chromatography from the fundamental viewpoint. In chapter "Dissolution Assay", the dissolution assay process is described. This type of analysis is used to measure the dissolution amounts and characteristics for many kinds of metal components of materials.

On the other hand, Part IV describes the novel methods to analyze environmental characteristics of materials from biological viewpoints. Therefore, these methods are not so general yet, and some of them are still at the research level. Chapter "Gene Analysis for the Evaluation of the Effect of Environmental Factors" describes the application of gene analysis to evaluate the effect of environmental factors. Chapter "The Application of Mammalian Cells for Assessment of Metallic Products (In Vitro Process)" explains how mammalian cells could be applied to evaluate metallic toxicity. Chapter "Metal Ion Sensor of Pore-Forming Toxin for Environmental Evaluation" describes a metal ion sensor of pore-forming toxin and its applicability for environmental evaluations. In chapter "The Application of Phytoplankton", the author describes the use of phytoplankton to evaluate metallic toxicity. Chapter "The Application of Fish" discusses the application of fish to evaluate metallic toxicity from the viewpoint of embryology. In chapter "Life Cycle Assessment (LCA) of Surface Treatment Products", the author gives us a general sketch of Life Cycle Assessment (LCA) and describes its application for surface finishing products and engineering.

Part V mentions and describes concrete surface finishing processes that are environmentally friendly, their basic characteristics, and their future scopes. In chapter "What Is Environmentally Friendly Surface Finishing?", environmentally friendly surface finishing is described. Two main situations are presented in this chapter. One of them is where completely new processes would replace the previous harmful ones. The other possibility is the process where more environmentally friendly materials would substitute the harmful previous ones, even though the process would not change at all. Chapter "Substitution with More Friendly Elements" and chapter "Change from Metals to Nonmetals" describe examples that belong to the latter situation. Chapter "Substitution with More Friendly Elements" also talks about the hexavalent chromium process. It has been replaced with trivalent chromium and also with a non-chromium process. On the other hand, chapter "Change from Metals to Nonmetals" describes different examples of substitution. In this chapter, the complete abolishment of metal usages is explained. Resin coatings are examples. Chapter "Process Changes" describes process changes. Physical vapor deposition (PVD) is provided as an example. Chapter "The Application of Corrosion Protection" discusses the relation between corrosion protection and environmental friendliness. The environmental toxicity of metals that are harmful to humans and many organisms in nature is generally caused by dissolution. Therefore, surface finishing that prevents dissolution is often effective at increasing environmental friendliness for many metallic products. In this chapter, some practical cases are introduced. Chapter "Future Scope" describes the future scope for this field. The schematic organization of the book's parts and chapters is shown in Fig. 7 and the relationship of the book's parts and their concepts is displayed in Fig. 8.

Since the topics contained in this book are interdisciplinary and cover broad academic and practical fields, we needed many colleagues to contribute to this

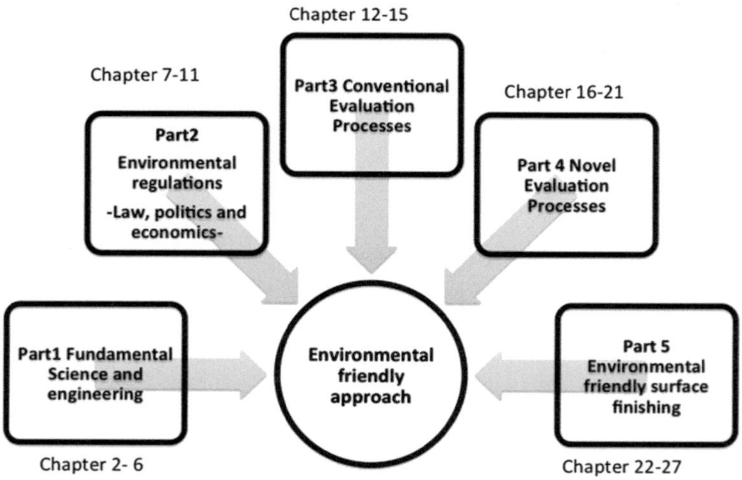

Fig. 7 The schematic organization of parts and chapters in this book

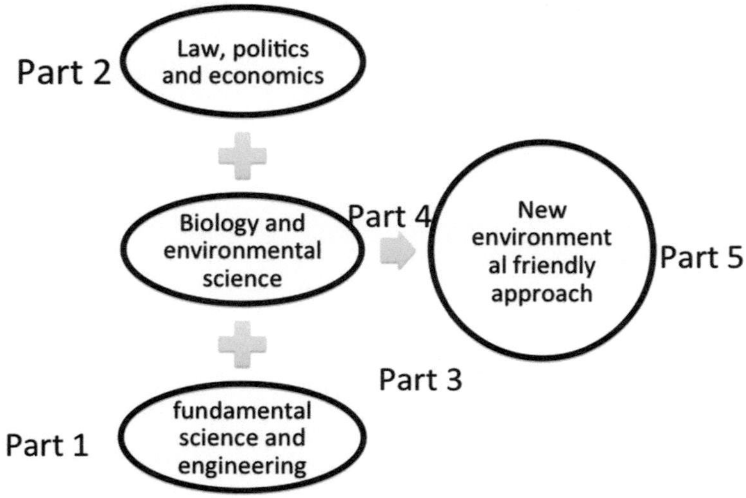

Fig. 8 The schematic positional relationship among parts in this book and their concepts. NOTE: Part 1 and Part 3 refer to the same concept

book. And some topics might still be at research levels, while others are very conventional and fundamental. However, the readers will have a concrete image and general sketch of the new concepts and academic adventures based on the fundamental conventional disciplines. Well, let's get started with the book's adventurous authors, for the new disciplines!

Part I
Fundamentals of Corrosion and Surface Finishing for Corrosion Control

The Fundamentals of Corrosion Science and Engineering: Equilibrium Theory and Its Meaning

Hideyuki Kanematsu and Dana M. Barry

Abstract Corrosion is basically the oxidation of metals, where electrons are transferred between oxidant and reductant. Therefore, corrosion is generally composed of redox reactions and should be analyzed from the viewpoint of electrochemistry. In this chapter, we describe the basic concept of electrochemistry and how various corrosion aspects can be explained by this discipline. We focus particularly on the equilibrium side because it could suggest possibilities that might be useful for corrosion prediction. The close relationship between redox reactions and corrosion are explained and stressed qualitatively and quantitatively.

1 Introduction

As mentioned in the introductory chapter, metals were originally combined with oxygen, sulfur, and other non-metallic components to form minerals. From the viewpoint of thermodynamics, the state of compounds has lower energy and is more stable than that of metals themselves. It means that we generally need to add energy in order to produce metals. Since we could consider corrosion to be a reverse reaction of metal production, it is a very natural chemical reaction that occurs spontaneously in nature (Fig. 1). At this point, we already know that corrosion is basically an oxidation reaction. Therefore, corrosion (as a redox reaction) can be analyzed from the viewpoint of electrochemistry. Here we can see the fundamental aspect of electrochemistry as it relates to corrosion.

H. Kanematsu, Ph.D. (✉)
National Institute of Technology, Suzuka College, Suzuka, Japan
e-mail: kanemats@mse.suzuka-ct.ac.jp

D.M. Barry, Ph.D.
Clarkson's Departments of Chemical & Biomolecular Engineering and Electrical & Computer Engineering, Clarkson University, Potsdam, NY, USA

Center for Advanced Materials Processing (CAMP), Clarkson University, Potsdam, NY, USA
e-mail: dmbarry@clarkson.edu

H. Kanematsu, D.M. Barry (eds.), *Corrosion Control and Surface Finishing*,
DOI 10.1007/978-4-431-55957-3_2

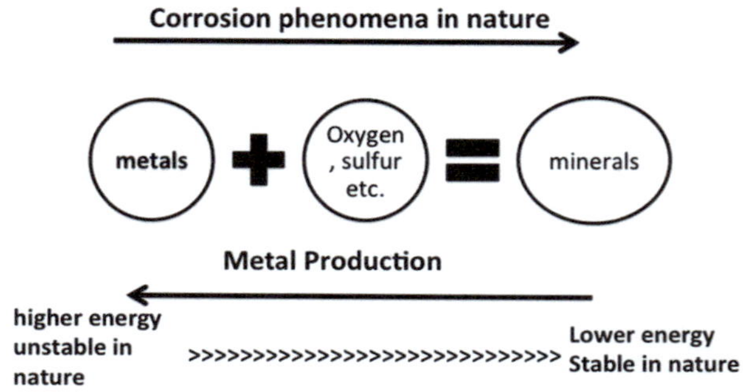

Fig. 1 Metal production and corrosion phenomena

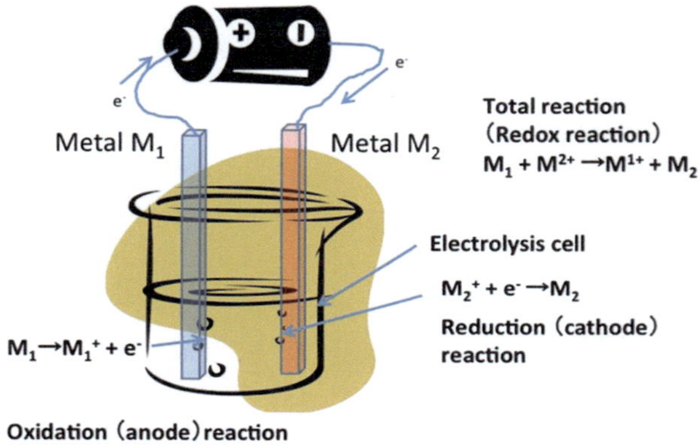

Fig. 2 Schematic illustration for electrolysis

2 Faraday's Law

Electrolysis is a fundamental phenomenon not only for corrosion, but also for electric cells, plating, electrolysis, synthesis, etc. Figure 2 shows the schematic illustration for electrolysis. Two electrodes are immersed in the electrolytic bath. One is an anode and the other is the cathode. At the anode, anodic reactions occur where electrons are released, as shown in Eq. (1).

$$M \rightarrow M^{n+} + ne^- \tag{1}$$

M: electrode metal, n: electric charge number, e: electron

On the other hand, electrons are bound with ions at the cathode. If the aqueous solution is acid, the main cathode reaction might be the reduction of hydrogen ion to produce hydrogen gas, shown in Eq. (2).

$$2H^+ + 2e^- \rightarrow H_2 \tag{2}$$

Electrolysis generally occurs this way. In 1833, Michael Faraday discovered that electrolysis obeys a certain law. Actually, it can be divided into two laws.

One of them is the relation between the amount of electrolyzed matter and electric charges consumed during electrolysis. The amount of matter produced by electrolysis is proportional to the electric charge used for it. The relation can be formulated as shown in Eq. (3).

$$m = K \times I \times t = K \times Q \tag{3}$$

m: the amount of matter produced or decomposed during electrolysis
K: electrochemical equivalent, I: current [A], t: time [s], Q: charge of electricity, electric charge [Q]

$$n = \frac{m}{M} = \frac{It}{zF} \tag{4}$$

n: the amount of substance [mol], m: mass [g], M: Molecular weight [g/mol], I: current [A], t: time [s], z: valence of ion,
F: Faraday constant $9.6485 \times 10^4 [C/mol]$

The law expressed mathematically in Eq. (4) is called Faraday's second law. This law makes certain that the electrochemical equivalent is compatible with the chemical equivalent. See Eq. (4).

It means that the amount of electrical charge needed to precipitate 1 g equivalent of a substance is always constant, regardless of the kind of matter. The constant value is usually denoted as F and called the Faraday constant. Faraday found the relations in the nineteenth century, when people did not know about the existence of electrons at all. However, the laws suggested it very strongly. From the viewpoint, it could open an important door to the quantum mechanics world in the twentieth century.

The equation could be changed in the following way.

$$v = \frac{dm}{dt} = \frac{n}{t} = \frac{I}{zF} \tag{5}$$

v means the production rate of a substance through electrolysis. Therefore, Eq. (5) suggests that the current could control the production rate of a substance during electrolysis.

3 The Electric Cell and Its Electric Motive Force

Figure 2 shows the case where the external electric force would drive the electrochemical reaction and produce the electric current as a result. On the other hand, we could think of a similar system without the external electric force. In this case, the electrochemical reactions would occur within the internal system and produce the flow of electrons as a result. We could take the electrical energy from the system and use it. The system is often called an electric cell. The system is shown in Fig. 3 schematically.

Historically, the Daniel Cell has been very well-known for the fundamental electric cell system. The schematic structure is shown in Fig. 4.

Two electrodes are immersed in aqueous sulfuric acid solution. One of the electrodes is zinc and the other one is copper. A barrier membrane is placed between the two electrodes. It could block the passage of cations, while it could allow anions to move through it freely. The copper is precipitated onto the zinc electrode. The color changes to dark brown, while the color of the solution (blue) gradually gets pale. This suggests that the copper ions in the aqueous solution precipitate onto the zinc electrode, as shown in Eq. (6).

$$Cu^{2+}(aq) + 2e^{-} \rightarrow Cu(s) \tag{6}$$

Here, (aq) means an aqueous solution and (s) means the element exists as a solid. At the other electrode (zinc electrode), the following reaction (7) occurs.

$$Zn(s) \rightarrow Zn^{2+}(aq) + 2e^{-} \tag{7}$$

The overall reaction could be written as the sum of the Eqs. (6) and (7) in the following way, Eq. (8).

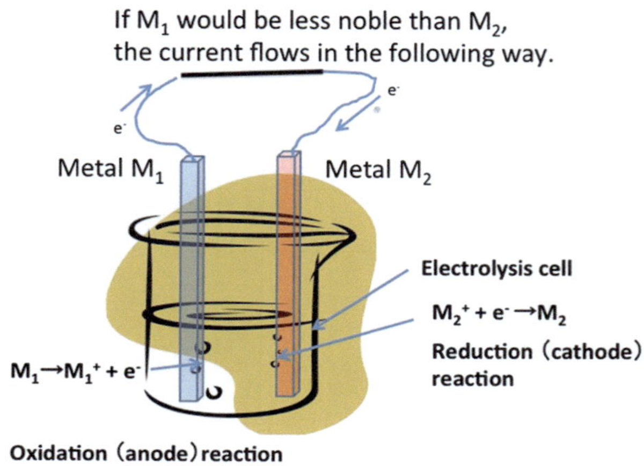

Fig. 3 Conversion of the energy from chemical reactions to electrical energy

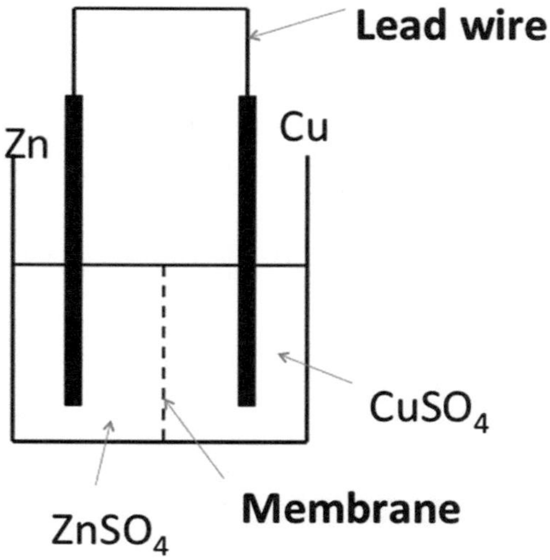

Fig. 4 Daniel Cell and its structure

$$Cu^{2+}(aq) + Zn(s) \rightarrow Cu(s) + Zn^{2+}(aq) \tag{8}$$

Electric Cells are usually expressed in the following way, Eq. (9), from the viewpoint of cell structure.

$$-Zn\left|Zn^{2+}\right|\left|Cu^{2+}\right|Cu+ \tag{9}$$

The symbol, "|", corresponds to the interface between two different phases, such as solid/liquid. The other symbol, "||", corresponds to the membrane.

At the interface, the potential difference generally appears. If the right potential difference would be E_r and the left one E_l, the electromotive force, E_e, can be defined as follows Eq. (10).

$$E_e = E_r - E_l \tag{10}$$

The left electrode, $Zn|Zn^{2+}$, is less noble and the right one, $Cu^{2+}|Cu$, is noble. The two electrodes (couple) constitute the Daniel cell.

As described above, the potential difference generally appears at the interface between the two different phases. Each interface where an anodic or a cathodic reaction occurs has its own potential defined by the Nernst Equation. The corresponding reaction is always a part of the whole Redox reaction. It is called a "half-cell reaction". Some examples are shown in Table 1. Using such a table, one can get the potential for a half-cell reaction. Then one can estimate the electromotive force for the couple of electrodes, using Eq. (10).

Table 1 Some representative half-cell reactions and their potentials [1]

Redox reaction	E (V)	Redox reaction	E(V)
$Li^{+} + e^{-} \leftrightarrow Li$	−3.05	$AgCl + e- \rightleftarrows Ag + Cl^{-}$	0.22
$K^{+} + e^{-} \leftrightarrow K$	−2.92	$Hg_2Cl_2 + e- \rightleftarrows Ag^{+}Cl^{-}$	0.27
$Ba^{2+} + 2e^{-} \rightleftarrows Ba$	−2.90	$Cu^{2+} + 2e^{-} \rightleftarrows Cu$	0.34
$Ca^{2+} + 2e^{-} \rightleftarrows Ca$	−2.76	$Ca^{+} + e^{-} \rightleftarrows Cu$	0.52
$Na^{+} + e^{-} \rightleftarrows Na$	−2.71	$I_2 + 2e^{-} \rightleftarrows 2I^{-}$	0.54
$Mg^{2+} + 2e^{-} \rightleftarrows Mg$	−2.38	$Fe^{3+} + e^{-} \rightleftarrows Fe^{2+}$	0.77
$Al^{3+} + 3e^{-} \rightleftarrows Al$	−1.67	$Ag^{+} + e^{-} \rightleftarrows Ag$	0.80
$Mn^{2+} + 2e^{-} \rightleftarrows Mn$	−1.03	$Br_2 + 2e^{-} \rightleftarrows 2Br^{-}$	1.09
$2H_2O + 2e^{-} \rightleftarrows H_2 + 2OH^{-}$	−0.83	$O_2 + 4H^{+} + 4e- \leftrightarrows 2H_2O$	1.23
$Zn^{2+} + 2e^{-} \rightleftarrows Zn$	−0.76	$MnO_2 + 4H^{+} + 4e^{-} \leftrightarrows Mn + 2H_2O$	1.28
$Cr^{3+} + 3e^{-} \rightleftarrows Cr$	−0.74	$Cr_2O7_2{}^{-} + 14H^{+} + 6e- \leftrightarrows 2Cr^{3+} + 7H_2O$	1.33
$Fe^{2+} + 2e^{-} \rightleftarrows Fe$	−0.44	$Cl_2 + 2e^{-} \leftrightarrows 2Cl^{-}$	1.36
$PbSO_4 + 2e^{-} \rightleftarrows Pb + SO_4^{2-}$	−0.36	$PbO_2 + SO_4{}^{2-} + 4H^{+} + 2e- \leftrightarrows PBSO_4 + 2H_2O$	
$Fe^{3+} + 3e^{-} \rightleftarrows Fe$	−0.04	$H_2O_2 + 2H^{+} + 2e- \leftrightarrows 2H_2O$	1.78
$2H^{+} + 2e^{-} \rightleftarrows H_2$	0	$F_2 + 2e- \leftrightarrows 2F^{-}$	2.87

4 Redox Potential and Chemical Reaction [2, 3]

In the previous section, we used electrode potential without any strict definition. However, it should be defined in advance, in order to better understand it. Originally, the electrode potential was closely related to Gibbs free energy change of a reaction at the interface (electrode). It is also related to chemical affinity. In this electrochemical case, we might say that the electrode reaction should be related to electrochemical affinity. So what is electrochemical affinity? To answer this question and others would lead to a complete understanding of what electrode potential is and what it really means. We would like to introduce the concept of inner potential for the explanation.

Let's think over the case shown in Fig. 5. A metallic electrode is placed in a vacuum. In the same vacuum phase, a point charge is placed. Here we postulate that the original distance of the point charge is separated enough from the electrode surface. When the point charge is moved from the infinite distance gradually ("gradually" means keeping the equilibrium state!), the work, W_t, can be expressed as follows.

$$W_t = W_\chi + W_i \tag{11}$$

Here W_χ is the work to move the point charge from the infinite distance to a very close point to the electrode surface. The special point (Point B) would be the limitation one where the image force available in the very close area to the electrode surface would not affect the moving of a point charge. On the other hand, the extra work would be required for the point charge to move into the inside phase (Point C) from point B. The work would be used for the point charge to overcome the very complicated potential barrier composed of dipole molecules, absorbed ones, etc. It corresponds to W_i. We can define the potential corresponding to the work. Therefore, the following Eq. (12) would be available.

$$\varphi^i = \psi^i + \chi^i \tag{12}$$

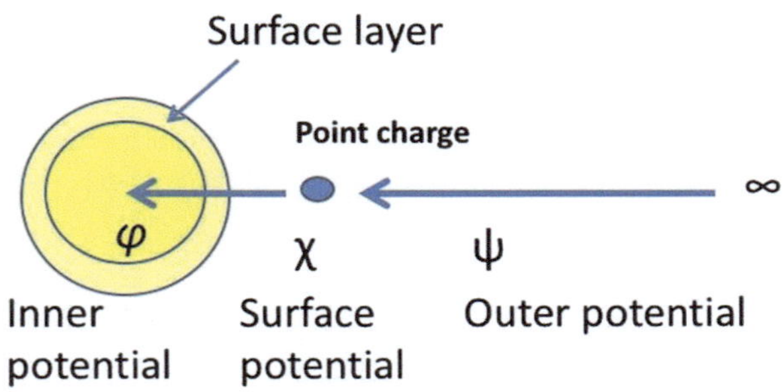

Fig. 5 Inner potential, surface potential, and outer potential

φ_i is called the inner potential of phase i,ψ_i is the outer potential of phase i andχi is the surface potential. Essentially the inner potential is closely related to the electric potential.

Now we postulate that the different two phases, α and β, would be in contact with each other. The inner potentials corresponding to each phase can be described as φ^α and φ^β, respectively. Here we can define two types of potential differences. One of them is called the Galvani potential difference and is shown in the following Eq. (13).

$$\Delta\varphi = \varphi^\beta - \varphi^\alpha \tag{13}$$

Another one is called the Volta potential difference or contact potential difference, as shown in Eq. (14).

$$\Delta\psi = \psi^\beta - \psi^\alpha \tag{14}$$

From (12) and (13), a Galvani potential difference can be described as follows (16).

$$\Delta\varphi = \left(\psi^\beta - \psi^\alpha\right) + \left(\chi^\beta - \chi^\alpha\right) = \Delta\psi + \Delta\chi \tag{15}$$

The Volta potential difference can be measured in fact, while the Galvani potential difference could not be measured due to the complicated states of surface potentials and their difficult estimation. The potential difference at the interface between two contact phases corresponds to the Volta potential difference.

The potential difference is closely related to the difference of the electrochemical potential based on the electrochemical affinity. If we could measure $\Delta\varphi$ directly, we could organize the table of electromotive forces based on the Galvani potential difference. However, $\Delta\varphi$ is generally hard to measure. We always need a reference electrode to measure the half cell potential at an electrode. When a certain electrode is coupled with a reference electrode, then the electromotive force can be measured. Since we usually use some reference electrodes as standards, the electromotive force is defined as the equilibrium potential of the reaction. The table was made in such a way and the hydrogen reference electrode was used to measure and calculate potentials for the half cell reactions.

Now, we postulate the following Redox reaction at equilibrium.

$$O(oxidated\ state) + ne^- = R(reduced\ state) \tag{16}$$

When the potential is measured, based on a reference electrode such as a hydrogen reference electrode, the potential E has a certain relation with the electrochemical potential as follows.

$$nFE = -(\mu_R - \mu_O) \tag{17}$$

Here, μ_R and μ_O are the electrochemical potentials for the oxidant and reductant, respectively. When we describe the activities for O and R as a_O and a_R, respectively, the electrochemical potentials for the two substances can be shown as follows.

$$\mu_O = \mu_O^0 + RTlna_O \tag{18}$$

$$\mu_R = \mu_R^0 + RTlna_R \tag{19}$$

μ_O^0 and μ_R^0 are the electrochemical potentials when the activities of the substances are the unit.

From the Eqs. (17), (18) and (19), the following equation can be obtained.

$$E = E^0 + \frac{RT}{nF} ln\frac{a_O}{a_R} \tag{20}$$

Here E_0 is the equilibrium potential when both a_0 and a_1 are the units.

The Eq. (20) is called Nernst Equation.

5 Potential: pH Diagram

In the previous section, we described how to estimate the electromotive force for a couple of electrodes. Actually, the estimation is often needed to evaluate the possibility, if a type of corrosion would occur or not. As explained in the next chapter, corrosion can occur by one or more local cells formed on a material's surface. Therefore, it would be important from the viewpoint of corrosion prediction and analysis to evaluate the electromotive force for a combination of some reactions.

In the same way, a potential – pH diagram, is often used for estimations. The diagram has a potential axis (vertical one) and a pH axis (horizontal one). This diagram is also called a Pourbaix diagram after the original creator's name.

Figure 6 is an example for the case of iron. The figure is a schematic one. Using the reactions and the Nernst equation, the figure can be obtained for any metals. The dotted lines correspond to the decomposition of water. Therefore, the area between the two dotted lines is the stable area for water. Each area is surrounded by lines in the diagram. The areas where divalent or trivalent iron ions are stable correspond to corrosion ones. On the other hand, the areas where iron hydroxides are stable suggest that the thin oxide layers form on the iron surface to protect the substrate from corrosion. This schematic figure indicates that iron would be vulnerable to corrosion in acidic solutions or environments, while it has stronger corrosion resistance in alkaline solutions/environments.

In this chapter, we focused on the electrochemical potential, E, rather than the current. The potential can be used to predict the possibility of electrical potential.

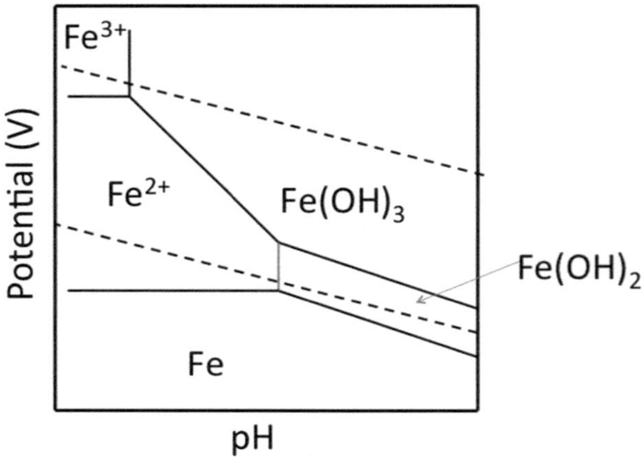

Fig. 6 Schematic potential – pH diagram for iron

As for the kinetic aspect based on the current, we will explain it in the following chapters.

References

1. Izumi S, Ishikawa M, Katakura K, Aoi Y, Nagao Y (2009) Fundamental electrochemistry. Morikita Publishing Co., Tokyo
2. Tamamushi R (1991) Electrochemistry, 2nd edn. Kagaku Doujin Co. Ltd., Tokyo
3. The Institute of Electrical Engineer of Japan (1976) Electrochemistry. Tokyo

Measurement and Evaluation for Corrosion

Hideyuki Kanematsu and Dana M. Barry

Abstract Since the corrosion phenomenon is basically an electrochemical redox reaction, it can be evaluated and analyzed by many electrochemical methods as well as non-electrochemical traditional ones. In this chapter, we describe some representative evaluation methods for corrosion from practical methods to scientific ones using electrochemical methods.

1 What Is the Corrosion?

Corrosion phenomenon is a natural one that occurs in nature. Gibbs free energy orders it as a spontaneous reaction. Basically, it is the oxidation reaction part of an electrochemical one. This suggests that corrosion could be explained from the viewpoint of electrochemistry.

According to ISO 8044 standard [1, 2], it is defined "Physicochemical interaction (usually of an electrochemical nature) between a metal and its environment which results in changes in the properties of the metal and which may often lead to impairment of the function of the metal, the environment, or the technical system of which these form a part".

The Japanese Industrial Standard (JIS Z 0103) also defines it as follows. Corrosion is the phenomenon where metals would be attached electrochemically or deteriorated by environmental factors [3].

Any definition indicates that corrosion is related to physical chemistry, particularly to electrochemistry. Figure 1 shows the photo and Fig. 2 shows the

H. Kanematsu (✉)
National Institute of Technology, Suzuka College, Suzuka, Japan
e-mail: kanemats@mse.suzuka-ct.ac.jp

D.M. Barry
Clarkson's Departments of Chemical & Biomolecular Engineering and Electrical & Computer Engineering, Clarkson University, Potsdam, NY, USA

Center for Advanced Materials Processing (CAMP), Clarkson University, Potsdam, NY, USA
e-mail: dmbarry@clarkson.edu

H. Kanematsu, D.M. Barry (eds.), *Corrosion Control and Surface Finishing*,
DOI 10.1007/978-4-431-55957-3_3

Fig. 1 A single metal immersed in an aqueous solution

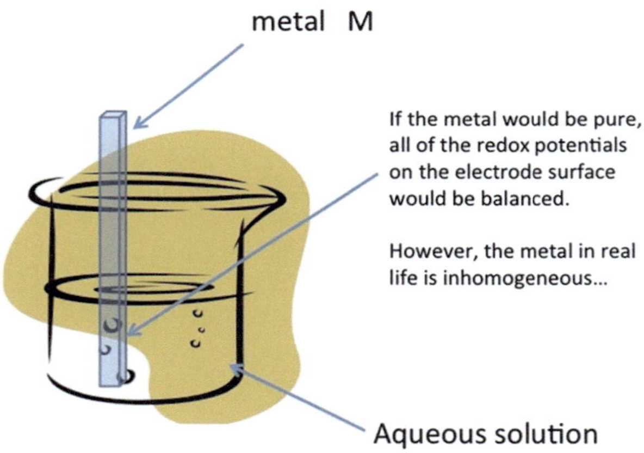

Fig. 2 Metal immersed in an aqueous solution

corresponding typical and simple example schematically. In the figure, not two metals, but only one metal is immersed in an aqueous solution. If the metal would be genuine, it would not produce any galvanic cells at all. However, the surface of the metal is not generally homogenous. Therefore, the potential energy state usually differs from place to place. This means that the electric potential also differs from one item to another. That is the reason why different electrochemical reactions occur, depending on the sites. Figure 3 shows an example. In this case, the metal M is immersed in an aqueous solution just like the situation displayed in Figs. 1 and 2.

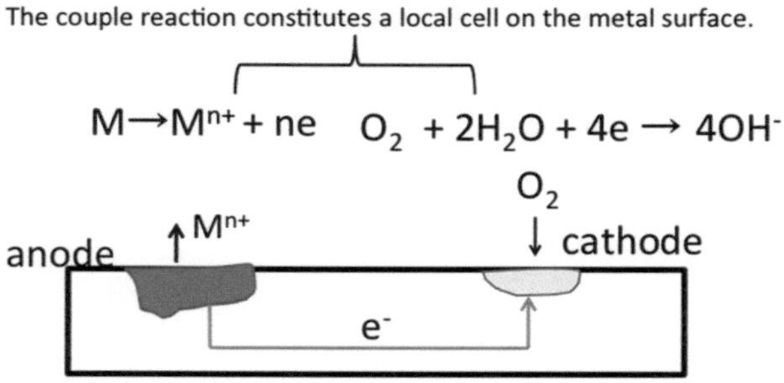

Fig. 3 Corrosion produced by a local cell in a neutral aqueous solution

If the specimen was rotated at 90° virtually, the cross section of the specimen could be observed in the figure.

If the aqueous solution was neutral and the specimen was iron, an anode reaction could be described by Eq. 1.

$$Fe \rightarrow Fe^{2+} \quad (1)$$

On the other hand, the corresponding reaction might be the following one in most cases.

$$O_2 + 2H_2O + 4e^- \rightarrow 4OH^- \quad (2)$$

The two half-cell reactions would be coupled and the corrosion phenomenon would proceed as a result. In fact, the plural complicated reactions proceed simultaneously. However, the simplified model is often the best for us to think about for the corrosion problem.

As mentioned above, the cathodic reaction is the reduction of dissolved oxygen. However, the cathodic reaction might be the reduction of hydrogen ion, when the solution would be an acid such as a sulfate solution.

$$2H^+ + 2e^- \rightarrow H_2 \quad (3)$$

As we can see in the example, the corrosion reaction generally changes from case to case and it depends on a combination of the environment and material.

2 Classification of Corrosion and the Basic Characteristics

Generally, corrosion is classified in various ways. Basically, the corrosion phenomena involve electrochemical reactions as shown above. Water is always related to them. However, we can tell the difference when corrosion occurs in aqueous solutions and when it occurs in the atmosphere, even though the latter contains a

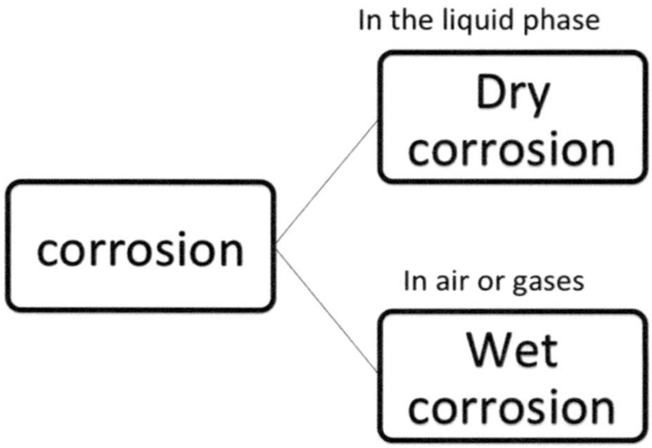

Fig. 4 The classification of corrosion based on the phase

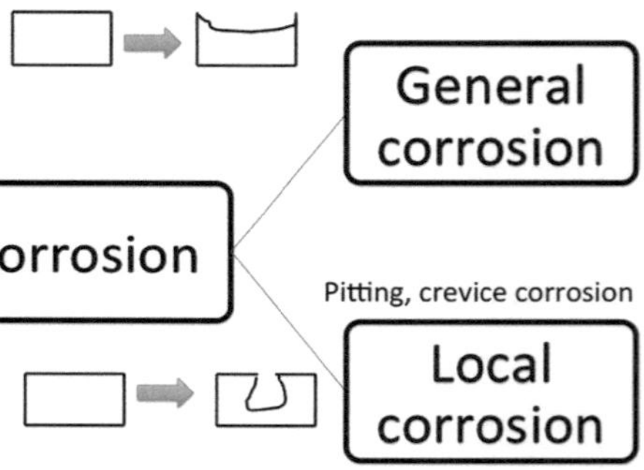

Fig. 5 The classification of corrosion based on morphology

water component (humidity) to some extent. The former is called wet corrosion and the latter dry corrosion (Fig. 4).

Corrosion can also be classified based on morphology. It is mainly classified into general corrosion and local corrosion, as shown in Fig. 5. For general corrosion, materials are corroded entirely and gradually. For example, iron immersed in an acid aqueous solution such as dilute hydrochloric acid can be mentioned. As can be seen in the Pourbaix diagram of chapter "The Fundamentals of Corrosion Science and Engineering: Equilibrium Theory and Its Meaning", iron is generally corroded and chemically unstable in acid solution. Therefore, the surface corrodes entirely. On the other hand, there are many cases where the material corrodes locally, even though most of the other areas might remain without any deterioration. Such corrosion is called local corrosion. Local corrosion generally leads to more dangerous situations due to many reasons. One of the main reasons is that one can't detect the local corrosion easily. Since the corrosion would be localized in a very small area, one can't observe and can't detect them inevitably. As a result, one may not be aware of the potential problems until the final fracture moment. The corrosion failure could be serious and destroy an entire structure. The tiny localized defect might lead to the formation of a penetration hole. The stress would be

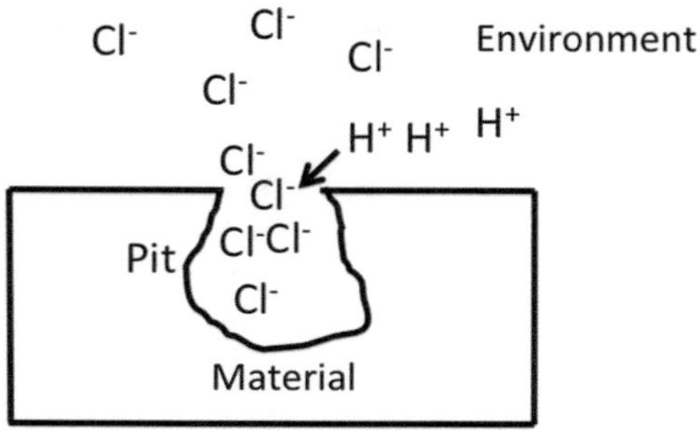

Fig. 6 Pitting and the concentration of chloride ion in pits

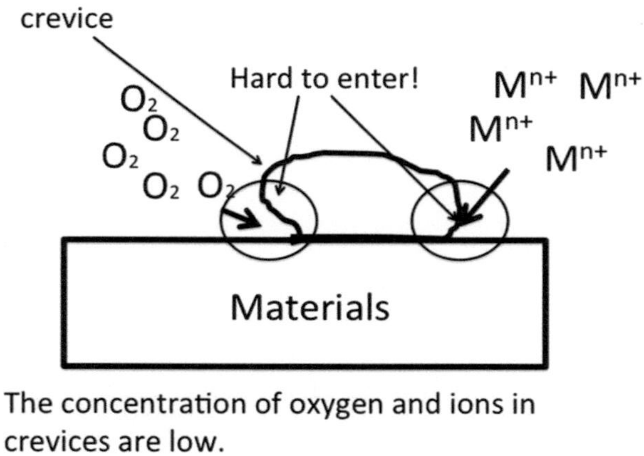

Fig. 7 The concentration of oxygen and ions in crevices

generally concentrated on this tiny part. Structures are often ruptured by the stress concentration on a very tiny part. Examples of localized corrosion are pitting and crevice corrosion.

Pitting corrosion is often observed by the naked eye, as small holes or impressions on metals. There are still discussions about the precise mechanisms for pitting corrosion. One mechanism of interest to the authors is that a special absorption of chloride ion onto a local area of a material's surface brings about pitting corrosion. As shown in Fig. 6, the absorption of chloride ion breaks the balance of electricity locally. To compensate for it, the hydrogen concentration increases in the area. This leads to the decrease of pH and the local part corrodes as a result. However, it is hard to demonstrate it scientifically. This is still a hypothesis, even though the pitting corrosion clearly depends on the concentration of chloride ion.

On the other hand, crevice corrosion is another type of local corrosion. As shown in Fig. 7 schematically, the surface area near a crevice is generally a special place where the surface energy is high and unstable. It might attract some substances into the place. If it attracts chloride ion, then pitting corrosion occurs. This corrosion might be accelerated due to the increase of concentration. Even when there would be no special matters, it could bring about corrosion with crevices. The reason might be attributed to the appearance of concentration cells.

Generally, it is hard for dissolved oxygen to reach the crevice area. Therefore, the inhomogeneous distribution of oxygen appears on a material's surface. If area 1 has a higher oxygen concentration than area 2, the activity of the place 1, a_1, is higher than that of the place 2, a_2. The two places having different potentials can constitute local cells. The electromotive force, E, could be described as follows.

$$E = E_{a1} - E_{a2} = \frac{RT}{nF} \ln\left(\frac{a_1}{a_2}\right) \tag{4}$$

Therefore, such a type of corrosion is also called a differential concentration cell. However, not only oxygen, but other elements dissolved in the solution can bring about the concentration cell.

In nature, "crevices" exist everywhere, not only on the macro-scale, but also on micro-scale. Actually, we could say that there are no material surfaces without crevices. This suggests that crevice corrosion is a very general corrosion, since crevices existing universally can always bring about the corrosion.

Corrosion can also be classified based on environments. For example, we can mention the following corrosion phenomena – atmospheric corrosion, fresh water corrosion, sea water corrosion, soil corrosion, high temperature corrosion, and gaseous corrosion.

Atmosphere corrosion is the general term for all of corrosion phenomena occurring in air. The vaporized water (humid component) forms very thin water films on materials' surfaces and the electrochemical reactions leading to corrosion proceed in the thin water film. There are many environmental factors existing in thin water films and they affect the corrosion mechanism.

Fresh water corrosion is the general term for all of the corrosion occurring in rivers, tap water, ground waters, etc. They usually do not contain so much chloride ion.

Sea water corrosion can be compared with fresh water corrosion. In the former, the water contains a lot of chloride ions and it accelerates the corrosion phenomenon very much.

Soil corrosion is the general term for the corrosion occurring in soil. The qualities of soils and the aeration property affect the corrosion characteristics in soil. Microbiologically influenced corrosion also occurs in soils characteristically.

High temperature corrosion occurs in high temperature water or vapor. One of the typical situations can be seen in boilers. However, not only oxygen, but also other gases might be involved in many cases.

Gaseous corrosion is the general term for the corrosion occurring in gases without liquids. It belongs to the dry corrosion category.

There are some corrosion cases where other factors, i.e. stress, or other mechanical factors are involved. For example, stress corrosion cracking, corrosion fatigue, hydrogen embrittlement, erosion corrosion, cavitation corrosion, fretting corrosion, etc. can be mentioned.

Stress corrosion cracking is brought about, when stress and corrosion reactions are applied to the materials simultaneously. Being compared with the case where each of the two factors is independent and separate, the life cycle of the material (with the combined condition) is generally reduced significantly. It often relates to local corrosion and therefore, stainless steel and aluminum alloys (that have higher resistance against corrosion otherwise) are often vulnerable to stress corrosion cracking.

Corrosion fatigue is the phenomenon where the strength would be reduced under the condition of corrosion and cyclic stresses. Also in this case, the life cycle of material is reduced significantly.

Hydrogen embrittlement is brought about, when the material absorbs evolved hydrogen during some processes. Some examples are acid pickling, electrolysis, and uses in various vaporized atmospheres. It sometimes leads to fractures. Usually, the phenomenon might occur after a long time. Therefore, it is often called “delayed fracture” or “delayed failure”.

Erosion corrosion occurs under the conditions of corrosion and wear by flowing water, the collision of sands, etc. The simultaneous application of both factors accelerates the corrosion and failure significantly also in this case.

Cavitation corrosion is brought about by cavitation damage. It can be seen when the repeated evolution and collapse of tiny air bubbles on material surfaces occurs. Cavitation breaks the stable passive films on materials’ surfaces, which would easily lead to corrosion.

Fretting corrosion is often seen when metals have contacts with other materials. The contact brings about microscopic slips in the vicinity of a material’s surface, which leads to the appearance of corrosion.

If these phenomena relate to other independent factors such as stress, etc. the result is accelerated corrosion with more complications. However, the phenomena in real life must be generally complicated and from the viewpoint, they may be universal corrosion types. Also for such corrosion, the basics and fundamentals of corrosion could be applied to the phenomena to solve technical and scientific problems.

3 Surface Finishing for Corrosion Control

In the previous section, we described many kinds of corrosion phenomena and the classification based on many concepts. As the description suggests, corrosion type and mechanism differ from a combination of materials and the environment. Therefore, an appropriate surface finishing for anti-corrosion purposes should change from case to case. Generally, non-metallic coatings such as polymers and ceramics have higher corrosion resistance. However, metal coatings are usually vulnerable to corrosive environments, even though some metals have pretty high corrosion resistance. As for metal plating, zinc, nickel, and chromium are versatile and practical metals for surface finishing that generally provides high corrosion resistance.

Fig. 8 The concept for environmentally friendly surface finishing in the future

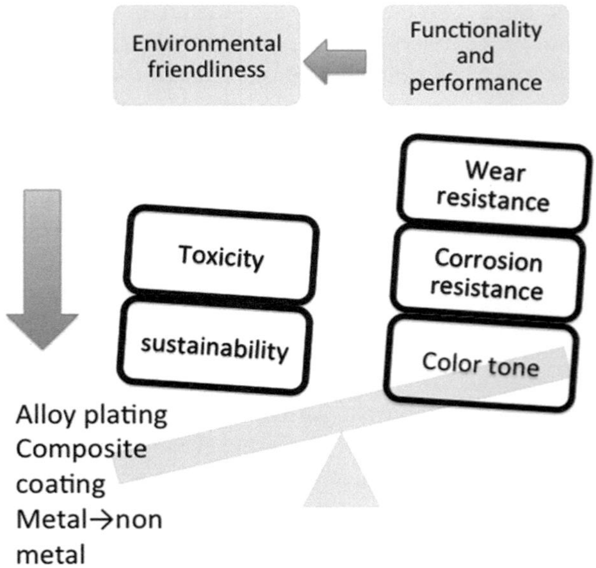

As described already, not only the property of corrosion resistance, but other desired functions are required for materials. Therefore, the best surface finishing processes need to be selected for specific materials in certain environments. Figure 8 shows the concept to design the surface finishing for corrosion control. In addition to the conventional concept, we have to take the possibility of environmental harmfulness into consideration. Corrosion often leads to the contamination of our environment, since the anodic dissolution is the essence in most cases. Therefore, sacrificial corrosion protection might be prohibited in some cases. From that perspective, the appropriate surface coating will be more versatile in the future. Chapter "The Application of Corrosion Protection" presents and discusses surface finishing processes.

4 Corrosion Evaluations

To design the appropriate surface finishing process against corrosion, we need to determine what kind of corrosion would cause an item to fail and how much material would be damaged. Therefore, we need appropriate and precise corrosion tests.

There are many kinds of corrosion tests. To make things simple, the two main types are shown in Fig. 9.

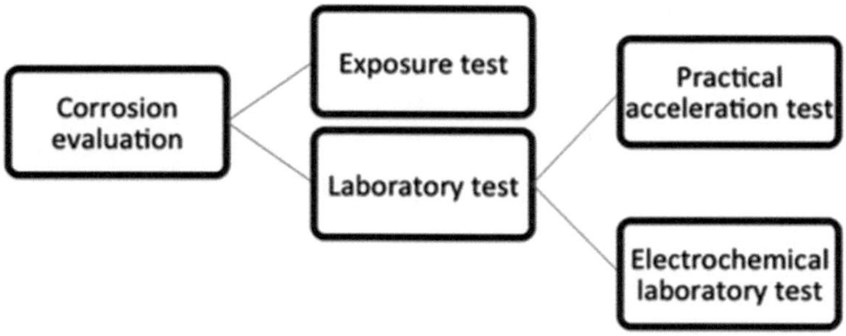

Fig. 9 Corrosion evaluation and its classification

4.1 Exposure Test in Practical Sites

One of them is the exposure test. For example, one finds a material in a certain environment. He/she may need the accelerating test on the laboratory scale. However, the importance of an exposure test in the practical environment is of importance, since the result in the exposure test is the closest simulation to the real phenomenon. Such an exposure test is carried out in various environments in which researchers and engineers are interested. Usually, the tests require a long time: from a couple of years to several tens of years. Even though the researchers and engineers in charge of the exposure tests may retire or quit their positions during the measurement periods, the measurements would still continue. The results obtained after a long term experiment would be precious and valuable. One can measure and estimate the material's life cycle from the results. Also the mechanism of failure for corrosion protection can be analyzed.

4.2 Practical Acceleration Test on the Laboratory Scale

Even though the exposure test in practical sites is still very important and useful, we still need laboratory tests. It would be very useful to accelerate the corrosion characteristics and to evaluate the results in a relatively short time. This category has some laboratory tests filling the gap between the exposure test and very fundamental electrochemical tests. To name a few, we can mention the salt spray test, a CASS (Copper accelerated acetic acid salt spray) test, and a combined cycle test [4].

The salt spray testing has been defined by some industrial standards such as JIS, ISO, MIL, etc. The specimen is set in a chamber and sprayed intermittently for a certain period of time. For example, neutral 5 % sodium chloride at 50 °C, is sprayed onto the specimen. The size of the specimen is fixed and the spraying amount is usually fixed.

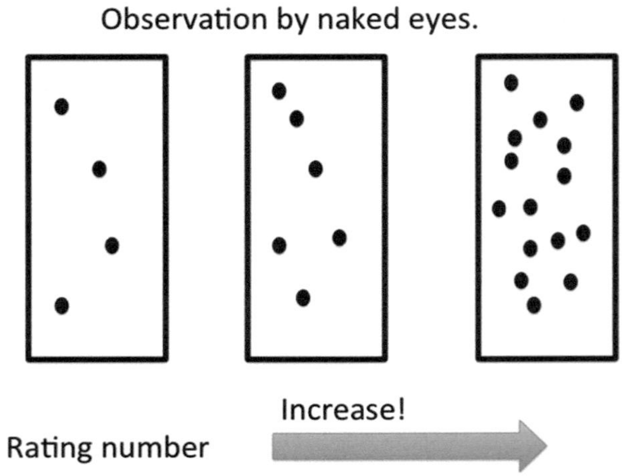

Fig. 10 Corrosion evaluation by rating number

On the other hand, the CASS test utilizes the sodium chloride – copper chloride mixed solution. The addition of copper ion accelerates corrosion much more than the salt spray testing. However, the results might not fit the real phenomenon so precisely in some cases.

For the combined cycle test, the spray testing is combined with drying and wetting. The cycle can be designed and changed flexibly. For example, the spraying will be carried out at 35 °C for several hours. Then the specimen is dried at 70 °C for a certain period and wetted at the same temperature for another certain period. The combination can simulate the weather conditions better. In such a way, the acceleration test can become closer to the real exposure test. In Japan, automobile companies usually require the related surface finishers to overcome a certain condition by the combined cycle test.

The specimen (after these tests) is evaluated for its extent of corrosion. If it would be genuine research, we could use various electron microscopies such as TEM, SEM, XPS, etc. However, we usually use a rating number method for the evaluation from the practical standpoint. Figure 10 shows schematic figures for the method. We can evaluate the distribution situation by the naked eye, by comparing the tested specimen to the standard samples. There are some other methods similar to the salt spraying test and the CASS test. However, the basic concept for the evaluation method is almost the same. The solution, temperature, and amount of spraying are varied from test to test.

4.3 Electrochemical Laboratory Test

On the other hand, electrochemical laboratory tests are closer to the basic fundamental theory. As described already, corrosion is basically the electrochemical

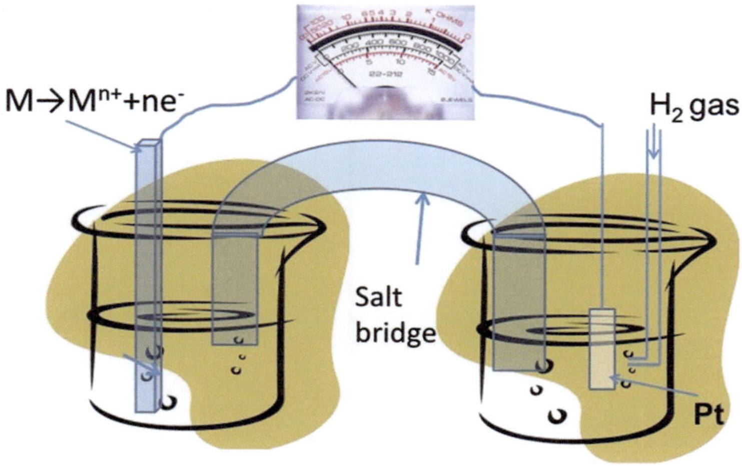

Fig. 11 The schematic principle for potential measurement

reaction composed of anodic and cathodic couple reactions. It means the electrochemical measurement would be the best way to evaluate corrosion phenomena on the laboratory scale. However, we always have the similar problem to confirm if the fundamental measurement could explain the real corrosion in practical sites.

The main parameters playing important roles for electrochemical measurements are potential and current. As described in the previous chapter (chapter "The Fundamentals of Corrosion Science and Engineering: Equilibrium Theory and Its Meaning"), the potential always suggests the corrosion possibility. One of the typical examples is the potential-pH diagram, as already described. Therefore, the measurement of potential is one of the basic corrosion measurements on the laboratory scale.

Figure 11 shows the schematic principle for potential measurement. It needs a reference electrode. The so-called normal hydrogen electrode is theoretically the best reference. However, it is hard to deal with it in practice. The calomel electrode and silver/silver chloride electrode are often used. As for the calomel electrode, the following reaction occurs.

$$Hg_2Cl_2 + 2e^- \rightleftarrows 2Hg + 2Cl^- \tag{5}$$

On the other hand, the reaction for a silver/silver chloride reference electrode is as follows.

$$AgCl + e^- \rightleftarrows Ag + Cl^- \tag{6}$$

At any rate, the reaction is reversible and very fast. Recently, the toxicity for mercury has been mentioned often. Therefore, the calomel electrode using a mercury compound is going to be out of service.

Fig. 12 Evans diagram and polarization

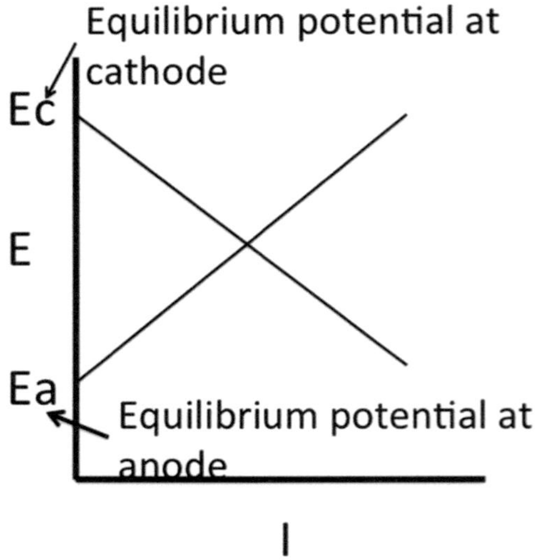

By such a simple system, the potential with time can be measured. The equilibrium potential (Usually, it is called mixed potential, since plural potentials are balanced at a value.) often shows a certain tendency. For example, the specimen's mixed potential generally moves in the noble direction, when the microbiologically influenced corrosion occurs. In such a case, only the potential measurement would provide us sufficient information to some extent.

When corrosion proceeds, the equilibrium is broken. The potential is shifted from the equilibrium state and the current flows as a result. The potential shift from the equilibrium state (Ee) is called overpotential (η) and the shift is called polarization. When the polarized potential is Ep, the following relation is established.

$$E_P = E_e + \eta \tag{7}$$

For example, we postulate the following corrosion reaction here.

$$M^{n+} + ne^- \rightarrow M \tag{8}$$

When the potential is polarized in the less noble (minus) direction, then the cathodic reaction proceeds. On the other hand, when the anodic reaction (corrosion reaction) proceeds, then the potential is polarized in the noble (plus) direction. The situation is shown in Fig. 12 schematically. The diagram is often called the Evans diagram and it is very helpful to analyze corrosion reactions.

In both polarization cases, current flows in the circuit. The current is equivalent to the reaction rate at the interface, according to Faraday's law. Therefore, it corresponds to corrosion rate.

When the corrosion occurs, the potential moves to corrosion potential where the anodic curve and cathodic one are intersected with each other on the Evans

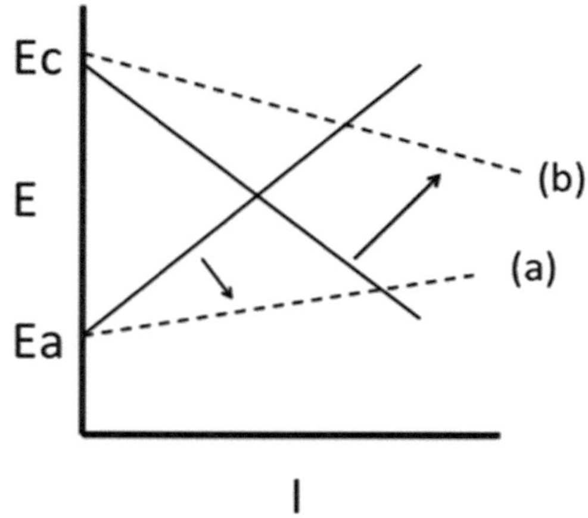

Fig. 13 Polarization changes on the Evans diagram

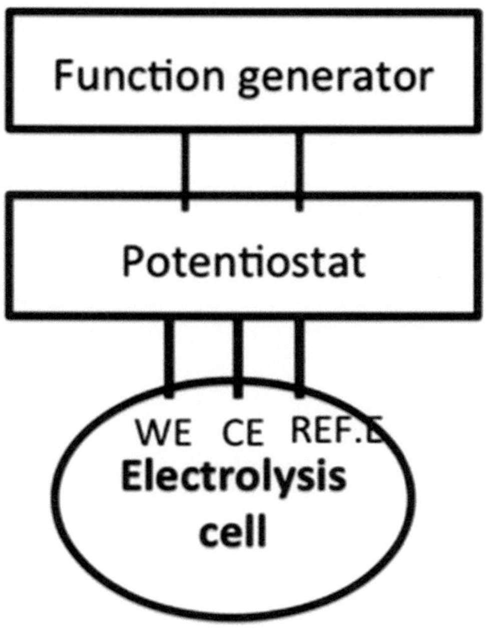

Fig. 14 Polarization measurement system

diagram. As anodic reaction occurs actively, the potential moves in the less noble direction (dotted line a). When the cathodic direction occurs actively, the potential moves in the less noble direction (dotted line b). Figure 13 shows the dotted lines and changes schematically.

The polarization curve is measured well, when one can use the polarization measurement system. A typical constituent of the system is shown in Fig. 14. A potentiostat is used for the system. It is the electronic hardware to control a three-electrode cell and run most electroanalytical experiments. When a function generator is combined with the potentiostat, the potential can be scanned and the potential-current curve can be obtained.

Even though there are many useful and advanced newly-developed techniques for the evaluation of corrosion, the polarization curve from the kinetic viewpoint and the potential-pH diagram from the equilibrium standpoint are still very important tools.

References

1. Javaherdashti R (2008) Microbiologically influenced corrosion. Springer, London, p 1
2. Mattson E (1989) Basic corrosion technology for scientists and engineers. Chater 3. Ellis Horwood Publishers, West Sussex
3. Japanese Industrial Standard (JIS) Z0103 (1996) Glossary of terms used in rust and corrosion preventive technology
4. Japanese Industrial Standard (JIS) 2371 (2000) Method of salt spray testing

Industrial Surface Treatments

Dana M. Barry and Hideyuki Kanematsu

Abstract Treatments to the surfaces of metals as well as other items are very important in most industries. They are used for many purposes including decoration, reflectivity, improving hardness, and preventing corrosion. This chapter introduces and describes various industrial surface finishing processes. It includes electroplating, electroless plating, spray coating, galvanization, painting, anodizing, physical vapor deposition (PVD), and chemical vapor deposition (CVD).

1 Introduction

Industrial treatments are made to material surfaces for various reasons. One is to improve the appearance of a finished product to attract potential buyers. Another is to improve the safety of products. For example, after surfaces are smoothed and sharp edges and burrs are removed, they become easier for individuals to handle and less likely to cause injuries. Other purposes for surface treatments are to improve corrosion resistance, wear resistance, mechanical properties, and electrical properties. This chapter introduces and describes various industrial surface finishing processes. It includes electroplating, electroless plating, spray coating, galvanization, painting, anodizing, physical vapor deposition (PVD), and chemical vapor deposition (CVD). Chapter 6 ("Corrosion and Surface Finishing") combines the concepts of corrosion and surface finishing and discusses how corrosion is controlled by a variety of surface treatments.

D.M. Barry (✉)
Clarkson's Departments of Chemical & Biomolecular Engineering and Electrical & Computer Engineering, Clarkson University, Potsdam, NY, USA

Center for Advanced Materials Processing (CAMP), Clarkson University, Potsdam, NY, USA
e-mail: dmbarry@clarkson.edu

H. Kanematsu
National Institute of Technology, Suzuka College, Suzuka, Japan
e-mail: Kanemats@mse.suzuka-ct.ac.jp

H. Kanematsu, D.M. Barry (eds.), *Corrosion Control and Surface Finishing*,
DOI 10.1007/978-4-431-55957-3_4

2 Electroplating

Electroplating can be defined as the application of a metal coating to a metallic or other conducting surface using an electrochemical process. The item to be plated serves as the cathode (negative electrode) and the anode is usually the metal to be plated on the item. Both components are placed in an aqueous (aq.) solution called an electrolyte containing dissolved metal salts and other ions to allow for a flow of electricity. A power source provides a direct current to the anode and oxidizes its metal atoms so they are able to dissolve in the solution. These dissolved metal ions are then reduced at the cathode and deposited there as a metal coating. Refer to the general example below where M represents the metal to be plated onto another metal.

At the Anode: $M(s) \rightarrow M^{2+}(aq.) + 2e^-$
At the Cathode: $M^{2+}(aq.) + 2\,e^- \rightarrow M(s)$

There are many industrial applications for electroplating. Articles are electroplated to enhance their appearance, prevent corrosion, and to improve their surface properties like hardness, thickness, and conductivity. The automotive industry uses electroplating to refurbish old chrome parts like bumpers, grills, and tire rims to make them look new. Electroplating is also used on plastic for chrome-plating lightweight parts of the modern automobile [1]. Electroplating is used for electrical parts and components too. Silver electroplating is used on copper or brass connections because of its superior conductivity. Gold and palladium plating are used for switchgear in the telecommunications industry [1]. In addition, electroplating is used to make pennies, small zinc coins covered with a layer of copper [2].

3 Electroless Plating

Electroless plating uses a redox reaction without an electric current to deposit metal on an object or workpiece. The electroless plating solution generally includes the following: a source of metal, a reducer, a complexing agent (which holds the metal in solution by forming coordinate bonds with the metal atom or ion), and buffers along with other chemicals to stabilize the bath. This type of plating process, which has been around for many years, was used to make mirrors. See the reaction below where a complex metal (silver-Ag) is reduced using formaldehyde as the reducing agent. As a result, shiny silver is plated on the workpiece [3].

$$H - CHO + 2[Ag(NH_3)_2]OH \rightarrow 2\,Ag(s) + HCOONH_4 + H_2O + 3\,NH_3$$
Formaldehyde + Silver metal complex → Silver + Ammonium Formate +Water + Ammonia

Electroless nickel plating is used to deposit a coating of nickel or nickel alloy on a substrate. It relies on the presence of a reducing agent that reacts with the metal ions to deposit metal. This process uses a unique chemical bath without any electrical current. The bath chemistry is constantly replenished during the plating process.

The most common form of electroless nickel plating produces a nickel phosphorus alloy coating [4]. For this process a reducing agent such as hydrated sodium hypophosphite ($NaPO_2H_2.\ H_2O$) reacts with the metal ions to deposit metal. The phosphorus content in these coatings can vary from about 2 to 14 %. It determines the alloys' metallurgical properties. These coatings are generally used on parts where hardness, corrosion protection, and wear resistance are required. Examples include door knobs, kitchen utensils, bathroom fixtures as well as electrical and mechanical tools.

There are advantages for using the electroless plating process. It does not require electrical power. Metal is deposited evenly along a part's edges and inside holes as well as over irregularly shaped objects. Also the process can deposit a conductive surface on a nonconductive item so that it can be electroplated.

4 Spray Coating

Spray coating is a process in which materials are sprayed onto a surface. It is widely known as the use of spray cans (or spray guns) by artists and as an industrial process for car body painting. For this method, the functional fluid or ink is atomized at the nozzle of the spray head, which makes a continuous flow of droplets. This system uses a stream of pressurized air or gas (such as argon or nitrogen) to break up the liquid into droplets at the nozzle. These droplets of ink (paint) are then carried onto the surface of a particular substrate. The spreading of the droplets on this surface is affected by the surface temperature, the fluid-surface interaction, and the kinetic impact of the droplets. Important variables for the atomization process include surface tension, viscosity, fluid density, gas flow properties, and the design of the nozzle. The quality of the coated layer depends upon a number of factors such as surface properties, coating speed, and droplet sizes.

Another form of this coating process is called Thermal Spraying, where melted (or heated) materials are sprayed onto a surface. This technique consists of an electrical (examples: plasma and arc) or chemical (ex. combustion flame) heat source and a coating material in the form of powder or wire which is melted into tiny droplets and sprayed at high velocity onto surfaces. These coatings can be thick (approximately 20 μm to several mm) over a large area and deposited at a high rate as compared to other coating processes such as electroplating [5]. Coating materials used for thermal spraying include metals, alloys, ceramics, plastics, and composites. They can be applied to some plastics and most metal substrates including aluminum, steel, stainless steel, copper, and bronze [6].

It should be mentioned that ceramic coatings are most often applied using Plasma Spray due to their high melting temperatures. On the other hand, Electric Wire Arc Spray (another type of Thermal Spraying) is generally used to apply coatings containing metals (examples: zinc and copper) and metal alloys. This process is cost effective and can be used to prepare a variety of coating textures.

5 Galvanization

Galvanization can be considered a process for coating iron and steel with zinc to prevent them from rusting. In addition to corrosion protection, the zinc coating provides other benefits such as high strength, light weight, aesthetics, and low cost. Zinc coated steel is referred to as galvanized steel.

The most common form of galvanizing is Hot-dip Galvanization. The metal parts must be free of grease, oil, and other surface contaminants prior to this process. For this method metals such as iron, steel, or aluminum, are coated with a layer of zinc by being submerged in a bath of molten zinc. The coating forms a barrier between the substrate and corrosive substances in the atmosphere. Its effectiveness depends on factors such as the coating thickness and coating corrosion rate which is influenced by the composition of the atmosphere as well as the type of coating.

Zinc also serves as a sacrificial anode (where there is a loss of electrons), so that if the coating is scratched the remaining zinc protects the exposed steel. Galvanized steel is widely used for applications requiring corrosion resistance without the cost of stainless steel. It can be identified by the crystal patterns on its surface which are often referred to as spangles. Galvanized steel is used in roofing, handrails, consumer appliances, automotive body parts, metal pails, and in most heating and cooling duct systems in buildings.

Another type of galvanization is called Dry Galvanizing because it does not involve any liquids. For this process, parts and zinc powder are tumbled in a rotating sealed drum at a specified temperature. The zinc evaporates and diffuses into the substrate to form a zinc alloy. This coating has good adhesion to paint, powder coatings, and rubber. It is used to coat small, complex-shaped metals.

Wires and fibers like thread are also galvanized to improve their strength, provide desired surface properties and to prevent corrosion. See the galvanized thread in Fig. 1 [7].

Fig. 1 Galvanized thread

6 Painting

Painting may be simply defined as the coating of a surface with paint (which includes pigment or color). This process is commonly used to decorate, color, and protect objects as well as provide them with texture. The paint coating can be applied by using brushes, rollers, and spray methods. (Refer to the Spray Coating Sect. 4 of this chapter). Paint is typically applied as a liquid and changes to a solid film after drying.

Industrial paint is used to protect metal, wood, and a variety of other materials. Its characteristics and properties are in part determined by the ingredients of the paint and the performance specifications required for specific applications of the paint.

Paint typically contains four major components. They are pigment, binder, liquid, and additives. The pigment is simply a powder in its unmixed form. It is used to determine the paint's color and appearance. Prime pigments are responsible for the paint's color or whiteness. Titanium dioxide is the main ingredient (prime pigment) in most white paints. Extender pigments are used to add bulk to the paint. Silica is an example of an extender pigment that is used to increase the paint's durability.

The binder is used to hold the pigment in place. Oil-based binders (used in oil-based paints) oxidize when exposed to air and harden along with the rest of the paint. The liquid component evaporates. Latex-based binders are generally used in water-based paints. In this case after the paint is applied, the water evaporates leaving behind a coating composed of the pigment and binder.

Three important binders for paints are mentioned [8]. Acrylic polymers (resins) are present in emulsion paints that are typically used to paint the inside and outside of buildings. Alkyd polymers (resins) are contained in decorative gloss paints. Epoxy polymers (resins) often serve as binders for industrial primer coatings. They give the paint great adhesion properties and high resistance to corrosion.

The liquid component of the paint is mainly used to transport the binder and pigment to the surface of the substrate. Oil-based paints tend to use paint thinner, while latex-based paints use water as their liquid.

Additives are used to change the properties of the dry film coating and/or the liquid paint. Several examples include thickeners (used to make the paint thicker and easier to apply), dispersants (used to separate pigment particles), and fungicides (used to protect exterior paint coatings from fungi).

7 Anodizing

Anodizing is an electrolytic passivation process that increases the thickness of natural oxide layers on the surface of metals. It basically forms an anodic oxide finish on a metal's surface to increase corrosion resistance. For the anodizing process, the metal to be treated serves as the anode (positive electrode) of an electrical circuit.

Consider the anodized aluminum layer. It is grown by having a direct current go through an electrolytic solution where an aluminum piece is the positive electrode (anode). Anodizing of aluminum usually takes place in an acid solution. Hydrogen is released at the negative electrode (cathode) and oxygen is released at the surface of the aluminum anode, forming a layer of aluminum oxide. Anodized aluminum can be found on smartphones, flashlights, cameras, and other products.

In addition to increasing corrosion resistance, anodizing improves surface adhesion for paint primers and glues. Thick anodic films tend to be porous so they can absorb dyes for cosmetic effects. On the other hand, thin transparent coatings can add interference effects to reflected light [9, 10].

Anodized films are most often applied to protect aluminum alloys. However there are processes for other metals such as titanium, zinc, and magnesium. Anodized titanium is used in dental implants and sometimes in art and costume jewelry because it generates various colors without dyes. Each color depends on a specific thickness of the oxide [9]. To ensure the preparation of a consistent oxide layer, one must control conditions such as electrolytic concentrations, acidity, and current. Also a sealing process is often needed to achieve corrosion resistance because thick coatings (oxide layers) are generally porous.

A variation of the typical anodized aluminum is Photosensitive Anodized Aluminum. This method uses the porous nature of unsealed anodized aluminum to create a sub-surface image. For this special case, the anodized aluminum is impregnated with a silver compound that creates an activated latent image when exposed to a source of light. After the developing and fixing processes take place

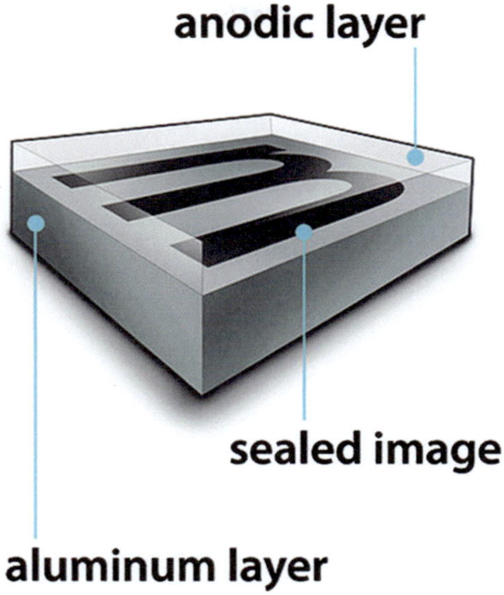

Fig. 2 Sub-surface image formed inside of the metal using the photosensitive anodized aluminum technique

(resembling those in black and white photography), then a black, silver-based image is formed inside of the metal. Finally the photosensitive anodized aluminum is sealed. See Fig. 2 [11]. This technique is used in applications where permanent product identification is required.

8 Physical Vapor Deposition

Physical Vapor Deposition (PVD) includes various methods for depositing thin films onto a substrate by the condensation of a desired vapor. For this process the coating material changes from the solid phase to the vapor phase and back to the solid phase (as it builds a film on the substrate). See Fig. 3 [12].

The PVD technique involves physical processes like the use of high-temperature vacuum evaporation of a coating material that is then condensed onto a substrate. It is carried out in a vacuum at temperatures between 150 and 500° C and provides coatings with an average thickness of about 2–5 μm [13]. Another example of this method is Plasma Spray-Physical Vapor Deposition. Here a ceramic powder is placed into a plasma flame where it is vaporized and then condensed onto a cooler substrate. Various compound coating compositions can also be created by introducing reactive gases such as nitrogen into the vacuum chamber during the metal deposition process.

Physical vapor deposition is used to deposit thin films onto items such as tools (examples: drills and saws). Several possible films include titanium nitride and zirconium nitride. Titanium nitride is very popular because it is extremely hard and

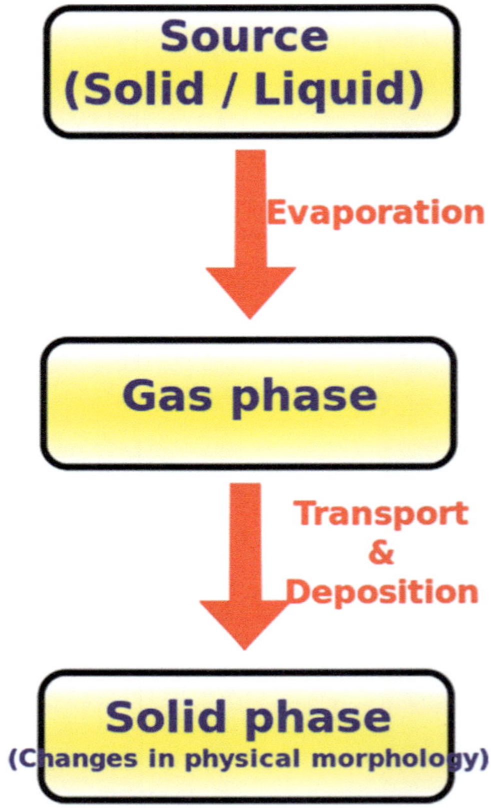

Fig. 3 Diagram of the PVD process

inert. Therefore, it is a desired coating for precision metal parts. The PVD method is also beneficial for decorative and equipment coating applications. This type of surface treatment, which improves hardness and helps prevent corrosion, allows for a variety of substrate materials to be coated with metals, alloys, ceramics, glass, polymers, etc. It is used in industries involving automobiles, optics, watches, firearms, and more.

9 Chemical Vapor Deposition

Chemical Vapor Deposition (CVD) is a chemical process to produce solid thin films on surfaces. Typically one or more precursor gases flow into a chamber containing heated objects that need to be coated. Chemical reactions take place on or near the hot surfaces. As a result, thin films are deposited on these surfaces. Volatile chemical by-products (which are frequently produced) can be removed by gas flowing through the reaction chamber.

Several benefits of the CVD process include the following. A variety of materials can be deposited with high purity. These films generally display uniform thickness on the insides and outsides of elaborately shaped objects. Also the CVD process has a relatively high deposition rate. On the other hand, the main disadvantage of this technique is that the precursors can be toxic or corrosive. Therefore, this coating method has some limitations.

CVD is used to produce coatings, powders, and fibers. This process is important for the semiconductor industry. Depending on the device, it can be used to deposit thin films of active semiconductor material like silicon dioxide (SiO_2), for use as an insulating dielectric, and silicon nitride (Si_3N_4), for use as an insulator and chemical barrier in making integrated circuits (IC).

Simplified equations are provided for forming silicon dioxide and silicon nitride.

$$SiH_4 + O_2 \rightarrow SiO_2 + 2H_2$$

Silane + Oxygen → Silicon dioxide + Hydrogen

$$3\ SiH_4 + 4\ NH_3 \rightarrow Si_3N_4 + 12H_2$$

Silane + Ammonia → Silicon nitride + Hydrogen

CVD is used to produce microelectromechanical structures (MEMS), very small devices. MEMS technology allows both electronic circuits and mechanical devices to be manufactured on a silicon chip. MEMS structures can be made from silicon wafers with CVD deposits of polycrystalline silicon (polysilicon) films and sacrificial silicon dioxide layers that are later removed by chemical etching [14].

In addition, chemical vapor deposition is used to make synthetic diamonds by creating the conditions required for carbon atoms in a gas to deposit on a substrate in crystalline form. This process needs sources for carbon and energy, and generally some hydrogen. The growth of diamond on the surfaces of various materials (substrates) imparts the diamond's qualities such as hardness and high scratch resistance to these materials. Such diamond coatings would be advantageous for industrial cutting tools and grinding machines.

10 Conclusion

This chapter introduced and briefly described a number of industrial surface treatments. These treatments are made to material surfaces for various reasons. One is to enhance the appearance of a finished product to attract potential buyers. Another is to improve the safety of products. Other purposes are to promote corrosion control and wear resistance, and to improve mechanical and electrical properties. The treatments discussed include electroplating, electroless plating, spray coating, galvanization, painting, anodizing, physical vapor deposition (PVD), and chemical vapor deposition (CVD).

References

1. http://www.sharrettsplating.com/electroplating-benefits.html
2. The Composition of the Cent (http://www.usmint.gov/about_the_mint/fun_facts/?action=fun_facts2)
3. http://antoine.frostburg.edu/chem/senese/101/redox/faq/electroless-plating.shtml
4. http://en.wikipedia.org/wiki/electroless_nickel_plating
5. http://en.wikipedia.org/wiki/Thermal_spraying
6. http://www.precisioncoatings.com/what-is-thermal-spray.html
7. Greiz TITV (2007) Galvanized thread Elitex, CC-BY- SA 3.0 DE. http://commons.wikimedia.org (Search Category: Galvanization)
8. http://www.essentialchemicalindustry.org/materials-and-applications/paints.html
9. http://en.wikipedia.org/wiki/anodizing
10. Lee W, Nielsch K (2007) Porous alumina. Max-Planck-Institut
11. Wainer S (2010) Photosensitive anodized aluminum cross-section, CC. BY- SA 3.0. http://en.wikipedia.org/wiki/photosensitive_Anodized_Aluminum
12. Gurulenінn PVD (2011) Process flow diagram, CC BY-SA 3.0. http://en.wikipedia.org/wiki/physical_vapor_deposition
13. http://www.corrosionpedia.com/definition/875/physical-vapor-deposition-pvd
14. Creighton JR, Ho P (2001) Introduction to chemical vapor deposition (CVD). Chapter 1, chemical vapor deposition (#06682G) published by ASM International. pp 1–10. www.asminternational.org

Surface Analysis

Nobumitsu Hirai

Abstract Surface analyses have been one of the key technologies for corrosion control and surface finishing. It is very important that the most appropriate apparatus for the purpose of the analyses should be selected from various analytical techniques. In this chapter, surface analytical methods for corrosion control and surface finishing, such as X-ray fluorescence analysis (XRF), X-ray diffraction analysis (XRD), X-ray photo-electron spectroscopy (XPS), scanning electron microscopy (SEM), transmission electron microscopy (TEM), Auger electron spectroscopy (AES), Secondary ion mass spectrometry (SIMS), Rutherford back-scattering spectrometry (RBS), Surface-enhanced Raman spectroscopy (SERS), Fourier-transform infrared spectroscopy (FTIR), and so on, are briefly introduced.

1 Introduction

Surface analyses have been one of the key technologies for corrosion control and surface finishing. It is very important that the most appropriate apparatus for the purpose of the analyses should be selected from various analytical techniques. In this chapter, surface analytical methods for corrosion control and surface finishing, mainly ex situ techniques, are described.

2 X-Ray Analyses

X-rays are kinds of electromagnetic waves. Wavelengths of X-rays range from 1 pm to 10 nm. They are shorter than those of ultraviolet rays and longer than γ-rays. Energy of X-rays, E, ranging from 0.1 keV to 1 MeV, is calculated by the following equation;

N. Hirai (✉)
National Institute of Technology, Suzuka College, Shiroko-cho, Suzuka 510-0294, Mie, Japan
e-mail: hirai@chem.suzuka-ct.ac.jp

H. Kanematsu, D.M. Barry (eds.), *Corrosion Control and Surface Finishing*,
DOI 10.1007/978-4-431-55957-3_5

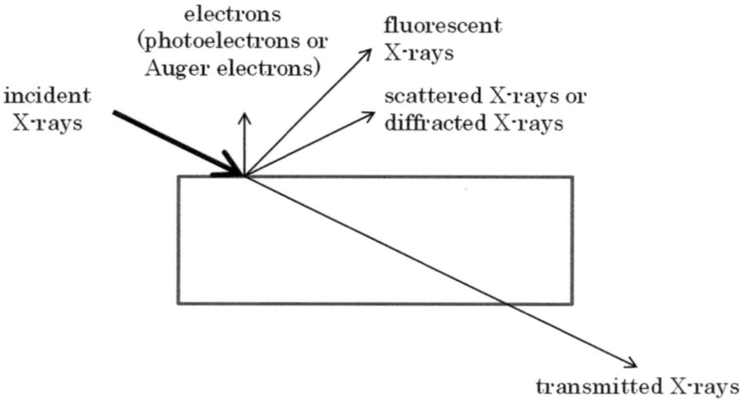

Fig. 1 Various interactions between X-rays and the samples

$$E = h\nu = \frac{hc}{\lambda}$$

h: Planck's constant, 6.626×10^{-34}/J sec;
ν: frequency;
c: speed of light, 2.998×10^{8}/m sec^{-1};
λ: wavelength.

Various interactions occur when X-rays are irradiated to the samples. Figure 1 illustrates various interactions between X-rays and the samples. One part of incident X-rays is scattered and/or diffracted, and another part is absorbed to generate fluorescent X-rays and/or electrons (photoelectrons or Auger electrons), and the rest is transmitted. In this section, operating principles and characteristics of X-ray fluorescence analysis (XRF), X-ray diffraction analysis (XRD), and X-ray photoelectron spectroscopy (XPS), which are often used for the analyses of corrosion and surface finishing, are described.

2.1 X-Ray Fluorescence Analysis (XRF)

Schematic illustrations for the principle of X-ray fluorescence are shown in Fig. 2. When X-rays having higher energies than the electrons in a K shell of atoms are irradiated, the electrons in the K shell receive enough energy from the irradiated X-rays to leave the atom, which results in electron holes. Then, electrons in L or M shells move to the electron holes in the K shell and X-rays, called "fluorescent X-rays", are generated. The energies of the fluorescent X-rays are equal to the difference between those of the electrons in the K and L (M) shells so that the

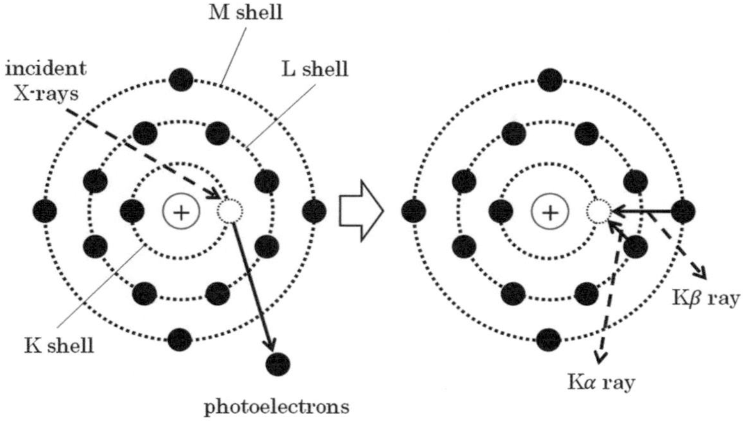

Fig. 2 Schematic illustrations for the principle of X-ray fluorescence

Table 1 Characteristics of WDS and EDS

	WDS (WDX)	EDS (EDX)
Merit	Higher precision	Higher detection efficiency
	Higher resolution	Smaller and cheaper apparatus
Demerit	Lower detection efficiency	Lower precision
	Larger and more expensive apparatus	Lower resolution

wavelength of the fluorescent X-rays is peculiar to the elements in the samples. Therefore, the elements in the samples can be analyzed qualitatively and quantitatively from the wavelength and the intensity of the X-rays, respectively.

In order to individually analyze the fluorescent X-rays from each element, the spectral separation of the X-rays is required. There are two types of spectrographic methods for XRF. They are wavelength dispersive X-ray spectrometry (WDS, WDX) and energy dispersive X-ray spectrometry (EDS, EDX). The characteristics of WDS and EDS are shown in Table 1. Please refer to the following reference for more detailed explanations of XRF [1].

2.2 *X-Ray Diffraction Analysis (XRD)*

Figure 3 shows schematic illustrations for the principle of X-ray diffraction. When the X-rays with the wavelength of about 0.1 nm, which is approximately equal to the lattice spacing of the samples, are irradiated, the X-rays are scattered by the electron in the atoms. The scattered X-rays interfere with each other, and mutually

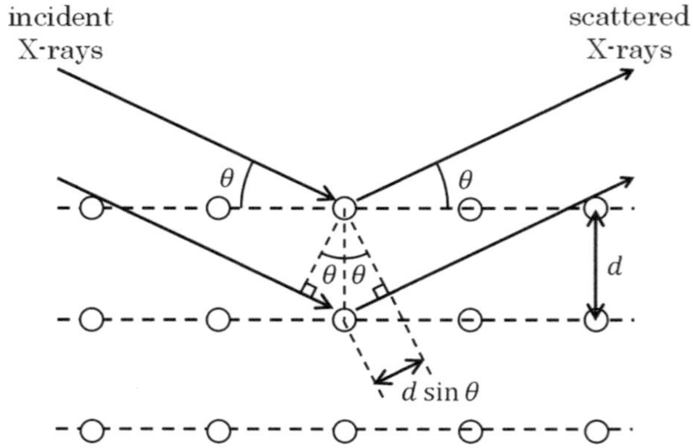

Fig. 3 Schematic illustrations for the principle of X-ray diffraction

intensify at certain angles. It is called “X-ray diffraction”. The angles where the scattered X-rays intensify each other are derived from the following equation (Bragg’s law);

$$2d \sin \theta = n\lambda$$

d: lattice spacing of the samples;
θ: irradiating angle of the X-rays;
n: a positive integer;
λ: wavelength of the X-rays.

Figure 4 schematically shows the diagrams for the principle of a typical source of X-rays. Thermo-electrons, generated from a heated filament, are accelerated by high voltage and then collide at the anode, called “target”, such as Cu, Fe, Co, Cr, Mo, and so on, to generate the X-rays. The generated X-rays consist of characteristic X-rays (Kα rays, Kβ rays, etc.) which are peculiar to the target materials. They are continuous X-rays having continuous energy dispersion. For XRD analyses, Kα rays from the target are usually used for the X-ray source. The wavelength of Kα rays of various targets is shown in Table 2. The CuKα rays are typically used for the XRD analyses.

There are two methods of XRD, out-of-plane XRD estimating the lattice spacing parallel to the material’s surface, and in-plane XRD estimating the lattice spacing perpendicular to the material’s surface. The incident angle of the X-rays is used for the in-plane XRD and is usually less than 1° so that the infiltrated depth of the incident X-rays to the material is about a few (10) nm. The diffracted XPD pattern can be detected with high accuracy without the effects of bulk structure. Thus, in-plane XRD is one of the most powerful tools for use on sample surfaces and thin films. Please refer to the following reference for more detailed explanations of XRD [1].

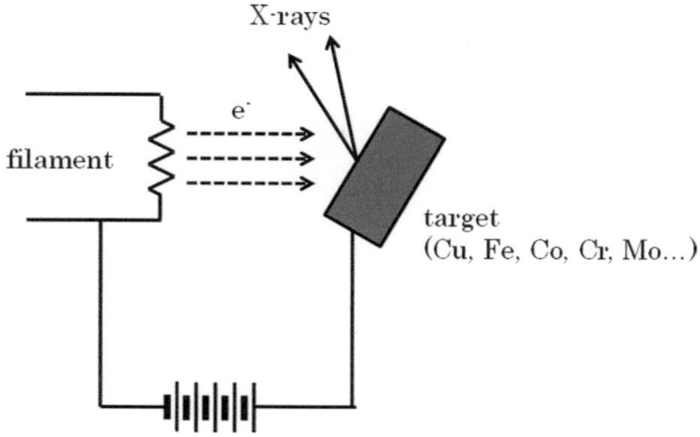

Fig. 4 Schematic illustrations for the principle of a typical source of X-rays

Table 2 Wavelength of Kα rays of various targets

Target	Wavelength of Kα rays [nm]
Cu	0.1542
Fe	0.1937
Co	0.1790
Cr	0.2291
Mo	0.0711

2.3 X-Ray Photoelectron Spectroscopy (XPS)

Schematic illustrations for the principle of X-ray photoelectron spectroscopy are shown in Fig. 5. As described in Sect. 2.1, some of the electrons receive the energies of the irradiated X-rays to jump out from the atom as photoelectrons when the X-rays having higher energies than the binding energy of the electrons (E_B) are irradiated. These photoelectrons have the following energies (E_P);

$$E_P = h\nu - E_B = \frac{hc}{\lambda} - E_B$$

h: Planck's constant, 6.626×10^{-34}/J sec;
ν: frequency of irradiated X-rays;
c: speed of light, 2.998×10^8/m sec^{-1};
λ: wavelength of irradiated X-rays.

The binding energies of the electrons (E_B) are peculiar to both the elements in the samples and their chemical bonding states so that the composition and chemical bonding states of the elements in the samples can be estimated when the energies of the photoelectrons (E_P) are measured. This analytical method is named X-ray photoelectron spectroscopy (XPS), or electron spectroscopy for chemical analysis (ESCA). Please refer to the following reference for more detailed explanations of XPS [2].

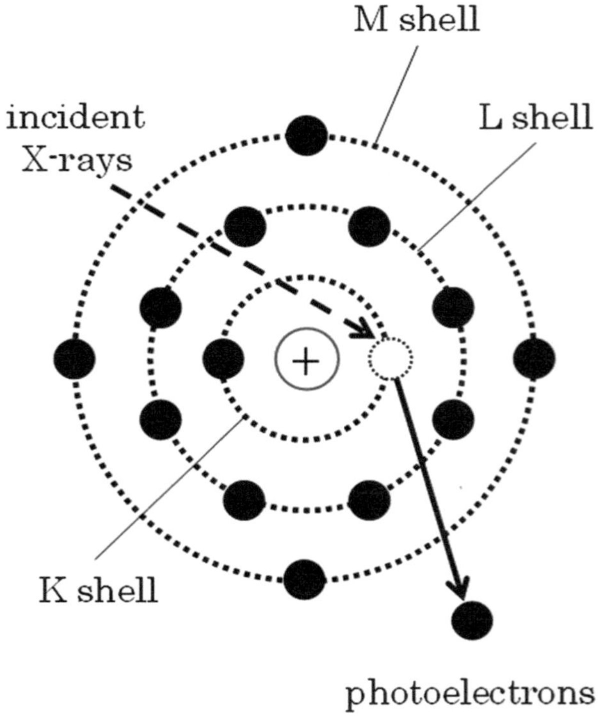

Fig. 5 Schematic illustrations for the principle of X-ray photoelectron spectroscopy

3 Electron Beam Analyses

Electron beams can be used for the probe which can analyze the micro portions of the sample's surface, ranging from nanometer to micrometer scales, because the electron beams can be easily narrowed by using the magnetic field. The infiltrated depth of the electron beams is usually smaller than that of the X-rays, suggesting that the electron beam analyses should be one of the best methods for surface analyses.

Various interactions between electron beams and the samples are schematically illustrated in Fig. 6. In this section, operating principles and characteristics of scanning electron microscopy (SEM), transmission electron microscopy (TEM), and Auger electron spectroscopy (AES), which are often used for the analyses of corrosion and surface finishing, are described.

3.1 Scanning Electron Microscopy (SEM)

Scanning electron microscopy (SEM) is one of the microscopes which can image the two-dimensional spatial distributions of secondary electrons, reflected

Fig. 6 Various interactions between electron beams and the samples

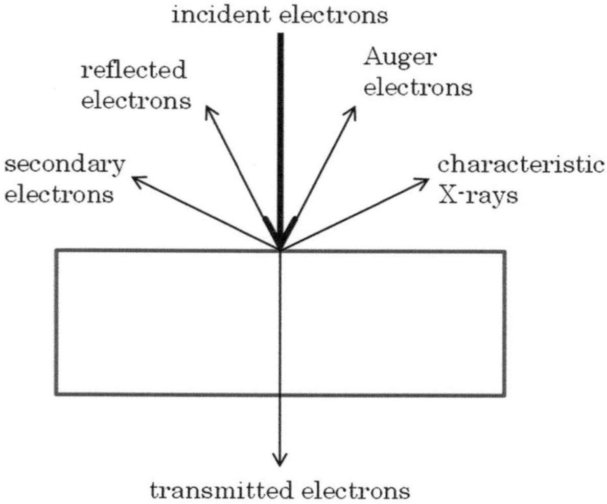

electrons, X-rays, transmitted electrons, and so on, generated from the samples, when the focused electron beams are scanned on sample surfaces. Secondary electrons are the vacancy electrons jumping out from the atoms in the samples when the electron beams are irradiated to the samples. Because of much lower energies of the secondary ions, the electrons generated only at the top surface can jump out from the samples so that spatial distribution images of the secondary ions are usually used to analyze the surface morphology of the samples.

The apparatus which generated and accelerated the electron beams is named an electron gun. Thermal electron emission type electron guns are used for general SEM, but a SEM equipped field emission type electron gun is called FE-SEM (field-emission scanning electron microscopy), which can obtain the higher resolution images with lower applied voltage.

When the surfaces of the samples are crystalline, electron backscatter diffraction (EBSD) patterns, called Kikuchi lines, generated from reflected electrons, are observed. EBSD patterns provide knowledge concerning crystal structures and orientations. Thus, the combination of SEM and EBSD is one of the powerful tools, which can tell us the microstructures of the sample surfaces and the orientations of the grains on the sample surfaces.

As described in Sect. 2.1, wavelength and intensity of the characteristic X-rays from a sample give knowledge concerning the species and amount of the elements, respectively. Therefore, a combination of SEM and EDX or WDX is also one of the powerful tools, which can tell us two-dimensional spatial distributions of elements on the sample surfaces.

SEM combined with focused ion beams (FIB), usually gallium ion beams, are used in order to obtain the cross sectional images near the sample surface. The sample surfaces can be etched in a nanoscale by spattering the surface atoms with

FIB, with a spot size from a few to a few hundred nanometers. Please refer to the following references for more detailed explanations of SEM [3, 4].

3.2 *Transmission Electron Microscopy (TEM)*

SEM images can be obtained by the signal from the sample surfaces when the focused electron beams are scanned, whereas transmission electron microscopy (TEM) images can be taken by transmitted electrons through a sample. Thus, thicknesses of these samples should be much thinner and the energies (the accelerating voltages) of the incident electrons should be much higher. However higher resolution images, for example, atomic resolution images, can be obtained with the TEM with higher accelerating voltages. Similar to SEM, the combination of TEM with EBSD, EDS, or WDS is one of the most powerful tools for surface analyses.

The thickness of the sample needs to be thin enough to transmit the electron beams through it. Therefore, the preparation of thin enough samples is one of the most important technologies of the surface analyses for corrosion and surface finishing. FIB, described in the Sect. 3.1, as well as chemical etching, electrochemical etching, ion milling, and using microtome, is one of the candidate tools to make the sample thin enough. Please refer to the following references for more detailed explanations of TEM [5, 6].

3.3 *Auger Electron Spectroscopy (AES)*

Figure 7 provides schematic illustrations for the principle of Auger electrons. Auger electrons are generated as follows;

1. Electron beams are irradiated for inner shell electrons to jump out from the atoms of the samples.
2. Photoelectrons, the inner electrons jumping out from the atoms, are generated, resulting in making electron holes.
3. Outer shell electrons move to the electron holes.
4. Other outer shell electrons accept the energy caused by the movement of the electrons from an outer shell to an inner shell.
5. Auger electrons (the outer shell electrons that accepted the energy caused by the movement of electrons from an outer shell to an inner shell) jump out from the atoms.

Similar to XPS, the energies of Auger electrons are peculiar to both the elements in the samples and their chemical bonding states so that the composition and chemical bonding states of the elements in the samples can be estimated when the energies of Auger electrons are measured. Please refer to the following reference for more detailed explanations of XPS [7].

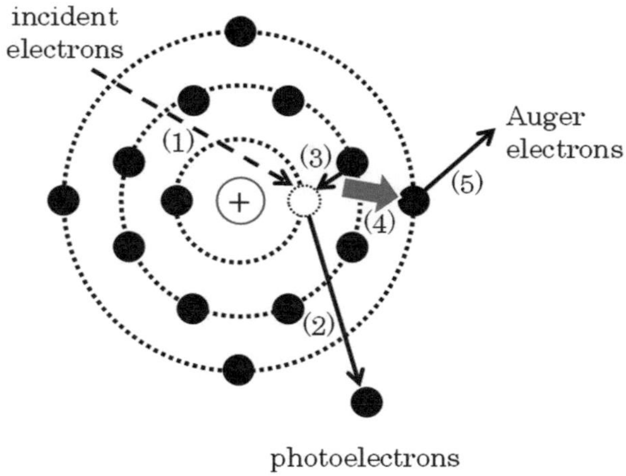

Fig. 7 Schematic illustrations for the principle of Auger electrons

4 The Other Techniques

When the ions (primary ions) having high energies are irradiated on the sample surfaces in a vacuum, some of the atoms of the samples are ionized (secondary ions). Secondary ion mass spectrometry (SIMS) [8] is the mass spectrometry of secondary ions. SIMS is not a nondestructive analysis, but a method for depth-directional distribution of a micro amount of an element.

When the hydrogen or helium ions with high energies are irradiated on the sample surface in a vacuum, some of the incident ions are subjected to elastic scattering (Rutherford scattering). The energies of spattered ions are dependent on the mass and depth of the atoms in the samples. This results in an analysis of a sample's chemical composition based on a depth profile from the energies and intensities of spattered ions. This type of analysis is called Rutherford backscattering spectrometry (RBS) [9].

Surface-Enhanced Raman spectroscopy (SERS) [10] is also one of the analytical tools for a sample's surface. When laser beams with frequency ν_0 are irradiated to the sample, some of the beams are scattered. Almost all of the frequency of the scattered beams is the same as that of incident beam (ν_0), but the frequency of some scattered beams ($\nu_0 \pm \nu_1$) is slightly different from that of the incident beam. This is called Raman scattering spectroscopy (RSS). The frequency of lattice vibration of the samples is ν_1 so that RSS gives us knowledge concerning molecular structure, crystal structure and residual stress. The combination of RSS with an optical microscope as well as an atomic force microscope (AFM) is also effective for spatial distribution analysis.

Fourier-transform infrared spectroscopy (FTIR) [11] is also one of the techniques for analyzing the surface of a sample by using infrared beams. Transmitted or refracted beams are detected when the infrared beams are irradiated to a sample's surface. FTIR is a method of measuring an infrared absorption spectrum. It can be used to analyze the molecular structure and functional group(s) of a sample.

References

1. Brittain HG (2006) X-ray diffraction and x-ray fluorescence. Compr Anal Chem 47:177–226
2. Tougaard S (2005) Surface analysis | X-ray photoelectron spectroscopy. In: Encyclopedia of analytical science, 2nd edn. Elsevier, Oxford, pp 446–456
3. Wells OC (2001) Scanning electron microscopy. In: Encyclopedia of materials: science and technology. Elsevier, Amsterdam, pp 8265–8269
4. Williams DB (2001) Electron microscopy: analytical. In: Encyclopedia of materials: science and technology. Elsevier, Amsterdam, pp 2569–2576
5. Williams DB (2001) Electron microscopy: transmission. In: Encyclopedia of materials: science and technology. Elsevier, Amsterdam, pp 2577–2584
6. Vecchio KS (2001) Electron microscopy: transmission. In: Encyclopedia of materials: science and technology. Elsevier, Amsterdam, pp 304–313
7. Unger WES (2005) Surface analysis | Auger electron spectroscopy. In: Encyclopedia of analytical science, 2nd edn. Elsevier, Oxford, pp 466–474
8. Williams P (1985) Secondary ion mass spectrometry. Annu Rev Mater Sci 15:517–548
9. Baumann SM (1992) RBS: Rutherford backscattering spectrometry. In: Encyclopedia of materials characterization, surfaces, interfaces, thin films. Elsevier, Grêenwich, pp 476–487
10. Littleford RE, Graham D, Smith WE, Khan I (2005) Raman spectroscopy | surface-enhanced. In: Encyclopedia of analytical science, 2nd edn. Elsevier, Oxford, pp 110–118
11. Parker SF (2005) Infrared spectroscopy | industrial applications. In: Encyclopedia of analytical science, 2nd edn. Elsevier, Oxford, pp 430–439

Corrosion and Surface Finishing

Dana M. Barry and Hideyuki Kanematsu

Abstract Surface finishing is very important to most industries for a variety of reasons. The finishing process could be used for reflectivity, improving hardness, or for decorating a finished product to attract potential buyers. However, from an engineering point of view, its main purpose is to prevent corrosion. Corrosion is actually a process whereby manufactured metals return to their natural oxidation states. Corrosion is of economic importance because it limits the life time of metal structures (example: bridges) and can result in accidents or incidents of pollution. This chapter describes corrosion and discusses how various surface treatments are used to prevent it. The topics covered include electroplating, electroless plating, spray coating, galvanization, painting, anodizing, physical vapor deposition (PVD), and chemical vapor deposition (CVD).

1 Introduction

Treatments to the surfaces of metals and other materials are very important in most industries. They are used for purposes such as decoration, reflectivity, improving hardness, and preventing corrosion. From an engineering point of view, the primary purpose of surface finishing is to prevent corrosion. Corrosion involves electrochemical reactions of a metal with an oxidizing agent in the environment. It is of economic importance because it limits the life time of metal structures, and can result in accidents or incidents of pollution. This chapter describes corrosion and discusses various surface treatments to prevent it. Plating and painting are identified

D.M. Barry (✉)
Clarkson's Departments of Chemical & Biomolecular Engineering and Electrical & Computer Engineering, Clarkson University, Potsdam, NY, USA

Center for Advanced Materials Processing (CAMP), Clarkson University, Potsdam, NY, USA
e-mail: dmbarry@clarkson.edu

H. Kanematsu
National Institute of Technology, Suzuka College, Suzuka, Japan
e-mail: Kanemats@mse.suzuka-ct.ac.jp

H. Kanematsu, D.M. Barry (eds.), *Corrosion Control and Surface Finishing*,
DOI 10.1007/978-4-431-55957-3_6

as common coating methods to prevent corrosion. They provide a protective film between the substrate and its surrounding environment. Galvanization is another technique that is mentioned. It is the process of applying a protective zinc coating to steel or iron to prevent rusting. Individuals need to select an appropriate coating and coating process for specific metals and their applications. In order to properly do this, various factors must be considered. These include the composition and properties of the substrate and the film that is to be applied to it. Also the coating process and the corrosive environment (where the metal item will be located and used) must be taken into account.

2 Corrosion

Corrosion can be considered the deterioration of materials (especially metals) as a result of their chemical reactions with the surrounding environment. Corrosion destroys the useful properties of a material such as the strength and structure of iron in bridges. It can cause environmental pollution resulting from a leak in an oil pipeline or a large loss of transported water from corrosion leaks in buried pipes for water distribution. Corrosion is a process whereby manufactured metals return to their natural oxidation states. A reduction-oxidation reaction takes place in which the metal is oxidized by its surroundings (example: oxygen in air). A common form of corrosion is rust, a reddish-brown compound, known as iron oxide (Fe_2O_3). See Fig. 1 [1]. This reaction is possible when iron (Fe) is in the presence of water and oxygen.

Fig. 1 Iron rust sample

Corrosion can affect an entire surface of a metal or just local spots. Uniform corrosion of the complete surface usually only happens in acidic conditions [2]. This generally results in overall thinning and causes no major damage. On the other hand, a very detrimental form of corrosion is pitting. This type is found at a single location on the surface and creates a pit or cavity which is difficult to prevent and often hard to detect. It can result in structural failure (example: a cracked pipe).

Corrosion occurs on exposed surfaces. Therefore, these surfaces should be coated to provide a protective barrier between the metal and its surroundings. Surface finishing is applied to protect metals from corrosion. Some industrial treatments are presented and described in this chapter. In order for them to be successful, one needs to recognize the type of metal and corrosive environment involved and to determine the appropriate coating process for that situation. In all cases, surface preparation is an essential first step before any coating is applied. This is because the performance of a coating is greatly influenced by its ability to properly adhere to a material's surface. Consider the corrosion of steel, an alloy containing mostly iron and about 1 % carbon. Steel stored inside a dry, clean, heated building has a low risk of corrosion. However, a steel structure exposed to a harsh environment of moisture, oxygen, air pollution, etc. needs to be protected so that the iron isn't oxidized to rust.

3 Electroplating to Control Corrosion

Electroplating can be defined as the application of a metal coating to a metallic or other conducting surface using an electrochemical process. The item to be plated serves as the cathode (negative electrode) and the anode is usually the metal to be plated on the item. Both components are placed in an aqueous (aq.) solution, called an electrolyte, containing dissolved metal salts and other ions to allow for a flow of electricity. A power source provides a direct current to the anode and oxidizes its metal atoms so they are able to dissolve in the solution. These dissolved metal ions are then reduced at the cathode and deposited there as a metal coating.

Electroplating is widely used in various industries to coat metal objects with a thin layer of another metal. Since chromium is very corrosion resistant, it is electroplated on to car parts, wheel rims, and other items. Nickel plating, tin plating, and their alloys are used for corrosion protection on nuts, bolts, brackets, and other metal components [3]. Gold electroplating is also very good for corrosion and tarnish protection, but it is rather expensive. In addition, a sacrificial coating can be electroplated to a metal surface. For example, zinc serves as a sacrificial coating for iron and is used up in the reaction. This protects the iron from rusting (corrosion).

4 Electroless Plating to Control Corrosion

Electroless plating uses a redox reaction without an electric current to deposit metal on an object or workpiece. The electroless plating solution generally includes the following: a source of metal, a reducer, a complexing agent (which holds the metal in solution by forming coordinate bonds with the metal atom or ion), and buffers along with other chemicals to stabilize the bath [4]. One form of this process is electroless nickel plating. It can be used to produce a nickel phosphorus alloy coating, which helps prevent corrosion. For this process a reducing agent such as hydrated sodium hypophosphite ($NaPO_2H_2$. H_2O) reacts with the metal ions to deposit metal. The phosphorus content in these coatings can vary from about 2 to 14 %. Low phosphorous electroless nickel provides excellent resistance to alkaline, corrosive environments [5]. On the other hand, high phosphorous electroless nickel finishing is used in industries (like oil drilling and coal mining) requiring resistance to strongly acidic corrosive environments and to those industries needing a pore free finish [6].

5 Spray Coating to Control Corrosion

Spray coating is a process in which materials are sprayed onto a surface. It is widely known as the use of spray cans (or spray guns) by artists and as an industrial process for car body painting. For this method, the functional fluid or ink is atomized at the nozzle of the spray head, which makes a continuous flow of droplets. This system uses a stream of pressurized air or gas (such as argon or nitrogen) to break up the liquid into droplets at the nozzle [7]. These droplets of ink (paint) are then carried onto the surface of a particular substrate. Spray coatings protect metals from corrosion by providing them with a barrier from the atmosphere and surrounding environment. Drawbacks to this process are that it is a line-of-sight technology (which makes it impossible to coat small cavities, etc.) and the coatings have pores/cracks that need to be sealed for certain applications [7].

Another form of this coating process is called Thermal Spraying, where melted (or heated) materials are sprayed onto a surface. This technique consists of an electrical (examples: plasma and arc) or chemical (ex. combustion flame) heat source and a coating material in the form of powder or wire which is melted into tiny droplets and sprayed at high velocity onto surfaces [8, 9]. Thermal spraying can produce thick coatings over a large area at a high deposition rate as compared to other coating processes. This technique also allows for versatile substrates and coating materials. Substrates include metals like aluminum, steel, stainless steel, copper, and bronze, as well as some plastics. The coating materials can be metals, alloys, ceramics, plastics, and composites [10].

6 Galvanization to Control Corrosion

Galvanization is a process for coating iron and steel with zinc to prevent them from rusting. The most common form of galvanizing is Hot-dip Galvanization [11]. For this method, metals such as iron and steel are coated with a layer of zinc by being submerged in a bath of molten zinc. The coating forms a barrier between the substrate and corrosive substances in the atmosphere. Its effectiveness depends on factors such as the coating thickness and coating corrosion rate which is influenced by the composition of the atmosphere as well as the type of coating.

Zinc also serves as a sacrificial anode (where there is a loss of electrons), so that if the coating is scratched the remaining zinc protects the exposed steel. Galvanized steel is widely used for applications requiring corrosion resistance without the cost of stainless steel. Galvanized steel is used in roofing, handrails, consumer appliances, automotive body parts, metal pails, and in most heating and cooling duct systems in buildings.

7 Painting to Control Corrosion

Painting can be defined as the coating of a surface with paint (which includes pigment or color). The paint coating can be applied by using brushes, rollers, and spray methods (which are described in chapter "Industrial Surface Treatments"). Industrial paint is used to protect metal, wood, and a variety of other materials. It provides a barrier between the substrate's surface and its surroundings.

Paint and galvanized steel (steel coated with zinc) can be used together to provide superior corrosion control than either method used alone. This technique, known as a "duplex system," consists of painting steel that has been hot-dip galvanized after fabrication [12]. The paint serves as a barrier protecting the galvanized steel from the atmosphere. It slows down the rate at which zinc is consumed and therefore extends the life of the galvanized steel.

In order for this method to be successful, the metal's surface must be properly cleaned and prepared, before an appropriate paint is applied. To start, the galvanized surface is cleaned to remove any dirt, grease, or oils, as well as other surface contaminants and zinc corrosion products. Then the surface is profiled to allow for a good adhesion of the paint. Therefore, it must be free of protrusions and slightly roughened. In any case, remove the least amount of zinc as possible during the cleaning and profiling steps. Otherwise, the corrosion protection (of the item) will be reduced. Finally, an appropriate paint should be selected and applied. Individuals are encouraged to contact paint manufacturers to obtain specific information regarding the suitability of paints for use on galvanized steel.

8 Anodizing to Control Corrosion

Anodizing is an electrolytic passivation process that increases the thickness of natural oxide layers on the surface of metals [13]. It basically forms an anodic oxide finish on a metal's surface to increase corrosion resistance. For the anodizing process, the metal to be treated serves as the anode (positive electrode, where electrons are lost) of an electrical circuit. Anodized films are most often applied to protect aluminum alloys. An aluminum alloy is seen on the front bicycle wheel in Fig. 2 [14]. For these alloys, aluminum is the predominant metal. It typically forms an alloy with the following elements: copper, magnesium, manganese, silicon, tin, and zinc [15]. Two main classifications for these alloys are casting alloys and wrought alloys, both of which can be either heat treatable or non-heat treatable.

Almost 85 % of aluminum is used for wrought products such as rolled plates, foils, and extrusions [16]. Cast aluminum alloys provide cost-effective products but usually have lower tensile strengths than those of wrought alloys. Aluminum alloys are used in engineering, the aerospace industry, and for components where corrosion resistance or light weight is needed. If left unprotected their surfaces are anodized, which means that they form a white protective layer of the corrosion product, aluminum oxide.

Fig. 2 Aluminum alloy displayed in the front bicycle wheel

9 Physical Vapor Deposition (PVD) to Control Corrosion

Physical Vapor Deposition (PVD) includes various methods for depositing thin films onto a substrate by the condensation of a desired vapor. For this process the coating material changes from the solid phase to the vapor phase and back to the solid phase (as it builds a film on the substrate). The PVD technique involves physical processes like the use of high-temperature vacuum evaporation of a coating material that is then condensed onto a substrate. It is carried out in a vacuum at temperatures between 150 and 500° C and provides coatings with an average thickness of 2–5 μm [17]. This method can be used to deposit thin films onto items for corrosion control. Some of these films are titanium nitride (TiN), titanium carbon nitride (TiCN), zirconium nitride (ZrN), and chromium carbide (CrC). They are hard, have a low coefficient of friction, and are chemically resistant, which makes them good chemical and thermal barriers when applied to surfaces [18]. ZrN provides a decorative, corrosion resistant coating for brass. TiN coatings extend the life of cutting tools and surgical instruments used in the medical industry.

PVD is an alternative to the processes of electroplating and to some painting applications because it generates less hazardous waste and uses smaller amounts of hazardous materials since no plating baths, etc. are involved [19]. Also in some cases, PVD coatings are harder and more corrosion resistant than those of the electroplating process. Most of these coatings do not require protective topcoats because they are durable and have good impact strength. The PVD method allows many different substrate materials such as metals, alloys, ceramics, glass, polymers, etc. to be coated. Also the coating properties such as hardness and adhesion can be accurately controlled. Therefore, these coatings have a wide range of applications. They are used in the aerospace, medical, automotive, optics, firearm industries, just to name a few.

10 Chemical Vapor Deposition (CVD) to Control Corrosion

Chemical Vapor Deposition (CVD) is a chemical process to produce solid thin films on surfaces. Typically one or more precursor gases flow into a chamber containing heated objects that need to be coated. Chemical reactions take place on or near the hot surfaces. See Fig. 3 [20]. As a result, thin films are deposited on these surfaces. The coatings serve many purposes but can also provide protective barriers from the surrounding atmosphere.

Several benefits of the CVD process include the following. A variety of materials can be deposited with high purity. These films generally display uniform thickness on the insides and outsides of elaborately shaped objects. Also the CVD process has a relatively high deposition rate.

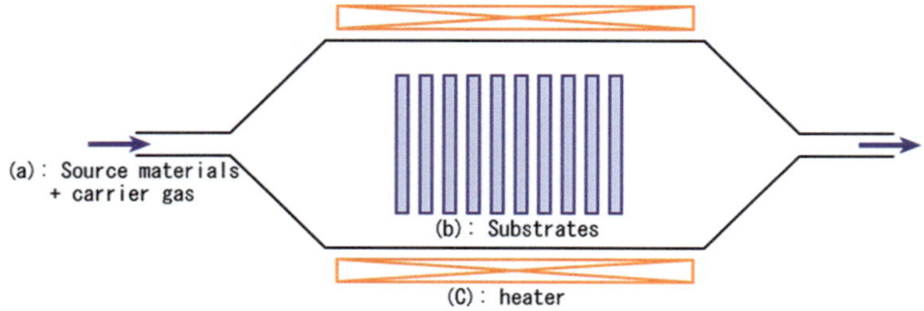

Fig. 3 A schematic diagram for the chemical vapor deposition process

In contrast, the physical vapor deposition method (PVD) generally requires a line of sight between the source and the surface to be coated. Also the PVD process often requires a higher vacuum than the CVD process.

Disadvantages exist for the CVD technique [21, 22]. The precursors can be toxic or corrosive. They can be expensive too. Also as a result of using this method, most of the films are deposited at high temperatures. This restricts the type of substrates that can be coated. Substrates with different thermal expansion coefficients can cause the deposited films to have mechanical instabilities. Therefore, this coating method has some limitations in regards to corrosion control.

11 Conclusion

This chapter described corrosion and discussed various surface finishing treatments to prevent it. It mentioned that corrosion involves electrochemical reactions of a metal with an oxidizing agent in the environment. Corrosion is actually a process whereby manufactured metals return to their natural oxidation states. We were also reminded of the economic importance of corrosion because it limits the life time of metal structures (example: bridges) and can result in accidents or incidents of pollution.

Industrial treatments for corrosion control were discussed too. Plating and painting were identified as common coating methods to prevent corrosion. They provide a protective film between the substrate and its surrounding environment. Galvanization is another technique that was described. Galvanization is the process of applying a protective zinc coating to steel or iron to prevent rusting. Individuals need to select an appropriate coating and coating process for specific metals and their applications. In order to properly do this, various factors must be considered. These include the composition and properties of the substrate and the film that is to be applied to it. Also the coating process and the corrosive environment (where the metal item will be located and used) must be taken into account.

The topics that were covered in this chapter include electroplating, electroless plating, spray coating, galvanization, painting, anodizing, physical vapor deposition (PVD), and chemical vapor deposition (CVD).

References

1. Benjah-bmm27-own work: sample of iron (III) oxide, public domain, Wikimedia Commons. 16 May 2007. http://en.wikipedia.org/wiki/Iron(III)_oxide
2. Shaw BA, Kelly, RG (2006) What is corrosion?. Electric Soc J Interface 24–26
3. http://www.sharrettsplating.com/electroplating-benefits.html
4. http://antoine.frostburg.edu/chem/senese/101/redox/faq/electroless-plating.shtml
5. http://en.wikipedia.org/wiki/electroless_nickel_plating
6. http://www.coastlinemetalfinishing.com/plating/electroless-nickel
7. http://plasticphotovoltaics.org/lc/lc-fabrication/lc-coating/lc-spray.html
8. http://en.wikipedia.org/wiki/Thermal_spraying
9. Fauchais P, Vardelle A (2012) Thermal sprayed coatings used against corrosion and corrosive wear. In: Jazi HS (ed) Advanced plasma spray applications. Published under CC By 3.0 license. doi:10.5772/34448. www.intechopen.com
10. http://www.precisioncoatings.com/what-is-thermal-spray.html
11. http://en.wikipedia.org/wiki/Hot-dip_galvanization
12. Dunham K (2002) Preparing hot-dip galvanized steel surfaces for painting or powder coating: a primer. Met Finish 100(9):20. doi:10.1016/S0026-0576(02)807-46-2
13. http://www.anodizing.org/Anodizing/definitions.html
14. Conollyb-own work: bootie bicycle front wheel balloon tire bootie bike, CC – By – SA 3.0. 13 Aug 2013. http://en.wikipedia.org/wiki/Aluminium_alloy
15. http://en.wikipedia.org/wiki/anodizing
16. http://en.wikipedia.org/wiki/Aluminium_alloy
17. http://www.corrosionpedia.com/definition/875/physical-vapor-deposition-pvd
18. Podob M (1995) Physical vapor deposition and wear resistant hard-coatings. Presentation at Finishing'95 conference and exposition, Cincinnati, 97–108. http://infohouse.p2ric.org/ref/28/27102.pdf
19. http://www.engineershandbook.com/MfgMethods/PVD.htm
20. S-Kei – own work: thermal CVD, public domain, Wikimedia Commons 2008 Sept 20. http://en.wikipedia.org/wiki/chemical_vapor_deposition
21. Sathiyanarayanan S, Rajagopal G, Palaniswamy N, Raghavan M (2005) Corrosion protection by chemical vapor deposition: a review. Corros Rev 23(4–6):355–370
22. Creighton JR, Ho P (2001) Introduction to chemical vapor deposition (CVD), Chapter 1. Chemical vapor deposition (#06682G). Published by ASM International, pp 1–10. www.asminternational.org

Part II
Environmental Regulations for Metallic Materials

Background of Environmental Regulations

Hideyuki Kanematsu and Dana M. Barry

Abstract Now in the twenty-first century, we are facing serious environmental problems on the global scale. Global warming by carbon dioxide, an increase of ultraviolet light due to the breakdown of the ozone layer, etc. can be mentioned as examples. The concept of environmental protection has been strengthened and has prevailed through many unfortunate episodes and conferences about environmental protection. Several include the United Nations Framework Convention on Climate Change in Rio de Janeiro, Brazil and Kyoto, Japan. In this chapter, we summarize the historical development of the concept and also the current situation.

1 Introduction

Nowadays, harmony with our natural environment is very important. In the eighteenth century, the industrial revolution began. Since then, we have developed a society based on bulk productions, inevitably followed by a mass consumer society. In this society natural resources have been exhausted and environmental pollution became remarkable. We are beginning to realize that our irreplaceable Earth is now facing difficulty in surviving. Figure 1 shows the schematic illustration for the current situation.

Table 1 shows the reserve-production ratio of some representative metal resources at the beginning of the twenty-first century [1–3]. The reserve-production ratio could be defined as the value of confirmed reserve (CR) per annual production (AP, output).

H. Kanematsu (✉)
National Institute of Technology, Suzuka College, Suzuka, Japan
e-mail: kanemats@mse.suzuka-ct.ac.jp

D.M. Barry
Clarkson's Departments of Chemical & Biomolecular Engineering and Electrical & Computer Engineering, Clarkson University, Potsdam, NY, USA

Center for Advanced Materials Processing (CAMP), Clarkson University, Potsdam, NY, USA
e-mail: dmbarry@clarkson.edu

H. Kanematsu, D.M. Barry (eds.), *Corrosion Control and Surface Finishing*,
DOI 10.1007/978-4-431-55957-3_7

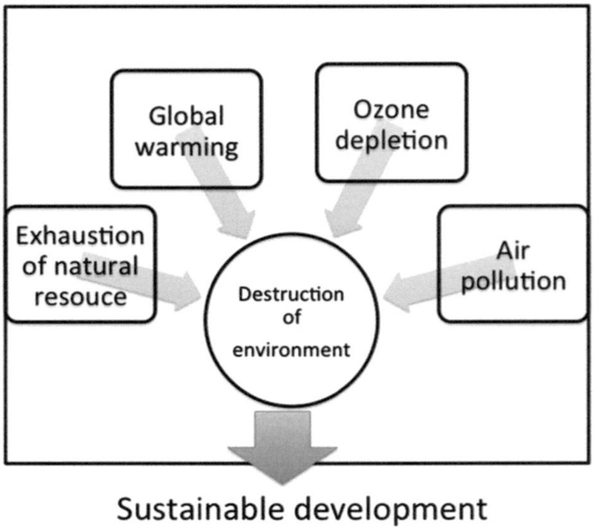

Fig. 1 Environmental problems nowadays

Table 1 Confirmed reserves for various metals (collected from various sources)

Substance	Fe	Pb	Sn	Ag	Au	Ti
CR (year)	117	20	18	19	20	128
Substance	Mn	Cr	Ni	Co	Nb	W
CR (year)	56	15	50	106	47	48
Substance	Mo	Ta	In	Bi	Cd	Zr
CR (year)	44	95	18	85	39	41

$$\text{The reserve-production ratio} = \text{CR}/\text{AP} \tag{1}$$

Confirmed reserve can change, depending on technologies and on a yearly basis. Therefore, the value cannot be absolute. However, the table shows clearly that our resources would never be inexhaustible.

On the other hand, Fig. 2 shows the reserve production ration for some fossil fuels [4].

On the other hand, environmental pollution has been serious, where industrial activities were active. In the mid-eighteenth century, the London areas were suffering from serious pollution. Similar situations were also found in Berlin, Germany and in some cities in the USA, like Chicago, etc. Recently Beijing in China has developed the same situation. The pollution was not restricted to local areas.

It caused serious problems such as ozone depletion and global warming too.

The ozone layer exists 20 km or 30 km above our Earth's surface and protects us from harmful ultraviolet light. If the ozone layer was severely damaged, then human beings would suffer from cataracts, skin cancer, compromised immune

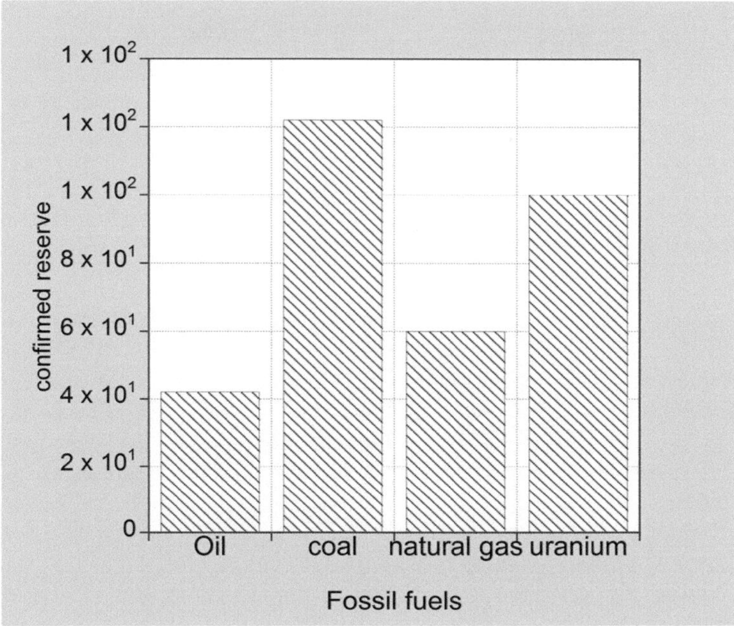

Fig. 2 Fossil fuels and their confirmed reserves

systems, etc. The breakdown of this layer is now occurring because of the use and release of chlorofluorocarbons (CFCs) on the Earth's surface.

Global warming is considered the phenomenon brought about by excessive production of so called greenhouse gases (GHG). The gases produced from the Earth's surface surround our Earth. As a result, it is difficult to release heat from the Earth's surface so the temperature of Earth's surface could increase. The greenhouse gases include carbon dioxide, methane, etc. Humans have produced too many of them.

All of these negative and pessimistic situations and values show that our natural resources must have limits. Also they show that our development would never be one directional for a better world in the future. However, we have to continue making our globe better for our descendants.

The concept of "sustainability" emerged. This concept was originally born from the World Commission on Environment and Development (WCED) established by the United Nations in 1984. Since the chairperson of the commission was Ms. Brundtland from Norway (the former prime minister of the country), the commission was called Bruntland Commission after her name. A report was summarized after several meetings and produced the concept of sustainable development. It can be defined as the development to satisfy the needs for current generations without violating any needs for future generations. We can say this committee became a great landmark to drastically change our idea about development. Also the concept for environmental protection has gradually been strengthened by it.

Now all of our technologies should be very sound from the viewpoint of sustainable development. Environmental friendliness must be added as an important innovation factor.

2 Global Warming

As for global warming, we should not forget the United Nations Framework Convention on Climate Change: UNFCCC or FCCC. This treaty fixed the international framework about global warming. As described above, greenhouse gases such as carbon dioxide, methane gas, etc. may violate the balance of nature. Therefore, this treaty was established for us so that we could clearly understand the potential causes for global warming and to maintain a safe level of global warming gases. The initial manuscript was made in New York in 1992. And it was approved among 155 nations in Rio de Janeiro in 1992. The conference became very famous as an Earth summit. However, it is called UNCED which stands for United Nations Conference on Environment and Development. The treaty was based on three principles.

#1: The signatory countries should have common, but different responsibilities, respectively.
#2: The consideration into special situations for each signatory country, particularly developing countries.
#3: Carrying out countermeasures rapidly and effectively.

Based on these three principles, the treaty obligated signatory countries to decrease the amount of emission for greenhouse gases down to the level at the beginning of 1990 within the 1990s.

The third UNFCCC was held in Kyoto in 1997. The characteristics in the Kyoto conference, Japan were the so-called Kyoto Protocol (Kyoto Protocol to the United Nationals Framework Convention on Climate Change). In this protocol, the reduction rates of some greenhouse gases such as carbon dioxide, methane, nitrous oxide, fluorocarbons and sulfur hexafluoride were set. It required about 5 % reduction in the term from 2008 to 2012, as compared with that in 1990. And the aim of reduction was also set for advanced countries, respectively: 8 % for the EU, 7 % for the United States and 8 % for Japan. It required the implementation of measures by all countries. The signatory countries total 83 and the contracting states are 192. However, the United States, the second largest country from the viewpoint of CO_2 production, refused ratification due to their internal circumstances. And at that time, the Russia Federation also rejected the acceptance. Therefore, the protocol did not become effective until 2004. Even though the United States did not join it, it became a great landmark for the collaborative effort to stop global warming in the world.

The example strengthened the tendency (for environmental problems) that emission values should be measured and that countermeasures for the reduction should be investigated based on the measured values.

3 EU Environmental Regulations and Their Effects on the World [5–8]

From the background, it was very natural that environmental regulations have been set in various countries. The United States, the EU, Japan, and many other countries set environmental regulations, depending on their situations and conditions. Particularly, the EU has led the global tendency so far. Their standards are going to become global ones. Therefore, the standards must inevitably contain somewhat political and geo-political meanings.

The characteristics of the EU standards are the regulations for products and matters involved in the production processes.

The law system of the EU is composed of three categories. The first one is the primary law (treaty) and corresponds to basic treaties that should be negotiated and decided among member countries. The other one is the secondary law set, based on the primary laws. They are classified into "Regulation", "Directive", "Decision", "Recommendation" and "Opinion". The law companies should follow the secondary law. From the viewpoint of environmental laws, we have to differentiate between Regulation and Directive. Regulations are the laws to consolidate many different laws of the EU member countries. A regulation is the law to control business and industrial activities directly and corporations must keep it as a top priority. REACH is one of the regulations, for example.

On the other hand, the "Directive" is the law for member countries to arrange and negotiate with their existing laws. WEEE and RoHS belong to the directive category. The directive is basically not applied to the member countries until they are able to adjust it to their existing laws. Therefore, the member countries have to rewrite the former laws by adjusting them to the directives. It means directives inevitably involve a "time lag" between the proposal and effect.

As representative environmental laws in the EU, four items can be mentioned: RoHS directive, WEEE directive, ELV directive and REACH regulation.

RoHS directive stands for Restriction of the use of certain Hazardous Substances in Electrical and Electronic Equipment. The purpose of the directive is to control the use of certain hazardous chemical substances such as mercury, cadmium, lead, hexavalent chromium, polybrominated biphenyl (PBB) and polybrominated diphenyl ether (PBDE) for electrical and electronic equipment. In the EU, no one is allowed to place electric and electronic equipment on the market that would include those hazardous substances exceeding a certain limit. The regulation values are mentioned in Table 2.

Table 2 Regulation values for RoHS

Hazardous substance	Limitations
Lead	<1,000 ppm
Mercury	<1,000 ppm
Cadmium	<100 ppm
Hexavalent chromium	<1,000 ppm
PBB	<1,000 ppm
PBDE	< 1,000 ppm

WEEE stands for Waste Electrical and Electronic Equipment. This directive requires the collection and recycling of electrical and electronic equipment in the following categories: major appliances, compact appliances, IT and communication devices, consumer durables, illumination items, electric and electronic tools, play-things, leisure and sporting instruments, medical instruments, monitors and control equipment and automatic dispensers. Companies that sell electric and electronic equipment in those categories within the EU have to make arrangements for their wastes and have product liabilities of manufacturers for collecting and recycling them. They may pay the fee to collect and recycle the wastes in many cases. RoHS and WEEE simultaneously became effective in 2003.

ELV stands for End of Life Vehicle and the directive to control the use of some hazardous substances mainly for automobiles. Lead, mercury, cadmium, and hexavalent chromium were the main targets for this directive. Particularly, the impact of controlling hexavalent chromium was very strong in plating industries, since a lot of hexavalent chromium was used for the surface finishing of automobiles. Conventionally, bolts and nuts used for automobiles were surface finished by chromate conversion coating.

Chromate conversion coating can be defined as the process where works are surface finished in baths containing hexavalent chromium [9]. Zinc plated steels, zinc, magnesium, aluminum, etc. in a chromate bath dissolve into the solution as ions and react with hexavalent chromium to form trivalent chromium. When the case of zinc plated steel is taken as an example, zinc on the surface dissolves as follows.

$$Zn + 2H^+ \rightarrow Zn^{2+} + H_2 \qquad (2)$$

Chromium acid is reduced by zinc and trivalent chromium ion forms.

$$2HCrO_4^- + 3Zn + 14H^+ \rightarrow 2Cr^{3+} + 3Zn^{2+} + 8H_2O \qquad (3)$$

And chromium hydroxide is formed and deposited on metal surfaces.

$$Cr^{3+} + 3OH^- \rightarrow Cr(OH)_3 \qquad (4)$$

In practice, the chromate film is not the simple chromium hydroxide, but gel-like composite hydroxide film composed of chromium acid, sulfate, chloride, fluoride, zinc ion, etc. Depending on the formation conditions (bath components, immersion times, temperatures, etc.) they show some different kinds of color. Particularly, black chromate has been used a lot for the automobile industries. This type of chromate film has very high corrosion resistance. However, it generally contains lots of hexavalent chromium in the film. The chromate conversion coating has two serious problems from the viewpoint of environmental friendliness. One of them is that the bath contains hexavalent chromium. A typical bath component is 20 g/L chromium acid, 15 g/L sulfuric acid, 10 g/L phosphate acid, 50 g/L acetic acid and 0.3 g/L silver particles. The bath temperature is at room temperature and the immersion time is about 2 min. By the immersion treatment, zinc plated steel shows black color and high corrosion resistance. However, the effluent caused environmental problems in the past. The hexavalent chromium, mixed into ground water, made habitants in the neighborhood of factories suffer from diseases. This broke the balance of the ecological system. The other problem is that hexavalent chromium is contained in the film to some extent. This has an excellent performance for the chromate film itself. When the film physically breaks to some extent, the contained hexavalent chromium dissolves into the solution around the film and is able to regenerate the chromate film by itself. This phenomenon is called self-repairing action of chromate film. However, it means the dissolution would simultaneously increase the content of hexavalent chromium in the ambient solution. See Fig. 3.

Based on the reasons just described, the directive of controlling hexavalent chromium has placed strong impact on the surface finishing industries. Industries have made great efforts to change the hexavalent chromium process to a trivalent chromium one. And finally, it will move to a completely non-chromium process for conversion coatings. This example suggests that environmental laws and regulations might be a great chance for the innovation, while it would enhance our sustainable development.

REACH appeared. It stands for Registration, Evaluation, Authorization and Restriction of Chemicals. The purpose and intention for the law is to protect

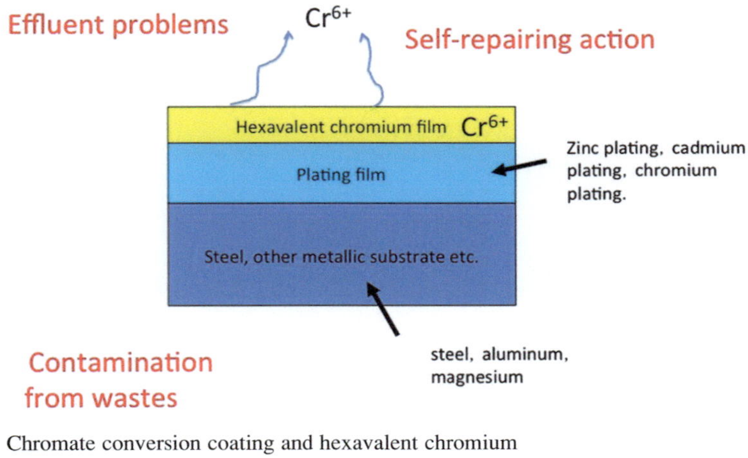

Fig. 3 Chromate conversion coating and hexavalent chromium

Fig. 4 Effects of EU directives and regulations on the world

human health and the environment from potential risks by harmful chemicals and to promote the competitiveness of chemical industries in the EU. The law should be applied to all chemical substances not only for industries, but also for commodities in our daily lives. Therefore, this law tremendously affects all industries and citizens within the EU. Since REACH requires corporations the burden of proof, corporations have to control and manage the risk of substances that would be sold in the EU markets. It means that they have to certify how to use the substances safely for potential users and at the same time let them know the proper procedure. If they can't control and manage the risk by themselves, they have to replace the substance with safer substances that they could control and manage more easily.

Since those directives and regulations appeared, the EU seems to lead environmental policies in the world. In many other countries outside the EU, similar laws have been established, obviously affected by the EU laws (directives and regulations). From the viewpoint, we can conclude that environmental laws of the EU became global standards of environmental protection policy, even though the US, Japan, and some other countries sometimes opposed parts of their laws. Figure 4 shows the effect of the EU environmental laws on the global scale schematically.

4 Future Scope

For many decades, we have seen how our ideas about the relation between civilization and environment have changed through many historical events and human efforts to seek harmony with the environment. The conventional idea in European citizenship around the birth of the industrial revolution was "the idea of progress" that we would always make progress to improve our future. However, the conventional idea has gradually changed to the new idea, "sustainable development." And the drastic change over the long and historical run will change our industries and innovations in the future.

References

1. US geological survey (2002) http://www.usgs.gov
2. Japan Environmental Ministry (2011) The white paper of environment, circulating society and life's society. http://www.env.go.jp/policy/hakusyo/h23/pdf.html
3. Yamauchi C (2012) Human beings and resources. Fubai Co., Nagoya
4. Report of Resources and Energy Agency, Ministry of Economy, Trade and Industry, Japan (2011) http://www.enecho.meti.go.jp/statistics/
5. European commission (2015) RoHS directive. http://ec.europa.eu/environment/waste/rohs_eee/index_en.htm
6. European commission (2015) WEEE directive. http://ec.europa.eu/environment/waste/weee/index_en.htm
7. European commission (2015) ELV directive. http://ec.europa.eu/environment/waste/elv/
8. European commission (2015) REACH – registration, evaluation, authorization and restriction of chemicals. http://ec.europa.eu/enterprise/sectors/chemicals/reach/index_en.htm
9. Morikawa T (2004) The current situation and the future for chromate conversion coating of zinc plating. The seminar paper of Technology Research Institute of Osaka Prefecture

World Health Organization's Standards from the Viewpoint of Health Risks

Dana M. Barry and Hideyuki Kanematsu

Abstract The World Health Organization (WHO) is a special agency of the United Nations (UN). It was established in 1948 and is concerned with international public health. Its objective is for all people to attain the highest level of health. In order to do this, the organization must control diseases, injuries, environmental pollution, etc. This chapter describes the World Health Organization and discusses its standards (acceptable limits of various contaminants present in the air, water, etc.), especially in regards to the heavy metals and several other metallic materials.

1 Introduction

The World Health Organization (WHO) was established on April 7, 1948. It is a special agency of the United Nations (UN) and is very concerned with international public health. Since it started, the organization has had a leading role in the eradication of smallpox. Its current focus includes communicable diseases like HIV/AIDS, Ebola, tuberculosis, and malaria. Other areas of interest include non-communicable diseases, development and aging, nutrition, food safety, substance abuse, and occupational health (which can be affected by contaminants at the workplace). This chapter describes the World Health Organization and discusses its standards (acceptable limits of various contaminants present in the air, water, etc.), especially in regards to the heavy metals and several other metallic materials.

D.M. Barry (✉)
Clarkson's Departments of Chemical & Biomolecular Engineering and Electrical & Computer Engineering, Clarkson University, Potsdam, NY, USA

Center for Advanced Materials Processing (CAMP), Clarkson University, Potsdam, NY, USA
e-mail: dmbarry@clarkson.edu

H. Kanematsu
National Institute of Technology, Suzuka College, Suzuka, Japan
e-mail: Kanemats@mse.suzuka-ct.ac.jp

H. Kanematsu, D.M. Barry (eds.), *Corrosion Control and Surface Finishing*,
DOI 10.1007/978-4-431-55957-3_8

2 World Health Organization

The World Health Organization (WHO) is a special agency of the United Nations (UN) that is very concerned with international public health [1]. It was established on April 7, 1948, and celebrates this date each year as World Health Day. The organization is led by Margaret Chan and has its headquarters in Geneva, Switzerland [2]. Dr. Margaret Chan is the Director-General of the WHO. She is from the People's Republic of China and earned her medical degree from the University of Western Ontario in Canada. She was elected to the position of Director-General in November 2006 for a 5-year term. After completing this service, the Assembly appointed her to a second term which continues until June 2017.

The overall goal of the WHO is that all people attain the highest possible level of health [3]. Therefore, the organization is interested in everything that affects one's health including communicable diseases, non-communicable diseases, mental-health, injuries, nutrition, environmental pollution, and more. One of its roles is to direct and coordinate international health within the United Nations' system. As an example, it assists countries as they work (with sectors of their government, partners, funding sources, etc.) to attain their health objectives and support their national health policies and strategies.

The World Health Assembly, located in Geneva, is the legislative body of the WHO. It appoints a Director-General every 5 years, and votes on issues such as policy and finance (including the budget). The Assembly also elects members to serve on the Executive Board to carry out the policies and decisions of the Assembly. In addition, the WHO contains regional divisions. Each region has a Regional Committee and a Regional Office headed by a Regional Director (elected by the Regional Committee). The six major regions and their websites are provided: Africa (http://www.afro.who.int), Europe (http://www.euro.who.int), South-East Asia (http://www.searo.who.int), Eastern Mediterranean (http://www.emro.who.int), Western Pacific (http://www.wpro.who.int), and The Americas (http://www.paho.org) [1].

The proposed 2016–2017 program budget for the WHO is about 4400 million U.S. dollars [4]. Therefore, this organization relies on financial contributions from member states (such as the United States, Japan, Germany, etc.) and voluntary donors. To enhance the monetary situation, it works hard to network and increase collaborations with international non-government organizations, private foundations, etc.

The WHO also has a number of publications including the *World Health Report*, a leading international publication about health. The publications provide reliable information and advice for human health. They also support national health strategies, and address the most important public health concerns. In addition, the organization has an online database (WHO MINDbank) that brings together country and international resources. MINDbank provides free access to resources for strategies, policies, and laws that relate to mental health, general health, non-communicable diseases, substance abuse, and other topics [5].

Humans need water, food, and a continuous supply of air. Recognizing the human need for clean air, the WHO's Regional Office for Europe published some air quality guidelines for Europe in 1987. These guidelines contain health risk assessment for many chemicals considered to be air contaminants [6].

Chemicals are released into the air from both natural and man-made sources. Inhalation of these pollutants can have adverse effects on human health. Also if these chemicals are taken up by plants and animals, then they enter the food chain and cause additional sources for human exposure. Over the years, major efforts have been made to reduce air pollution. This is a complicated task. It starts with an analysis to determine what chemicals are present in the air, their concentrations, and if exposure to them is hazardous to human health. Then control strategies and human exposure guidelines are prepared for each hazardous chemical. In addition, the WHO continues to carry out similar work for controlling contaminants and preparing safety guidelines for human exposure to pollutants in air, water, food, herbal medicines, etc.

3 Heavy Metals and Health Risks

Generally speaking, heavy metals are considered to be metallic elements with an atomic mass greater than 40 [7]. These metals and others occur naturally in the Earth's crust. Heavy metals enter our environment as a result of industrial discharge, weathering of the Earth's crust, air pollution, and other ways. Some individuals are exposed to these contaminants at the workplace, while others come into contact with them through food and drinks (ex. water). The threat to human health depends on factors such as the concentration of the contaminants and the amount of time an individual is exposed to them. The WHO's general description of human exposure is the amount of a substance in contact with a person over time and space. Human exposure to contaminants can be assessed by using environmental monitoring and biomarkers [8]. Atomic absorption spectrometry (AAS) and atomic emission spectrometry (AES) are techniques often used for quantitative analysis of heavy metals in environmental samples. Potential biomarkers include DNA and mutations (ex. in cells and organs). Direct approaches for assessing exposure to hazardous materials include the evaluation of human skin damage and the testing of biological fluids such as urine and blood. The WHO provides safety guidelines and standards for acceptable intake amounts of specific contaminants by humans. It also sets limits on the amount/concentration of various contaminants allowed in the air, water, etc.

The heavy metals lead, cadmium, and mercury are introduced and described in this section.

Fig. 1 Lead sample

3.1 Lead

Lead is an element with the symbol Pb and the atomic number 82. See Fig. 1 [9]. It is soft, malleable, and has a high density. Lead was previously used in paint and solder, pipes, pipe fittings, and for making bullets. Today it is mainly used in sheets for roofing and for radiation shields.

Lead is toxic to humans [10]. The main sources of lead exposure are food, air (examples: factory exhaust and lead dust from petro), and drinking water. Plants and animal food that we eat may be contaminated with lead as a result of their surrounding air, water, and soil. Also humans may ingest lead through the use of lead-containing cups, plates, and other utensils. Ingested lead can cause damage to the nervous system. It is a neurotoxin that accumulates in soft tissues and the bones. The levels of lead in the bones, teeth, and hair seem to increase with age.

A provisional tolerable weekly intake, PTWI, of lead was determined (by a Joint Food and Agriculture Organization/World Health Organization Expert Committee on Food Additives, JECFA) to be 0.025 mg/kg body weight in 2004 [11]. Also in 2004, the WHO's standard for lead in drinking water was 0.01 mg/L [11]. In addition as mentioned in a 2007 WHO publication, the WHO recommendation for the limit of lead in herbal medicines and products was 10 mg/kg [12].

3.2 Cadmium

Cadmium is an element with the symbol Cd and atomic number 48. Cadmium has an average concentration in the Earth's crust of 0.1–0.5 parts per million (ppm) [13]. It is a minor component in most zinc ores and is a zinc production byproduct.

Fig. 2 Cadmium sample next to a penny

See Fig. 2 [14]. This element is a soft, malleable metal that is resistant to corrosion. Therefore, when deposited on other metals it serves as a protective layer.

Cadmium was previously used for batteries, electroplating, pigments, and alloys. However, applications of cadmium have declined over the years because of its toxicity to humans. Acute exposure to cadmium fumes can cause flu-like symptoms (examples: chills, fever, etc.).

The inhalation of cadmium dust can lead to respiratory tract and kidney problems, which may be fatal. Also ingesting large amounts of cadmium causes liver and kidney damage.

Cadmium is naturally present in the environment and is sometimes emitted from factories. It can be transferred from the soil to the plants that we eat and to tobacco used in cigarettes. The largest single source of cadmium exposure for humans is tobacco smoke, especially because the lungs absorb this toxic element [13].

As for plants, cadmium is taken up through their roots to the edible fruits and leaves. People are also exposed to cadmium when they consume contaminated animal food.

In 2004 the Joint Food and Agriculture Organization/World Health Organization Expert Committee on Food Additives (JECFA) recommended the provisional tolerable weekly intake of cadmium as 0.007 mg/kg body weight [15]. Also the WHO (in 2004) adopted the provisional guideline of 0.003 mg/L for cadmium in drinking water [15]. In addition, a 2007 WHO publication recommended the limit of cadmium in herbal medicines and products to be 0.3 mg/kg [16].

3.3 Mercury

Mercury is an element with the symbol Hg and atomic number 80. It is a heavy, silver metal that exists as a liquid at standard conditions for temperature and pressure. See Fig. 3 [17].

Mercury has been known for a long time because it was found in Egyptian tombs that date back to 1500 B.C. [18]. It has had applications in electronics, amalgams, thermometers, barometers, and other devices. This element's use has decreased over the years because exposure to it can result in serious health problems.

Mercury occurs naturally in the air, water, and soil. However, it is mostly found in deposits as cinnabar (mercuric sulfide). Mercury is extracted by heating cinnabar (in air) and condensing the vapor. See the following equation.

$$HgS + O_2 \rightarrow Hg + SO_2$$

It is also present in other rocks including coal. When coal is burned, mercury is released into the environment. Once present in the air, it can eventually settle onto land and be washed into sources of water. Here microorganisms can change it to a highly toxic form known as methyl mercury. This form of mercury tends to accumulate in fish (like tuna and halibut) that are consumed by humans. In addition to eating contaminated food (especially marine life), mercury can be absorbed through our skin and mucous membranes. Also its vapors can be inhaled. Exposure to high levels of mercury cause harm to the human brain, heart, kidneys, lungs, and immune system.

In 2004 the Joint Food and Agriculture Organization/World Health Organization Expert Committee on Food Additives set a provisional tolerable weekly intake of mercury to be 1.6 μg/kg body weight [19]. Also in 2004 the WHO set the following standard for mercury in drinking water: 0.001 mg/L [19].

Fig. 3 Drop of mercury

4 Other Metallic Materials of Concern

This section includes several other metallic materials that pose health risks. It introduces and describes chromium and arsenic.

4.1 Chromium

Chromium is an element with the symbol Cr and atomic number 24. This metal is highly valued because of its hardness and corrosion resistance. See Fig. 4 [20]. This metal is added to steel to form stainless steel. It is also electroplated onto other materials to protect them from corrosion. Chromium compounds are found in the environment as a result of erosion of chromium-containing rocks, volcanic eruptions, and the dumping of chromium wastes from facility production (in landfills). Chromium is mined as chromite ($FeCr_2O_4$) in places like South Africa, India, Russia, and Turkey [21].

Chromium, especially chromium VI (with a charge of positive 6) is toxic. Exposure to chromium can cause skin inflammation or a rash. In the past, workers in the chrome ore industry who breathed large amounts of chromium dust developed asthma-like conditions. Also scientists found chromium VI to be mutagenic when inhaled and to have a possible link to cancer [22].

According to the Regional *Air Quality Guidelines for Europe* (published by the WHO in 2000), the concentration of chromium (VI) in air associated with an excess lifetime risk of 1:10,000 was 2.5 ng/m^3 at that time [23]. Also a provisional guideline value for total chromium in drinking water was listed as 0.05 mg/L in a 2011 WHO publication [24].

Fig. 4 Chromium sample

4.2 Arsenic

Arsenic is an element with the symbol As and the atomic number 33. It can occur as a pure element but is most often found in minerals containing sulfur and metals. Arsenic can exist in different structural forms (allotropes). However, gray arsenic is the most common. It is a metalloid that is brittle and a bit shiny. See Fig. 5 [25]. This form has metallic properties and has been used in industry to strengthen alloys of copper and lead. Arsenic is also a common n-type dopant in semiconductor electronic devices (example: gallium arsenide is a semiconductor). Over the years arsenic and its compounds were used in the production of products like pesticides, insecticides, and treated wood items. However, because of its toxicity and harmful effects to humans, arsenic's applications have decreased.

In addition to being in the Earth's crust, arsenic is present in natural water sources. It can exist at higher concentrations in ground waters where there are sulfide mineral deposits, etc. from volcanic rocks. Arsenic is found in drinking water and in our food (examples: fish, seafood, and rice). A WHO Advisory Conference held in 2014 considered setting the limits for arsenic in rice to 200–300 ppb [26].

Arsenic and its compounds are toxic [26]. The inorganic form is considered to be carcinogenic and is related to lung, kidney, bladder, and skin disorders. Due to a great concern about arsenic-related health problems, the WHO set the provisional guidelines value (listed in a 2011 publication) for arsenic in drinking water to 0.010 mg/L [27]. Also a provisional tolerable weekly intake (PTWI) of inorganic arsenic was determined (by a Joint Food and Agriculture Organization/World Health Organization Expert Committee on Food Additives, JECFA) to be 0.015 mg/kg body weight in 2004 [28].

Fig. 5 Gray arsenic sample

5 Conclusion

The World Health Organization (WHO) is a special agency of the United Nations (UN). It was established in 1948 and is concerned with international public health. Its objective is for all people to attain the highest level of health. In order to do this, the organization must control diseases, injuries, environmental pollution, etc. This chapter described the World Health Organization and discussed its standards (acceptable limits of various contaminants present in the air, water, etc.) for the heavy metals lead, cadmium, and mercury. The metallic materials chromium and arsenic were also introduced.

References

1. http://en.wikipedia.org/wiki/world_Health_Organization
2. http://www.who.int/dg/chan/en/
3. http://www.who.int/about/mission/en/
4. World Health Organization Sixty-Eighth World Health Assembly "Proposed Program Budget 2016–2017:" http://www.who.int/about/finances-accountability/budget/en/. A 68/7, 30 Apr 2015)
5. http://www.who.int/mental_health/mindbank/en/
6. (2000) Air quality guidelines for Europe, 2nd edn. WHO Regional Publication, European Series, No. 91, pp 1–288
7. Morais S, Costa FGe, Pereira MdL (2012) Heavy metals and human health (Chapter 10). In: Oosthuizen J (ed) Environmental health – emerging issues and practice. Published by InTech, p 227. http://cdn.intechopen.com/pdfs-wm/27687.pdf
8. Morais S, Costa FGe, Pereira MdL (2012) Heavy metals and human health (Chapter 10). In: Oosthuizen J (ed) Environmental health – emerging issues and practice. Published by InTech, pp 227–246. http://cdn.intechopen.com/pdfs-wm/27687.pdf
9. Lead-2, creative commons attribution 3.0 attribution: http://images-of-elements.com/lead.php. Also available at: http://en.wikipedia.org/wiki/Lead
10. http://en.wikipedia.org/wiki/Lead
11. Morais S, Costa FGe, Pereira MdL (2012) Heavy metals and human health (Chapter 10). In: Oosthuizen J (ed) Environmental health – emerging issues and practice. Published by InTech, p 228. http://cdn.intechopen.com/pdfs-wm/27687.pdf
12. (2007) WHO guidelines for assessing quality of herbal medicines with reference to contaminants and residues. Published by the World Health Organization, p 24. http://advanced3.com/w/who-guidelines-for-assessing-w16950
13. http://en.wikipedia.org/wiki/cadmium
14. Cadmium metal USGOV: a work of the U.S. Federal government that is in the public domain. http://commons.wikimedia.org/wiki/file:CadmiumMetalUSGOV.jpg
15. Morais S, Costa FGe, Pereira MdL (2012) Heavy metals and human health (Chapter 10). In: Oosthuizen J (ed) Environmental health – emerging issues and practice. Published by InTech, p 229. http://cdn.intechopen.com/pdfs-wm/27687.pdf
16. (2007) WHO guidelines for assessing quality of herbal medicines with reference to contaminants and residues. Published by the World Health Organization, p 24
17. Hg mercury: Wikimedia commons in the public domain. http://commons.wikimedia.org/wiki/File:Hg_Mercury.jpg
18. http://en.wikipedia.org/wiki/Mercury_(element)

19. Morais S, Costa FGe, Pereira MdL (2012) Heavy metals and human health (Chapter 10). In: Oosthuizen J (ed) Environmental health – emerging issues and practice. Published by InTech, p 230. http://cdn.intechopen.com/pdfs-wm/27687.pdf
20. Chrom 1: Wikimedia Commons in the public domain. Author: Tomihahndorf http://commons.wikimedia.org/wiki/File:Chrom_1.jpg
21. http://en.wikipedia.org/wiki/chromium
22. http://dartmouth.edu/~toxmetal/toxic-metals/more-metals/chromium-faq.html
23. (2000) Air quality guidelines for Europe, 2nd edn. WHO Regional Publication, European Series, No. 91, p 141
24. (2011) Guidelines for drinking water quality, 4th edn. Published by the World Health Organization, p 178
25. Native arsenic: Wikimedia Commons in the public domain. Author: Aram Dulyan. http://commons.wikimedia.org/wiki/File:Native_arsenic.jpg
26. http://en.wikipedia.org/wiki/Arsenic
27. (2011) Guidelines for drinking water quality, 4th edn. Published by the World Health Organization, p 315
28. Morais S, Costa FGe, Pereira MdL (2012) Heavy metals and human health (Chapter 10). In: Oosthuizen J (ed) Environmental health – emerging issues and practice. Published by InTech, p 231. http://cdn.intechopen.com/pdfs-wm/27687.pdf

European Union (EU) Directives and Regulations

Dana M. Barry and Hideyuki Kanematsu

Abstract This chapter introduces the European Union (EU) and discusses some of its directives and regulations. Today the EU has some of the most progressive environmental policies in the world that extend to all areas of environmental protection. Some of these areas are air pollution control, water protection, waste management, and the control of chemicals. This chapter includes information about the End of Life Vehicles Directive (ELV), the Restriction of Hazardous Substances Directive (RoHS), and the Waste Electrical and Electronic Equipment Directive (WEEE). The EU regulation called Registration, Evaluation, Authorization and Restriction of Chemicals (REACH) is described too.

1 Introduction

The European Union (EU) contains 28 member states that are mainly located in Europe. It is a political-economic union that operates through a system where negotiated decisions are made by the member states. This system is comprised of various state governments and institutions like the European Commission, the European Council, the European Central Bank, the European Parliament, etc. The European Parliament is elected every 5 years by the EU citizens. The European Union was formally established in 1993 by the Maastricht Treaty [1]. Its start is traced back to the European Coal and Steel Community (ECSC) and the European Economic Community (EEC).

D.M. Barry (✉)
Clarkson's Departments of Chemical & Biomolecular Engineering and Electrical & Computer Engineering, Clarkson University, Potsdam, NY, USA

Center for Advanced Materials Processing (CAMP), Clarkson University, Potsdam, NY, USA
e-mail: dmbarry@clarkson.edu

H. Kanematsu
National Institute of Technology, Suzuka College, Suzuka, Japan
e-mail: Kanemats@mse.suzuka-ct.ac.jp

H. Kanematsu, D.M. Barry (eds.), *Corrosion Control and Surface Finishing*,
DOI 10.1007/978-4-431-55957-3_9

Today the EU has some of the most progressive environmental policies in the world that extend to all areas of environmental protection. One of the top priorities of the environmental policy of the EU is to combat climate change. In 2007 member states agreed that in the future the European Union would use 20 % renewable energy and would reduce carbon dioxide emissions in 2020 by at least 20 % as compared to levels in the 1990s [1]. This also means that by 2020, 10 % of the overall quantity of fuel used by cars and trucks in the EU should consist of renewable energy like biofuels [1].

The body of EU environmental law contains more than 500 directives, regulations, and decisions, according to the Institute for European Environmental Policy [1, 2]. A directive is a goal set by a legislative act that all EU countries must achieve. However the individual countries can decide how to achieve this goal. On the other hand, a regulation is a law, a binding legislative act. It applies to the entire EU. Other acts include decisions and recommendations. A decision is binding only to those whom it addresses (example: a specific EU country). Recommendations are not binding but merely suggestions.

This chapter provides a general description of the following European Directives: the End of Life Vehicles Directive (ELV), the Restriction of Hazardous Substances Directive (RoHS), and the Waste Electrical and Electronic Equipment Directive (WEEE).

2 End of Life Vehicles Directive

The End of Life Vehicles (ELV) Directive is the European Union's way of addressing the problem of the end of life for automotive products. Each year motor vehicles that reach the end of their useful lives generate about 8 million tons of waste in the European Union [3, 4]. Figure 1 displays some end of life vehicles [5].

This directive, which is basically limited to passenger cars and light commercial vehicles, aims to reduce the amount of waste resulting from end of life vehicles. Therefore, it covers aspects of a vehicle's life cycle and its treatment operations. In this case treatment refers to activities (such as depollution and dismantling) after the end of life vehicle has been handed over to a facility, where operations are carried out for the recovery and/or disposal of vehicles and their components.

This directive involves four major stakeholders: the producer, the recycling industry, the last holder or owner of the vehicle, and the authorities. It should be mentioned that customers are the primary decision makers in determining when a vehicle is at the end of its life. However, sometimes a vehicle is considered to be at the end of its life based on the condition that it is in.

In addition to reducing waste, the ELV Directive seeks to prevent pollution and make vehicle dismantling and recycling more environmentally friendly. Under this directive, manufacturers need to reduce the use of hazardous substances when designing and making vehicles. They must ensure that mercury, hexavalent

Fig. 1 Crushed and stacked end of life vehicles

chromium, cadmium, and lead, are not included in vehicle components placed on the market after July 1, 2003, unless special exemptions have been granted [6, 7]. Also the ELV Directive requires that producers use the International Organization for Standards (ISO) guidelines for labeling and identifying vehicle components [6]. This assists workers in identifying vehicle components that are suitable for recovery and reuse.

As part of this directive, EU member states are required to establish collection systems for the end of life vehicles. They must also ensure that all vehicles are transferred to authorized treatment facilities. Since the treatment facilities are to depollute and remove fluids and specific compounds, etc., the automotive industry provides them with information about the vehicles and their components.

In order to actually measure the performance of the countries, targets were defined with the ELV Directive. They are calculated based on the average weight of a single vehicle per year. As of January 1, 2015, 85 % of ELV (end of life vehicles) by weight needed to be reused or recycled, with a total recovery of 95 % [3]. Reuse means to use for a similar purpose for which it was originally intended. Recycling is the reprocessing in a production process of the waste materials for the original use or for other purposes [8].

Details about the achievements of various countries (over the years) are available at the Eurostat website listed below.

http://epp.eurostat.ec.europa.eu/portal/page/portal/waste/data/wastestreams/elvs

3 Restriction of Hazardous Substances Directive

The Restriction of Hazardous Substances Directive (RoHS) is the shortened form for Directive on the restriction of the use of certain hazardous substances in electrical and electronic equipment. It was adopted by the European Union in February 2003 [9]. In July 2006, the RoHS 1 directive took effect and was required to be enforced in each member state [10]. Each member state used the directive as a guide to establish its own enforcement and implementation policies. This directive restricts (with some exceptions) the use of certain hazardous materials in the manufacturing of various types of electronic and electrical equipment. The hazardous materials include lead (Pb), mercury (Hg), cadmium (Cd), hexavalent chromium (Cr^{6+}), polybrominated biphenyl (PBB), and polybrominated diphenyl ether (PBDE), [10].

Lead is found in paints, pigments, solder, etc. Mercury has been used in lighting applications and automotive switches, etc. Cadmium is in plastic pigmentation and photocells for nightlights, etc. Hexavalent chromium is used in metal finishing for corrosion prevention, while polybrominated biphenyls and polybrominated diphenyl ethers are flame retardants for several plastics.

Figure 2 shows the structure of polybrominated biphenyl [11]. Figure 3 shows the structure of polybrominated diphenyl ether [12]. Both structures consist of halogenated aromatic rings.

The maximum permitted concentrations in non-exempt products are 0.1 % (1,000 ppm), except for cadmium which is limited to 0.01 % (100 ppm) by weight [9]. The restrictions are on each homogeneous material in the product. For example if a radio (which consists of various components) has a case made of plastic with 2,200 ppm (0.22 %) PBB, then the entire radio would not meet the requirements of the directive.

Fig. 2 Structure of polybrominated biphenyl (PBB)

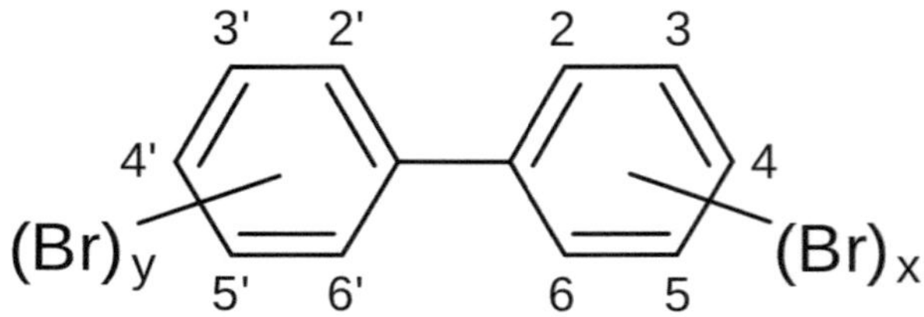

Fig. 3 Structure of polybrominated diphenyl ether (PBDE)

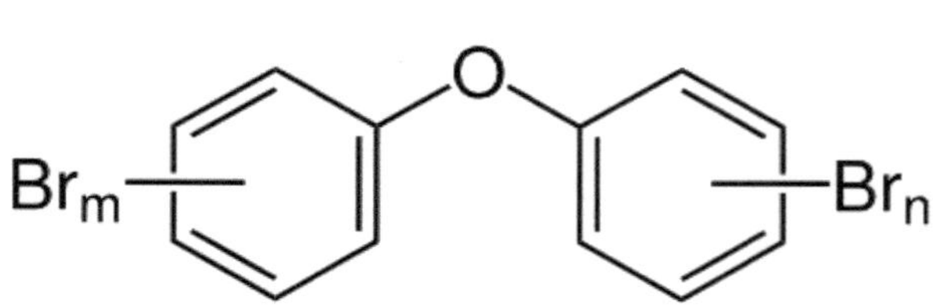

The Restriction of Hazardous Substances Directive applies to items such as household appliances, IT and telecommunications equipment, electrical tools, medical devices, monitoring and control instruments, toys, and lighting equipment.

The RoHS 2 directive is a modified form of the original directive and took effect in January 2013. It addresses the same substances as the original one with improved conditions. This newer directive requires periodic reevaluations to ease a gradual broadening of its requirements to cover additional electronic and electrical equipment. Products within the RoHS 2 directive must display the CE mark (a logo indicating compliance), the manufacturer's name and address, and a batch or serial number. Manufacturers have to also keep various forms of documentation.

As for the country of Japan, it does not have any direct legislation dealing with the RoHS materials. However, its recycling laws have encouraged Japanese manufacturers to move to a lead-free process that follows RoHS guidelines [9].

RoHS provides health benefits. It helps reduce damage to people and the environment, especially to those in third-world countries where much of today's high technology wastes end up. It also ensures a healthier environment for workers in the electronics industry.

4 Waste Electrical and Electronic Equipment Directive

The Waste and Electronic Equipment Directive (WEEE) sets collection, recycling, and recovery targets for electrical goods. It is closely linked to the RoHS directive (previously described), which (with exceptions) restricts the use of certain hazardous materials in the manufacture of electronic and electrical equipment.

Waste of electrical and electronic equipment such as computers and televisions, etc. is quickly growing in the EU and expected to be more than 12 million tons by 2020 [13]. Figure 4 displays old and defective electronic equipment [14].

Because of the hazardous contents of these items and other electronic waste, they can cause health and environmental problems if not managed properly. Therefore, to address these problems the RoHS and WEEE Directives were put in place. The first WEEE Directive was enforced in February 2003. It provided collection plans, where consumers could return their electronic and electrical wastes free of charge. These plans aimed to increase the recycling and /or reuse of electronic equipment (that had been discarded).

The WEEE Directive has undergone a number of revisions since 2003. It was updated in 2006 and 2009. The changes, which affected the method for calculating collection rates, were made in order to tackle the fast growing amount of waste. The collection rates were previously 4 kg per inhabitant per year, but in 2012 they became 20 kg per inhabitant per year [15].

The directive categorized the wastes as being historic and non-historic. Equipment placed on the market before 2005 is considered historic and the owner had to make provisions for its recycling. Electronic equipment placed on the market after

Fig. 4 A collection of old and defective televisions, monitors, and computers

2005, is considered non-historic, and is the responsibility of the producer/distributor to make provisions for its collection and recycling.

Some of the electronic waste groups include large and small household appliances, IT and telecommunications equipment, lighting equipment, toys, electrical tools, medical devices, and more.

5 Registration, Evaluation, Authorization and Restriction of Chemicals

Registration, Evaluation, Authorization and Restriction of Chemicals (REACH) is a European Union regulation. It came into force June 1, 2007 and replaced many European Directives and Regulations [16]. REACH addresses the production and use of chemical substances and their impact on human health and the environment. It is the strictest law so far to regulate chemical substances and will affect industries throughout the world. This regulation established the European Chemicals Agency to manage its technical, scientific, and administrative aspects.

REACH applies to substances manufactured or imported into the EU in quantities of 1 ton or more per year [17]. It generally applies to individual chemical substances but may include some substances contained in objects (articles in REACH terms). The manufacturing and importing companies must register these chemical substances with the new European Chemicals Agency (located in Helsinki, Finland). This agency has set three major deadlines for registration of chemicals. The dates are usually determined by the number of tons manufactured or imported.

The European Commission supports businesses affected by REACH. It provides them with free software (IUCLID) which simplifies managing and submitting data about chemical properties and effects. The submission is a mandatory part of the registration process.

A major component of the REACH regulation is the requirement to communicate information about chemicals throughout the supply chain. This ensures that manufacturers, importers, and their customers, are aware of information relating to the health and safety of the products. Also by having detailed information about the substances present in the products helps workers to find substitutes and/or remove harmful chemicals from them. Since the list of harmful chemicals continues to grow, organizations are required to monitor additions to the REACH list.

6 Conclusion

This chapter introduced the European Union (EU) and discussed some of its directives and a regulation. Keep in mind that a directive is a goal set by a legislative act that all EU countries must achieve. On the other hand, a regulation is a law, a binding legislative act.

The European Union is a political-economic union that operates through a system where negotiated decisions are made by member states, many of which are located in Europe. Today the EU has some of the most progressive environmental policies in the world that extend to all areas of environmental protection.

A brief summary is provided for the EU's directives and regulation described in this chapter. The End of Life Vehicles Directive (ELV) addresses the problem of the end of life for automotive products. Each year these products generate tons of waste. In addition to reducing waste, the ELV Directive seeks to prevent pollution and make vehicle dismantling and recycling more environmentally friendly.

The Restriction of Hazardous Substances Directive (RoHS) restricts (with some exceptions) the use of certain hazardous materials in the manufacturing of various types of electronic and electrical equipment. These hazardous materials include lead (Pb), mercury (Hg), cadmium (Cd), hexavalent chromium (Cr^{6+}), polybrominated biphenyl (PBB), and polybrominated diphenyl ether (PBDE).

The Waste Electrical and Electronic Directive (WEEE) set collecting, recycling, and recovery targets for electrical goods. It is closely linked to the RoHS directive.

Finally the European Union's regulation for Registration, Evaluation, Authorization and Restriction of Chemicals (REACH) was described. It focuses on the production and use of chemical substances and their impact on human health and the environment.

Since the list of harmful chemicals continues to grow, the European Union's directives and regulations will remain of high importance. In order to ensure that human health and the environment are protected, new laws and directives may appear in the future. Also existing ones will continue to be broadened and modified.

References

1. http://en.wikipedia.org/wiki/European_Union
2. http://europa.eu/eu-law/decision-making/legal-acts/index_en.htm
3. https://en.wikipedia.org/wiki/End_of_Life_Vehicles_Directive
4. http://ec.europa.eu/environment/waste/elv/
5. IFCAR, File: Auto scrapyard 1.jpg –Wikimedia Commons. Date: January 18, 2008. This work is in the public domain. https://commons.wikimedia.org/wiki/File:Auto_scrapyard_1.jpg
6. http://www.epa.gov/oswer/international/factsheets/200811_elv_directive.htm
7. http://www.rsjtechnical.com/WhatisELV.htm
8. Directive 2000/53/EC on end of life vehicles: guidance document by the European Commission, Version 1 (2005). http://ec.europa.eu/environment/waste/pdf/guidance_doc.pdf
9. https://en.wikipedia.org/wiki/Restriction_of_Hazardous_Substances_Directive
10. Implementation of the Restriction of Hazardous Substances in Electrical and Electronic Equipment (RoHS) Directive 2011/65/EU: Consultation. by the Department for Business Innovation & Skills (2012). http://product-industries-research.hktdc.com/business-news/article/Electronics-Electricals/BIS-Consultation-on-UK-Implementation-of-the-Recast-RoHS-Directive-2011-65-EU/psls/en/1/1X3VBFM5/1X07WHST.htm
11. Dschanz L (2007) File: Polybrominated biphenyls.svg. This work is the public domain. https://commons.wikimedia.org/wiki/File:Polybrominated_biphenyls.svg. https://en.wikipedia.org/wiki/Polybrominated_biphenyl
12. Leyo R (2007). File: Polybrominated diphenyl ether.svg. This work is in the public domain. https://commons.wikimedia.org/wiki/File:Polybrominated_diphenyl_ether.svg. https://en.wikipedia.org/wiki/Polybrominated_diphenyl_ethers
13. http://ec.europa.eu/environment/waste/weee/index_en.htm
14. Av Wijk (2008) File: Ewaste-pile.jpg. This work is in the public domain. https://commons.wikimedia.org/wiki/File:Ewaste-pile.jpg. https://en.wikipedia.org/wiki/Electronic_waste
15. https://en.wikipedia.org/wiki/Waste_Electrical_and_Electronic_Equipment_Directive
16. http://www.hse.gov.uk/reach/whatisreach.htm
17. https://en.wikipedia.org/wiki/Registration,_Evaluation,_Authorisation_and_Restriction_of_Chemicals

Regulations by the Environmental Protection Agency in the US

Dana M. Barry and Hideyuki Kanematsu

Abstract The United States Environmental Protection Agency (EPA) is an agency of the US federal government. Its purpose is to protect the environment and human health. It carries out this duty by writing and enforcing regulations that are based on laws passed by Congress. There are many aspects to the EPA. Therefore, this chapter provides a general description of the agency and discusses its regulations, especially for heavy metal contaminants in the water, air, and soil.

1 Introduction

The United States Environmental Protection Agency (EPA) was formed on December 2, 1970. Its mission is to protect human health and the environment. The EPA is led by an administrator, one appointed by the president and approved by Congress. Gina McCarthy is the current leader of the agency. She was appointed by President Barack Obama and approved by Congress.

The EPA has its headquarters in Washington, D.C. It includes regional offices for each of the agency's ten regions (which contain US states and territories) and many laboratories. It also has about 15,000 employees, many with jobs as scientists, engineers, specialists in environmental protection, as well as experts in the areas of legal advice, finance, and information technology [1].

As mentioned above, the EPA's mission is to protect human health and the environment. Therefore it needs to ensure that federal laws for protecting human health and the environment are enforced and that all members of society (examples:

D.M. Barry (✉)
Clarkson's Departments of Chemical & Biomolecular Engineering and Electrical & Computer Engineering, Clarkson University, Potsdam, NY, USA

Center for Advanced Materials Processing (CAMP), Clarkson University, Potsdam, NY, USA
e-mail: dmbarry@clarkson.edu

H. Kanematsu
National Institute of Technology, Suzuka College, Suzuka, Japan
e-mail: Kanemats@mse.suzuka-ct.ac.jp

H. Kanematsu, D.M. Barry (eds.), *Corrosion Control and Surface Finishing*,
DOI 10.1007/978-4-431-55957-3_10

individuals, and state, local, and tribal governments) have access to appropriate information so they can contribute to the management of human health and environmental risks. The United States also works with other nations to protect the global environment.

In order to carry out its mission, the EPA performs various tasks. It writes and enforces regulations that are based on environmental laws passed by Congress. The EPA's enforcement powers include fines and sanctions. Sometimes the agency sets national standards that states and tribes enforce on their own. If the states and tribes are not able to meet these standards, then the EPA assists them. The agency helps companies understand the requirements. It also works with industry and all levels of government in a variety of voluntary pollution prevention and energy conservation projects.

The EPA gives grants to various organizations and environmental programs for scientific studies and for advice in making decisions about community cleanups. Also it identifies and tries to solve environmental problems at laboratories throughout the US. In addition, this agency informs the public about its activities through websites and written materials.

The EPA does not work on its own but forms partnerships with businesses, and state and local governments, etc. Actually everyone is responsible for protecting the environment. Individuals should conserve materials and energy, re-use items, and recycle things whenever possible.

2 Laws and Regulations

The Environmental Protection Agency is a regulatory agency because Congress gives it the authority to write regulations for implementing laws. These regulations are mandatory requirements that can apply to individuals, groups, and various organizations. The EPA is responsible for enforcing many of the environmental laws and regulations of the United States. Sometimes the EPA needs to prepare guidance and policy documents [2]. The documents for guidance are used to clarify and provide assistance for implementing environmental regulations. On the other hand, the policies represent the EPA's views for particular topics of concern. The EPA also has a number of offices that deal with specific issues [1]. Examples of these offices are as follows: Office of Air and Radiation, Office of Chemical Safety and Pollution Prevention, Office of Environmental Information, and the Office of Water. For instance, the Office of Air and Radiation includes the Air Quality Modeling Group. This group provides direction for air quality models and simulation techniques used to assess air pollution control strategies.

The next section introduces and briefly describes the Clean Water Act, the Clean Air Act, the Toxic Substances Control Act, and the Resource Conservation and Recovery Act.

2.1 Clean Water Act

The first major US law to address the important issue of water pollution was the Federal Water Pollution Control Act of 1948. It was greatly expanded and reorganized in 1972, when it became known as the Clean Water Act (CWA) [3]. The CWA set up a basic structure for regulating pollutant discharges into the waters of the US and for regulating surface water standards.

The EPA has implemented pollution control programs under the CWA. For example, it set wastewater standards for industry and water quality standards for contaminants in surface water.

Over the years, various laws have changed parts of the Clean Water Act. Title 1 of the Great Lakes Critical Programs Act of 1990 put into place parts of the Great Lakes Water Quality Agreement of 1978, which was signed by the US and Canada [4]. Both nations agreed to reduce certain toxic pollutants in the Great Lakes. This law required the EPA to establish water quality standards for the Great Lakes. It addressed 29 toxic pollutants and determined the maximum levels (concentrations of each in the water) that are safe for human and aquatic life.

2.2 Clean Air Act

The first federal legislation relating to air pollution was the 1955 Air Pollution Control Act. However, the first actual law pertaining to air pollution control was the Clean Air Act (CAA) of 1963. Important amendments, that required regulatory controls for air pollution, were passed for this law in 1970, 1977, and 1990 [5]. The 1970 amendments required federal and state regulations for stationary pollution sources (example: industry) and mobile sources such as vehicles. Note that the Environmental Protection Agency (EPA) was established on December 2, 1970. Its main purpose was to ensure environmental protection by consolidating federal research, standard-setting, monitoring, enforcement activities, etc. into one agency.

The Clean Air Act of 1970 gave the EPA the authority to regulate the emissions of hazardous air pollutants and to establish the National Ambient Air Quality Standards (NAAQS) [6]. A goal of this Act was to set and achieve these Standards in every state by 1975. Therefore, each state was directed to implement plans appropriate for its own sources of air pollution.

The 1977 amendments to the CAA were mainly concerned with preventing deterioration of the air quality in areas that attained the NAAQS.

Additional amendments were made to the CAA in 1990 [7]. For example, they authorized a program to control about 200 toxic pollutants. They also established permit program requirements and a program to phase out the use of chemicals that deplete the ozone layer.

2.3 Toxic Substances Control Act

The Toxic Substances Control Act (TSCA) of 1976 addresses the production, importation, use, and disposal of certain chemicals [8]. This Act provides the Environmental Protection Agency (EPA) the authority to identify potentially dangerous chemicals in the US commerce and to gather and disseminate information about them. The EPA may require the manufacturers, etc. of these chemicals to carry out certain tests and report the results. Based on the test results and other information, the EPA makes decisions on how to regulate specific chemicals. This Agency has a variety of regulatory tools available, which range from a total ban on the production, etc. of a chemical to just requiring a warning label on the product. Also the TSCA requires pre-market screening for chemicals new to the US commerce.

To summarize, it can be said that the TSCA provides the EPA with the power to require testing, record keeping, reporting, and enforced regulations for specific chemicals [9]. The EPA evaluates chemicals and their risks in order to prevent and reduce pollution in the environment.

2.4 Resource Conservation and Recovery Act

The Resource Conservation and Recovery Act (RCRA) of 1980 gave the EPA the authority to control hazardous waste [10]. This includes hazardous waste generation, transportation, treatment, storage, and disposal. Under this Act, generators of hazardous waste are required to register and get an EPA identification number and to abide by its rules. There are also requirements for employee training, reporting, emergency procedures, etc. RCRA's hazardous waste rules apply to substances on the EPA list of hazardous wastes and to substances that through testing, display toxicity, reactivity, ignitability, and characteristics of corrosion. Some hazardous waste produced at metal finishing facilities includes solvents and wastewater treatment sludge from plating and other operations.

Even though RCRA is a Federal law, many States implement their own programs because their waste management rules are approved by the EPA.

3 Heavy Metals and EPA Regulatory Limits

Generally speaking, heavy metals are considered to be metallic elements with an atomic mass greater than 40 [11]. These metals and others occur naturally in the Earth's crust. Heavy metals enter our environment as a result of industrial discharge, weathering of the Earth's crust, air pollution, and other ways. Some individuals are exposed to these contaminants at the workplace, while others come into contact with them through food and drinks (ex. water). The threat to human health depends on factors such as the concentration of the contaminants and the amount of time an individual is exposed to them.

Some heavy metals that affect human health are presented. Also their EPA regulatory limits (for drinking water) are provided. The metals that are described include arsenic, cadmium, chromium, lead, and mercury.

3.1 Arsenic

Arsenic can exist in different structural forms (allotropes). However, gray arsenic is the most common. It is a metalloid that is brittle and a bit shiny. This form has metallic properties and has been used in industry to strengthen alloys of copper and lead. Arsenic is also a common n-type dopant in semiconductor electronic devices (example: gallium arsenide is a semiconductor). Over the years arsenic and its compounds were used in the production of products like pesticides, insecticides, and treated wood items. However, because of its toxicity and harmful effects to humans, arsenic's applications have decreased.

In addition to being in the Earth's crust, arsenic is present in natural water sources. It can exist at higher concentrations in ground waters where there are sulfide mineral deposits, etc. from volcanic rocks. Arsenic is found in drinking water and in our food (examples: fish, seafood, and rice).

Arsenic and its compounds are toxic [12]. Lower level exposures can cause nausea and vomiting and damage blood vessels. Ingesting very high levels of arsenic can possibly result in death. The inorganic form is considered to be carcinogenic and is related to lung, kidney, bladder, and skin disorders.

Due to a great concern about arsenic-related health problems, the EPA set the regulatory limit to 0.01 parts per million (ppm) in drinking water [13].

3.2 Cadmium

Cadmium has an average concentration in the Earth's crust of 0.1–0.5 parts per million (ppm) [14]. It is a minor component in most zinc ores and is a zinc production byproduct. Cadmium, which is naturally present in the environment, is sometimes emitted from factories. It can be transferred from the soil to the plants that we eat. As for plants, cadmium is taken up through their roots to the edible fruits and leaves. People are also exposed to cadmium when they consume contaminated animal food. In addition, cadmium is found in tobacco that is used in cigarettes. The largest single source of cadmium exposure for humans is tobacco smoke, especially because the lungs absorb this toxic element [14].

Cadmium was previously used for batteries, electroplating, pigments, and alloys. However, applications of cadmium have declined over the years because of its toxicity to humans. Acute exposure to cadmium fumes can cause flu-like symptoms (examples: chills, fever, etc.). The inhalation of cadmium dust can lead to respiratory tract and kidney problems, which may be fatal. Also ingesting large amounts of cadmium causes liver and kidney damage.

The EPA set the regulatory limit of cadmium in drinking water to 5 parts per billion (ppb) [15].

3.3 Chromium

Chromium is highly valued because of its hardness and corrosion resistance. This metal is added to steel to form stainless steel. It is also electroplated onto other materials to protect them from corrosion. Chromium compounds are found in the environment as a result of erosion of chromium-containing rocks, volcanic eruptions, and the dumping of chromium wastes from facility production (in landfills). Chromium is mined as chromite ($FeCr_2O_4$) in places like South Africa, India, Russia, and Turkey [16].

Chromium, especially chromium VI (with a charge of positive 6) is toxic. Exposure to chromium can cause skin inflammation or a rash. In the past, workers in the chrome ore industry who breathed large amounts of chromium dust developed asthma-like conditions. Also scientists found chromium VI to be mutagenic when inhaled and to have a possible link to cancer [17].

The EPA set the regulatory limit of chromium to 0.1 parts per million (ppm) in drinking water [15].

3.4 Lead

Lead is soft, malleable, and has a high density. It was previously used in paint and solder, pipes, pipe fittings, and for making bullets. Today it is mainly used in sheets for roofing and for radiation shields.

Lead is toxic to humans [18]. The main sources of lead exposure are food, air (examples: factory exhaust and lead dust from petro), and drinking water. Plants and animal food that we eat may be contaminated with lead as a result of their surrounding air, water, and soil. Also humans may ingest lead through the use of lead-containing cups, plates, and other utensils. Ingested lead can cause damage to the nervous system. It is a neurotoxin that accumulates in soft tissues and the bones. The levels of lead in the bones, teeth, and hair seem to increase with age. In addition, exposure to high lead levels can cause severe damage to the brain and kidneys.

The EPA's regulatory limit for lead in drinking water is 15 parts per billion (ppb) [19].

3.5 *Mercury*

Mercury is a heavy, silver metal that exists as a liquid at standard conditions for temperature and pressure. It has been known for a long time because it was found in Egyptian tombs that date back to 1500 BC [20]. Mercury has had applications in electronics, amalgams, thermometers, barometers, and other devices. This element's use has decreased over the years because exposure to it can result in serious health problems.

Mercury occurs naturally in the air, water, and soil. However, it is mostly found in deposits as cinnabar (mercuric sulfide). It is also present in other rocks including coal. When coal is burned, mercury is released into the environment. Once present in the air, it can eventually settle onto land and be washed into sources of water. Here microorganisms can change it to a highly toxic form known as methyl mercury. This form of mercury tends to accumulate in fish (like tuna and halibut) that are consumed by humans. In addition to eating contaminated food (especially marine life), mercury can be absorbed through our skin and mucous membranes. Also its vapors can be inhaled. Exposure to high levels of mercury cause harm to the human brain, heart, kidneys, lungs, and immune system.

The EPA's regulatory limit for mercury in drinking water is 2 parts per billion (ppb) [21].

4 Coal and Heavy Metals

Coal contains heavy metals because it is made by compressed organic matter that includes a variety of elements. The mining and burning of coal release toxic metals like mercury, arsenic, and cadmium, into the environment. The largest source of toxic pollutants in the US, due to coal ash and coal waste, is the electric power sector [22]. Each year the burning of coal generates millions of tons of waste. Some of this waste is used in new products and the rest is stored in ponds or pits.

Coal ash refers to airborne compounds known as fly ash and bottom ash. These compounds are produced by burning coal. They can contain heavy metals that settle out of the atmosphere onto the land and/or into water sources. These heavy metals eventually accumulate in plants and animals that are consumed by humans. Toxicity from heavy metals can result in damage to the nervous system and to vital organs such as the lungs and kidneys.

A liquid form of coal waste is coal sludge, which is formed during mining activities. It contains heavy metals and is usually kept in impoundments that sometimes leak and contaminate an area.

Coal is the most abundant fossil fuel on Earth. Therefore, it is used in coal-fired power plants to produce electricity. The general process is that heat energy from the combustion of coal is converted into mechanical energy, which operates an electrical generator. See Fig. 1 [23]. It displays a typical coal-fired power plant process.

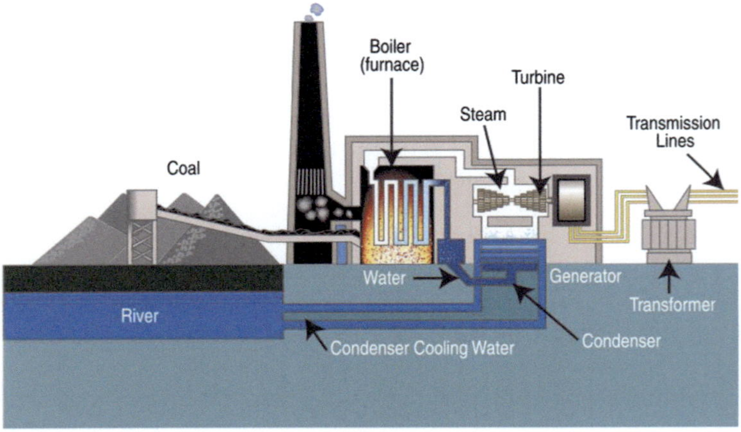

Fig. 1 A typical coal-fired power plant process

The Environmental Protection Agency finalized the very first national standards to reduce toxic air pollution (example: mercury) from coal and oil-fired power plants on December 16, 2011 [24]. This is especially important because the world's power demands are increasing. There were over 50,000 coal plants operating throughout the world in 2007 [25]. This number is expected to increase with time, so larger amounts of pollutants could be released into the environment.

5 Soil and Heavy Metals

Soil, which is important to both urban and rural communities, may contain heavy metals. This type of contamination is possible near mining, manufacturing, and old landfill sites, etc. Excess accumulations of heavy metals in soil are toxic to humans. This is especially true, when the pollutants are taken up by plants that are consumed by humans and animals (that are part of our regular diet). Heavy metals like lead, cadmium, and arsenic, cause damage to our central nervous system and our vital organs.

The best method for protecting soils from heavy metal contamination is prevention. It is very expensive and difficult to clean contaminated soils. Some treatments include high temperature methods to produce a non-leachable material, solidifying agents to make cement-like material, and a washing process to remove contaminants. Several other techniques involve soil and crop management by adjusting the pH of the soil and the use of plants to remove metals from soil and water.

To help protect soils from heavy metal contamination, the US Environmental Protection Agency set regulating limits for industrial waste and sludge (a semisolid material). Maximum concentrations for specific metals in sludge (mg/kg or parts per million, ppm) are provided: arsenic is 75, cadmium is 85, chromium is 3,000, lead is 420, and mercury is 840 [26].

6 Conclusion

This chapter introduced the United States Environmental Protection Agency (EPA). It described the Agency's main purpose which is to protect the environment and human health by writing and enforcing regulations based on laws passed by Congress. Also the EPA's regulations were discussed, especially for heavy metal contaminants in water, air, and soil.

References

1. http://en.wikipedia.org/wiki/United_States_Environmental_Protection_Agency
2. http://www2.epa.gov/laws-regulations/policy-guidance
3. http://www2.epa.gov/laws-regulations/summary-clean-water-act
4. http://www2.epa.gov/laws-regulations/history-clean-water-act
5. http://www.epa.gov/air/caa/amendments.html
6. http://www2.epa.gov/laws-regulations/summary-clean-air-act
7. https://en.wikipedia.org/wiki/Clean_Air_Act_(United_States)
8. Schierow L-J (2013) The Toxic Substances Control Act (TSCA): a summary of the act and its major requirements. CRS Report for Congress, 1–16. http://www.fas.org/sgp/crs/misc/RL31905.pdf
9. http://www2.epa.gov/laws-regulations/summary-toxic-substances-control-act
10. Chalmer P (2008) The future of finishing. National Center for Manufacturing Sciences (NCMS), Ann Arbor, p 32. www.nmfrc.org/futfinrev2.doc
11. Morais S, Costa FG e, Pereira MdL (2012) Heavy metals and human health (Chapter 10) In: Oosthuizen J (ed.) Environmental health – emerging issues and practice. Published by InTech, 227. http://cdn.intechopen.com/pdfs-wm/27687.pdf
12. http://en.wikipedia.org/wiki/Arsenic
13. Martin S, Griswold W (2009) Human health effects of heavy metals. Environmental Science and Technology Briefs for Citizens (Issue 15): 1. http://www.engg.ksu.edu/CHSR/outreach/resources/docs/15HumanHealthEffectsofHeavyMetals.pdf
14. http://en.wikipedia.org/wiki/cadmium
15. Martin S, Griswold W (2009) Human health effects of heavy metals. Environmental Science and Technology Briefs for Citizens (Issue 15): 2. http://www.engg.ksu.edu/CHSR/outreach/resources/docs/15HumanHealthEffectsofHeavyMetals.pdf
16. http://en.wikipedia.org/wiki/chromium
17. http://dartmouth.edu/~toxmetal/toxic-metals/more-metals/chromium-faq.html
18. http://en.wikipedia.org/wiki/Lead
19. Martin S, Griswold W (2009) Human health effects of heavy metals. Environmental Science and Technology Briefs for Citizens (Issue 15): 3. http://www.engg.ksu.edu/CHSR/outreach/resources/docs/15HumanHealthEffectsofHeavyMetals.pdf
20. http://en.wikipedia.org/wiki/Mercury_(element)

21. Martin S, Griswold W (2009) Human health effects of heavy metals. Environmental Science and Technology Briefs for Citizens (Issue 15): 3. http://www.engg.ksu.edu/CHSR/outreach/resources/docs/15HumanHealthEffectsofHeavyMetals.pdf
22. http://www.sourcewatch.org/index.php/Heavy_metals_and_coal
23. Coal fired power plant diagram. svg: A work of the U.S. Federal government that is in the public domain. http://en.wikipedia.org/wiki/Coal-fired_power_station
24. http://www.epa.gov/mats/powerplants.html
25. http://en.wikipedia.org/wiki/Coal-fired_power_station
26. Heavy metal soil contamination (2000) Soil Quality – Urban Technical Note, No. 3, 2. http://www.nrcs.usda.gov/Internet/FSE_DOCUMENTS/nrcs142p2_053279.pdf

Japanese Environmental Regulations

Hideyuki Kanematsu and Dana M. Barry

Abstract After World War Two (WW2), Japan sought for industrial development and entered a high-growth period with the growth of baby boomers. However, the Japanese focused too much on the development and neglected consideration for environmental protection. As a result, many unfortunate affairs relating to environmental pollution occurred in Japan. Then the Japanese government finally decided to establish the Environmental Basic Act of 1993. Since then, their environmental policy has been going in the right direction based on the viewpoint of protection. In this chapter, we describe the history for the development of environmental protection acts in Japan and explain some of the important ones.

1 Introduction: Historical Development of Environmental Protection [1, 2]

During WW2, Japanese land was destroyed and burned to the ground. The function as a unified nation was completely stopped and the economy slowed down very naturally and inevitably. However, with time it was restored to its former state under the protective umbrella of the USA. Moreover, Japan achieved a miracle of economic growth. Due to the very rapid economic and industrial growth, the Japanese society became distorted and the balance was broken in some fields. Destruction of the environment was one of them.

The worst environmental problem in Japan's human history occurred in Minamata, Japan in 1956 (Fig. 1) [3]. Quite a few people who ate fish in Minamata

H. Kanematsu (✉)
National Institute of Technology, Suzuka College, Suzuka, Japan
e-mail: kanemats@mse.suzuka-ct.ac.jp

D.M. Barry
Clarkson's Departments of Chemical & Biomolecular Engineering and Electrical & Computer Engineering, Clarkson University, Potsdam, NY, USA

Center for Advanced Materials Processing (CAMP), Clarkson University, Potsdam, NY, USA
e-mail: dmbarry@clarkson.edu

H. Kanematsu, D.M. Barry (eds.), *Corrosion Control and Surface Finishing*,
DOI 10.1007/978-4-431-55957-3_11

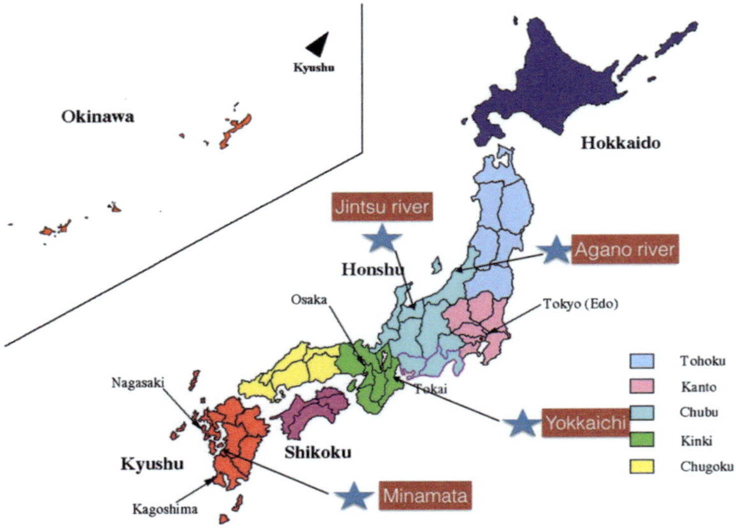

Fig. 1 The four worst environmental diseases in Japan

Bay began to notice abnormalities in their bodies. The symptoms varied from person to person. The main ones were perception disorders, ataxia, axipetal constriction of the visual field, language disorders, etc. At the beginning stage, the real reason was unknown. However, it could be attributed to methyl mercury concentrated in fish. The methyl mercury was discharged by Chisso, a chemical giant in Japan. The medical term for this problem is called the Hunter-Russel syndrome. However, the Japanese usually refer to it as the Minamata disease. Minamata disease is actually the starting point for all of the environmental destruction.

Around the same time, a different, but also a serious environmental problem was coming to a head along the Jintsu River at the boarder of the Toyama and Gifu prefectures (Fig. 1). It was called the Itai-itai disease and caused by cadmium from the mines in the area. "Itai" means aching in Japanese. It directly suggests that the patients had experienced severe pain. The patients suffered from the deterioration of their kidneys and displayed serious anemia. This illness has been mentioned as one of the four worst environmental diseases in Japan.

In 1965, a similar environmental problem and disease appeared near Showa Denko, another Japanese chemical giant. People living along the Agano River in the Niigata prefecture, Japan, suffered from symptoms like those caused by the discharged methyl mercury. This disease was called the second Minamata disease (Fig. 1).

The last of the four worst environmental diseases is asthma, which was found in Yokkaichi, Mie Prefecture (Fig. 1). Many chemical companies were concentrated in that area. They were referred to as the Yokkaichi industrial complex, which also included other related industrial companies. As a result of their gaseous exhausts (emissions), many people suffered from chronic and constant coughing. Their coughing was a result of the air pollution, particularly SO_X gas. At the beginning of the 1970s, the inhabitants won the court case against the industrial complex.

Based on this background information, etc., the Japanese people have gradually fostered their environmentally friendly minds. Also the Japanese government took action and made environmental protection acts and laws. Several tentative laws were enacted during those difficult times (the Basic Act for Environmental Pollution Control in 1967 and the Nature Conservation Act of 1972). In 1971, an environmental agency was formally established after some working groups were tentatively formed. This suggests that Japan decided to formally tackle with the environmental problems. Actually, this was the revolutionary starting point of a new era. In 1993, the Environmental Basic Act was established. However, the Environmental Pollution Control Act was abolished, while the Nature Conservation Act was revised according to the concept of the "new" Environmental Basic Act. In 1997 the third session for the conference of the parties to the United Nations Framework Convention (on Climate Change) was held in Kyoto, Japan. This event was not restricted to Japan, but was open to the world. However, the conference and the agenda affected the following Japanese environmental policies very much. At the turn of the century (2002), the Basic Act for the Promotion of the Recycling-Oriented Society was enacted. The law began to accelerate a change from the old non-sustainable society to the new sustainable one in the twenty-first century. In 2001, the environmental agency was reorganized and scaled up to the environmental ministry. In 2008, the Basic Act on Biodiversity was established. This law aimed to make our world a sustainable society, focusing on the balance among species. Both laws constitute the main components of the Environmental Basic Act.

The Japanese regulation (law) for hazardous substances that corresponds to REACH (a European regulation standing for registration, evaluation, authorization and restriction of chemicals) is the Act on the Evaluation of Chemical Substances and Regulation of Their Manufacture, etc. [4]. The law was established and developed by the Environmental Ministry, as compared with the former one (Fig. 2). It is generally called Kashinho in Japanese. The law was originally established because quite a few people suffered from polychlorinated biphenyl (PCB) that was unintentionally mixed with cooking oil by a company.

After the unfortunate event occurred 1968 in Japan, the people and the government had to drastically change the knowledge requirements for health hazards. Till then, people thought that human health would be protected from environmental pollution by controlling the use, production, and discharge of hazardous chemical substances.

However, in this case, people realized that the persistent bioaccumulation of toxins also seriously affected human health. At this point, a new law was required and Kashinho was finally established. Since then, mainly the Ministry of Economy, Trade and Industry and the Ministry of Health, Labor and Welfare have developed the law.

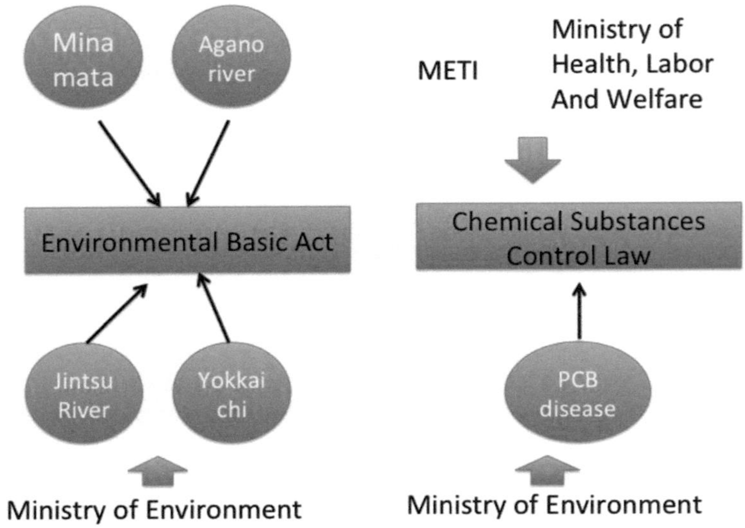

Fig. 2 Environmental Basic Act and Japanese Kashinho

Gradually, the range of hazardous chemical substances expanded and the concept of eco-system integrity was incorporated into them. The next section provides more information about the basic environmental law in Japan and Kashinho.

2 General Sketch for Environmental Basic Act [5, 6]

The basic concept of the act is composed of three factors.

One of them is enjoying the gifts from the environment and the devolution (Article 3). The second is the establishment of sustainable societies (Article 4). And the final is the implementation of international cooperation and the maintenance of the global environment (Article 8). The agents are the nation, the local government, the business operator and the citizen (defined from Articles 6–9). The law's basic guideline has three purposes. The first one is to keep the natural constituents of the environment well. The second one is to keep life's diversity. The third one is to keep the friendly contact between humans and the environment (defined in Article 14).

The concept of the design for the law is shown schematically in Fig. 3.

Concretely, it defines and controls the broad range of environmental items. Table 1 summarizes them, while corresponding to each article.

As described above, the two subordinate new laws that constituted the Environmental Basic Act, when it was developed, are provided: Basic Act for the Promotion of the Recycling-Oriented Society and Basic Act on Biodiversity. The former recognized the current situation as non-sustainable and outdated. Therefore, it was

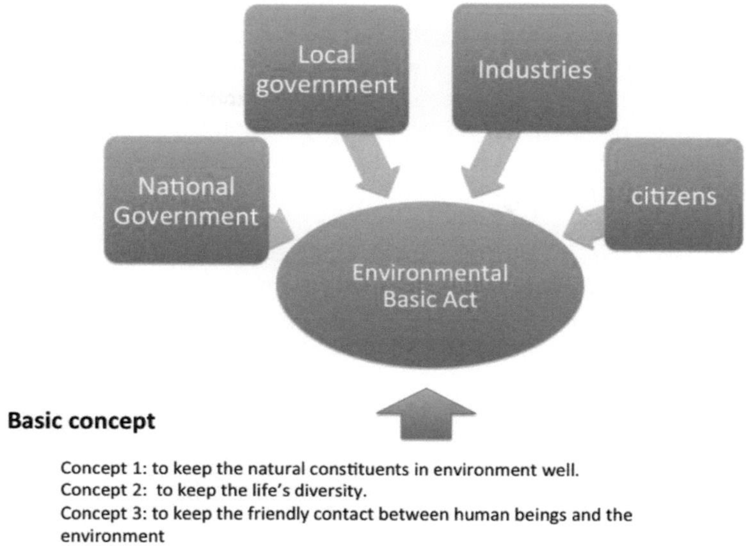

Fig. 3 Environmental Basic Act and its basic concepts

Table 1 Concrete policies for the Environmental Basic Act

Article number	Object
Article 16	Air pollution, water contamination, ground pollution, undesired sounds
Article 17 and 18	The plan for prevention of deterioration of environment. The driving of accomplishment
Article 19	Attention for environment. Enforcement of national policy
Article 20	The driving for the evaluation of environmental effect
Article 21	Regulation
Article 22	Economic steps, economic grants, economic burden share and its encouragement
Article 23	The positive use for the environmentally friendly products
Article 25	Education and learning for environmental conservation
Article 26	Promotion for voluntary activities by civil groups
Article 28 and 29	Investigations required for the establishment of laws and decisions
Article 30	Promotion for science and technology
Article 31	Processing to solve conflicts caused by environmental pollution
Article 32, 33, 34 and 35	International collaboration for protection of global environment

established to realize the recycling oriented society and also to solve the waste problems. A society was needed where people would make much of good products based on a comfortable life style. Industries would be required to make products with much longer life cycles and also to promote and introduce the lease/rental system much more. And for the flow of materials, a numerical goal was set from the viewpoints of input, output, and circulation.

On the other hand, the Basic Act on Biodiversity was established in 1995 [7]. It has been amended four times since then. The latest national strategy (according to the law) was the one for 2012–2020 at this point. Basically, the law aims for the protection of biodiversity and its sustainable utilization. The strategy incorporated the achievement of the 10th Meeting of the Conference of the Parties (COP10) to the Convention on Biological Diversity (CBD) that was held in Nagoya, Aichi Prefecture, Japan, on Oct. 30, 2010. It also took the big earthquake that occurred in the Tohoku region in Japan (2011) into consideration and tried to propose how human beings should live together with nature in the future. This law recognizes that the current situation would face four major crises – the negative impacts by humans' activities, a decreasing population, problems caused from alien invasive species/matters brought by human beings, and also by the change of our global environment. Also it arranged and summarized five problems to solve. It proposed the basic strategy constituted by five components based on seven fundamental standpoints.

The seven basic standpoints are:

#1: Scientific recognition for the current environmental situation and protective/adaptive attitudes.
#2: Approach for local areas and communities.
#3: World wide recognition.
#4: Cooperation and Coproduction
#5: Mainstreaming of biodiversity in socio-economics
#7: Long-term merits by sustainable utilization.

On the basis of those components, it proposed the following five problems to solve.

#1: Understanding and actions for biodiversity
#2: To keep the bearers and their collaboration
#3: Recognition of a natural symbiosis sphere connected by ecological service
#4: Maintenance engineering of a country from the standpoint of falling population
#5: Enhancement of scientific knowledge

According to the guideline mentioned above, the government established the roadmap and is now executing its policy on a full scale.

3 General Sketch for Kashinho [8]

Kashinho started due to the occurrence of the PCB pollution disease, as described above. Table 2 shows the organization of the law.

Due to the historical special situation when the law was established, it incorporates a Preliminary Review System for new unknown substances. For example, a very new substance such as PCB in the 1960s cannot be produced nor imported until the three Ministries of Health, Labor and Welfare, Economy, Trade and Industry and the Environment approve it.

What is the new chemical substance? The law defines it as the one that can't be applied to any of the following categories.

#1: The substance all three ministries judge not applicable for any of the following classes. Class 1 designated a particular chemical as a persistent and potentially long-term toxic substance and persistent and an ecological toxic substance.
#2: The substance already defined by Kashinho as a control subject. Class 1 designated particular chemicals. Class 2 designated particular chemicals and priority assessment of chemical substances.
#3: Chemical substances mentioned in the list of existing chemical substances.
#4: Chemical substances mentioned as Chemical Substances subject to Type II or Type III Monitoring before the amendment of 2009.

According to the basic principles shown in Table 2, three ministries have to classify the new substances into the following six categories for their judgments.

#1: Class 1 designated particular chemicals.
#2: Persistent and potentially long-term toxic substances. Not ecologically toxic.
#3: Persistent and ecological toxic substances. Not long-term toxic.
#4: Persistent, ecologically toxic and potentially long-term toxic substances.
#5: The case not available for #1, #2, #3 and #4.
#6: The unclear cases for which a category should be available.

The result of the judgment by the three ministries can generally be classified into three cases. If the substance would belong to case #1, then it would be defined as Class 1 (designated a particular chemical substance) and the rule for the category would be applied. If the substance would belong to cases #2–#4, then the ministries must listen to their evaluation by the advisory board about if the substance would be a priority assessment chemical substance or not. And if the final judgment would be

Table 2 Constitution of Kashinho

Article	Contents
Article 1 and 2	Purpose and definition
Article 3, 4, 5, 6 and 7	Judge and regulation for new chemical substances
Article 8	Notification for general chemical substances
Article 9, 10, 11 and 12	Measures for priority chemical assessment substances
Articles 13, 14, 15 and 16	Measures for chemical substances subject to monitoring
From Articles 17–34	Regulations for Class 1 designated particular chemical substances[a]
Articles 35, 36 and 37	Regulations for Class 2 designated particular chemical substances[b]
From Articles 38–56	Miscellaneous regulations
From Articles 57–63	Penal provisions

[a]Class 1 designated particular chemical substances – 28 chemical substances including PCB
[b]Class 2 designated particular chemical substances – 23 chemical substances including trichloroethylene

positive, then the substance would be officially announced as a priority assessment chemical substance. If the substance would belong to case #5, then the substance would be out of the regulations and judged officially as a non-new chemical substance by the three ministries. If the case would be true for #6, then the three ministries can require the subscriber to submit the report for the test (of the substance) if needed. According to the report, they would judge which category from #1 to #5 the substance would belong to once again.

The detailed and updated information is reflected on a website made by two ministries (Health, Labor and Welfare, Environment) and a governmental organization (Incorporated administrative agency: National Institute of Technology and Evaluation, NITE) [9]. The website is called Japan CHEmicals Collaborative Knowledge Database (J-CHECK) and everyone can refer to the site through the Internet.

References

1. Japanese Ministry of Environment, Environmental white papers, 1972–2006 http://www.env.go.jp/en/wpaper/
2. Japanese Ministry of Environment, White paper on environmental pollution, http://www.env.go.jp/en/wpaper/ 1969–1971
3. Ishimure M, Monnet L (2003) Paradise in the sea of sorrow: our Minamata disease (Michigan Classics in Japanese Studies). University of Michigan Center, Ann Arbor, MI
4. Bergkamp L (2013) The European Union reach regulation for chemicals: law and practice. Oxford University Press, Oxford, pp 13.89–13.121
5. Kaseno S (1994) From Basic Act for Environmental Pollution Control to Basic Environmental Act – the establishment of Basic Environmental Act and the meaning. Okayama University, Scientific Achievement Repository. http://escholarship.lib.okayama-u.ac.jp/Detail.e?id=2051220100519012046
6. The Ministry of Environment (1993) Japanese government, The Basic Environment Law, http://www.env.go.jp/en/laws/policy/basic/
7. The Ministry of Environment (2012) Japanese government, Biodiversity. http://www.biodic.go.jp/biodiversity/about/initiatives/
8. The Ministry of Economy, Trade and Industry (2011) Chemical substances and control law. http://www.meti.go.jp/policy/chemical_management/english/files/setsumeisiryou_eng.pdf
9. The Ministry of Health (2010) Labor and Welfare, Environment and National Institute of Technology and Evaluation, Japan CHEmicals Collaborative Knowledge Database (J-CHECK). http://www.safe.nite.go.jp/jcheck/top.action

Part III
Conventional Environmental Evaluation Processes for Surface Finishing

The Necessity and Meaning

Dana M. Barry and Hideyuki Kanematsu

Abstract It is necessary for environmental evaluation of surface finishing processes. From the viewpoint of current environmental protection acts and laws in various countries (which seem to be getting stricter) surface finishing industries are facing serious challenges. These industries tend to release heavy metals and other pollutants into our air, water, and soil. Therefore, to meet the demands of the environmental protection laws, they will need to modify current processes and propose some new ones to replace their traditional methods. In order for these new processes to be adequate and effective, appropriate environmental evaluations are needed to test them. This chapter describes several ways to evaluate and treat harmful contaminants in air and water (especially those from the surface finishing industry).

1 Introduction

Industrial treatments are made to materials' surfaces for various reasons. One is to improve the appearance of a finished product to attract potential buyers. Other purposes for surface treatments are to improve corrosion resistance, wear resistance, mechanical properties, electrical properties, etc. Surface finishing processes include electroplating, electroless plating, painting, physical vapor deposition, chemical vapor deposition, and more.

From an engineering point of view, a major purpose of surface finishing is to prevent corrosion. Corrosion is actually a process whereby manufactured metals

D.M. Barry (✉)
Clarkson's Departments of Chemical & Biomolecular Engineering and Electrical & Computer Engineering, Clarkson University, Potsdam, NY, USA

Center for Advanced Materials Processing (CAMP), Clarkson University, Potsdam, NY, USA
e-mail: dmbarry@clarkson.edu

H. Kanematsu
National Institute of Technology, Suzuka College, Suzuka, Japan
e-mail: Kanemats@mse.suzuka-ct.ac.jp

H. Kanematsu, D.M. Barry (eds.), *Corrosion Control and Surface Finishing*,
DOI 10.1007/978-4-431-55957-3_12

return to their natural oxidation states. It is of economic importance because it limits the life time of metal structures (example: bridges) and can result in accidents or incidents of pollution.

Surface finishing industries release heavy metals and other pollutants into our air, soil, and water. Therefore, they are facing serious challenges in order to meet the requirements of various environmental protection laws. This means that they will need to modify current processes and propose some new ones.

For example, the United States Environmental Protection Agency (EPA) protects human health and the environment. It carries out this duty by writing and enforcing regulations that are based on laws passed by Congress. The Clean Air Act gave the EPA authority to regulate the emissions of hazardous air pollutants. In the same way, the Clean Water Act required the EPA to establish water quality standards. The EPA's enforcement powers include fines and sanctions.

Environmental protection laws and various aspects of the surface finishing industry are discussed in other chapters. This chapter describes several ways to evaluate and treat harmful contaminants in the air and water (especially those from the surface finishing industry).

2 Air Pollutants

Air pollution is the introduction of harmful materials into the Earth's atmosphere. Indoor air pollution and urban air quality were listed as two of the world's worst toxic pollution problems in 2008 [1]. A pollutant in the air is a substance that can have adverse effects on humans and the environment. It exists as a solid particle, a liquid droplet, or a gas. A pollutant can be man-made or originate in nature (example: radon gas). The two main sources of pollutants are stationary ones like factories and industries and mobile sources such as motor vehicles and ships.

Pollutants are either primary or secondary ones. Primary pollutants are generally produced directly from a process while secondary pollutants result when primary pollutants react or interact. Primary pollutants produced by human activity include sulfur oxides, nitrogen oxides, carbon monoxide, volatile organic compounds, particulate matter (fine particles), toxic metals (examples: lead and mercury, etc.) and others.

3 Analyzing Air Pollution

The United States Environmental Protection Agency and similar organizations in other countries have published a list of air pollutant emission factors for many industrial sources. These emission factors are representative values that result from people's attempt to relate the amount of a pollutant released to the ambient air with an activity associated with the release of that pollutant. These factors help estimate emissions from various sources of air pollution. They are generally expressed as the weight of pollutant divided by a unit weight, volume, distance, or the length of time for the activity emitting the pollutant [1]. One example could be kilograms of pollutant per ton of coal burned.

Air pollution is generally analyzed through the use of a variety of mathematical models. These models help predict the transport of air pollutants in the lower atmosphere. The Point Source Dispersion Model is used for industrial sources. It contains simple mathematics and has been studied for a long time. It has been validated and calibrated with experimental data from all kinds of atmospheric conditions. This method uses a Gaussian dispersion model for continuous buoyant pollution plumes. The model helps predict air pollution isopleths, taking into account wind velocity, stack height, emission rate, and atmospheric turbulence [1]. Simply stated, an isopleth is a line on a map through all points having the same numerical value.

4 Controlling Air Pollution

A large amount of air pollution is due to the combustion of fossil fuels like coal. By reducing the use of fossil fuels, one can reduce the amount of air pollution. Therefore, an effective way to reduce air pollution is to switch over to clean power sources such as hydro power, solar power, and wind power.

In addition, certain devices are used for pollution control in industry. They either destroy contaminants or remove them (from an exhaust stream) so they are not emitted into the atmosphere. Several examples are provided. An electrostatic precipitator removes (collects) particles from a flowing gas, using the force of an induced electrostatic charge. It can remove fine particles such as dust and smoke from a stream of air.

Scrubbers are a form of pollution control technology. They can be used to remove some particulates and/or gases from industrial exhaust streams. Traditionally scrubbers referred to the use of liquid to wash unwanted pollutants from a gas stream. Today it also applies to the injection of a dry reagent or slurry into a dirty exhaust stream to clean out acid gases. Scrubbers are one of the main devices to control gaseous emissions, especially acid gases.

In a wet scrubber, the polluted gas stream is brought into contact with the scrubbing liquid. The design of wet scrubbers depends on the industrial process and the nature of the air pollutants involved. Scrubbers can be made to collect particulate matter and/or gaseous pollutants. Wet scrubbers remove dust particles by capturing them in liquid droplets. These droplets in the inlet gas must be separated from the outlet gas stream by another device called a mist eliminator. The resulting liquid must be treated before its final discharge or reuse in the plant. High efficiency removal of particles (1 μm or less) generally requires a high energy device such as a Venturi scrubber. Figure 1 shows a Venturi scrubber and another device called the mist eliminator [2]. One way to improve the efficiency of

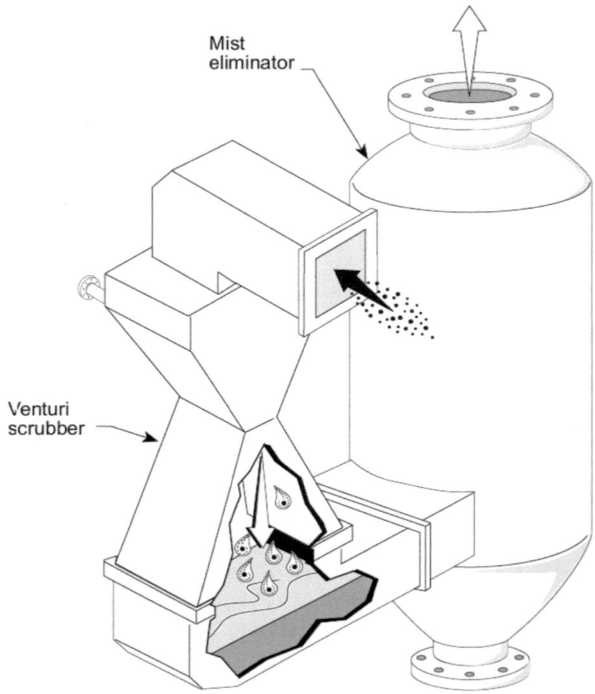

Fig. 1 This figure shows a Venturi scrubber and the device called a mist eliminator

removing pollutants is to increase their residence time in the scrubber. This can be accomplished by using packed towers, etc. Fig. 2 shows a packed tower scrubber [3]. The packed tower scrubber (often considered an absorber) is commonly used to remove gaseous pollutants. It should be noted that if a gas stream contains both gases and particle matter, wet scrubbers are generally the only single air pollution control device capable of removing both pollutants.

5 Industrial Wastewater Treatment

Industrial wastewater treatment includes the processes used to treat wastewater that is produced as a by-product of industrial activities. After appropriate treatment, the wastewater (effluent) may be reused or released to our environment's surface water.

Most industries produce some wastewater. For example, the production of iron from its ores involves reduction reactions in blast furnaces. The cooling waters end up with contaminants that include ammonia and cyanide. Also wastewater from metal mines and ore recovery plants is contaminated by minerals present in the original rock formations. Depending upon the industry, a variety of pollutants can potentially occupy the wastewater. Some of the contaminants include solids, oil and

Fig. 2 Display of a packed tower scrubber

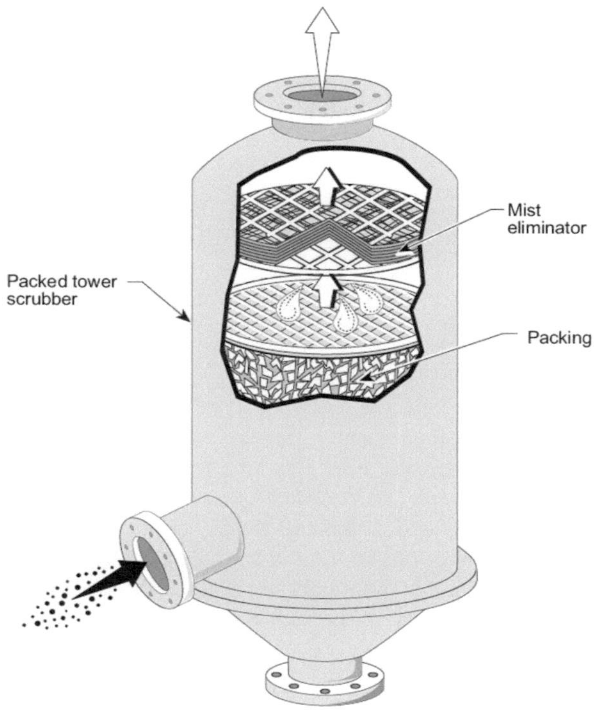

grease, biodegradable organics, acids and alkalis, and toxic materials such as heavy metals.

Solids can be removed by using sedimentation techniques, where they are recovered as slurry or sludge. Very fine solids may require filtration or ultrafiltration. Also flocculation may be used. Oil and grease can be recovered from open water surfaces by using skimming devices. Another technique is to use hydrocyclone oil separators. For this process, wastewater enters a cyclone chamber and is spun under extreme centrifugal forces up to 1000 times the force of gravity [4]. This results in the separation of water from the oil droplets. The oil leaves one end of the cyclone while the water travels through the other end for further treatment or discharge.

Activated sludge is a biochemical process for industrial wastewater and sewage that uses air and microorganisms to biologically oxidize organic pollutants [4]. This results in a waste sludge that contains the oxidized material. The activated sludge process includes an aeration tank and a settling tank. See Fig. 3 [5]. Air is injected into the aeration tank and mixed into the wastewater. The settling tank (clarifier) allows the waste sludge to settle. Some of the waste sludge is recycled to the aeration tank and the rest of it is removed for additional treatment and finally disposed of.

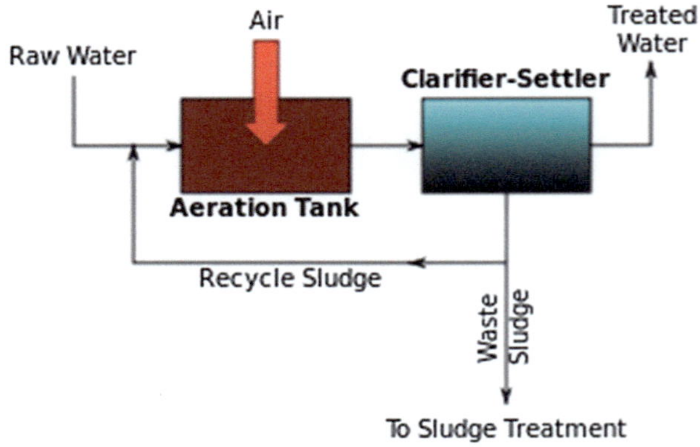

Fig. 3 General diagram of the activated sludge process

Biodegradable organic material can also be treated by using the trickling filter process. A trickling filter has a bed of rocks, gravel, etc., over which wastewater flows downward and contacts a layer of microbial slime covering the bed media. Air is forced through the bed to maintain aerobic conditions. For this process, the organic compounds in the wastewater are absorbed by the microbial slime layer. Thus the organic compounds are oxidized to form end products like carbon dioxide gas, water, etc. Fig. 4 shows a typical trickling filter system [6].

6 Treatment of Wastewater from Plating

An important category of the surface finishing industry is plating. This includes the process of electroplating which provides metals with surface coatings (usually by electrodeposition) for corrosion protection, wear resistance, improved electrical and mechanical properties, etc. Ferrous and non-ferrous materials can be electroplated with copper, nickel, chromium, zinc, lead, cadmium, etc. They can also be electroplated with precious metals such as gold and silver.

Various organic chemicals are added to plating solutions to improve deposit qualities and enhance smoothness, ductility, etc. However, due to the passage of direct current and electrolytic oxidation and reduction reactions, unwanted breakdown products accumulate in the solution. These products adversely affect the plating process and the physical properties of the deposited metal. Therefore, the use of activated carbon is very important to the metal finishing industry. Activated carbon is a major purification process for removing impurities and restoring the performance of plating solutions. Activated carbon is a form of carbon (usually derived from charcoal) that has small, low-volume pores that increase the surface area available for adsorption. It is usually used in water filtration systems. See Fig. 5 [7].

Plating baths include many chemicals so the plating wastewater contains solids, oil, grease, heavy metals, etc. The heavy metals are very toxic and the plating

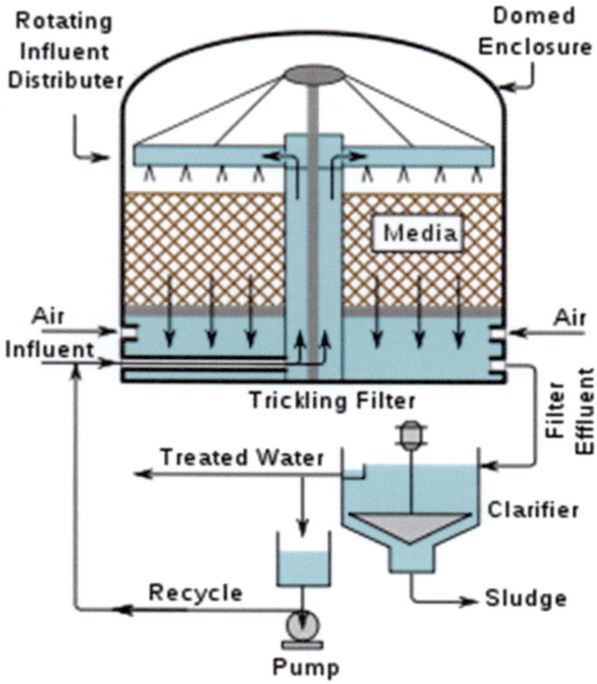

Fig. 4 A typical trickling filter system

wastewater has corrosive materials. Therefore, plating facilities must pretreat wastewater before discharging it in accordance with permits (of the National Pollutant Discharge Elimination System) as required by the Clean Water Act [8]. Large amounts of wastewater are generated due to the plating process and rinsing steps.

Conventionally, precipitation has been the method most often used to remove heavy metals. The hydroxide and sulfide precipitation methods are the main ones. Hydroxide precipitation is typically carried out between a pH of 8.5–9.5 for metals in the wastewater [8].

Chemistry is used for many aspects of wastewater treatment. Lime or caustic soda can be used to raise the pH, while sulfuric acid lowers it. Cyanide can be destroyed by introducing chlorine, as a gas or in liquid from (with sodium hypochlorite at a pH of 11) [9]. When enough chlorine is added the cyanide is converted to carbon dioxide and nitrogen. Hexavalent chromium can be reduced to trivalent chromium by using SO_2 gas or SO_2 releasing chemicals [9].

Surface finishing industries should determine and use the lowest concentration of chemicals needed to obtain a quality finish. Also whenever possible, they are encouraged to use less toxic alternatives to replace harmful chemicals.

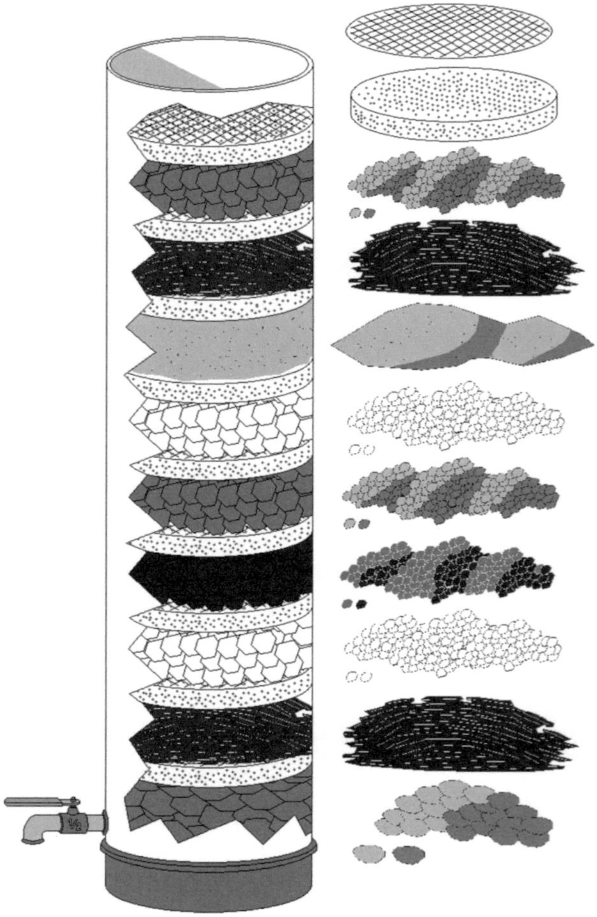

Fig. 5 Activated carbon used in a water filtration system. It resembles charcoal and is displayed in the fourth level from the bottom

7 Conclusion

It is necessary for environmental evaluation of surface finishing processes. From the viewpoint of current environmental protection acts and laws in various countries (which seem to be getting stricter) surface finishing industries are facing serious challenges. These industries tend to release heavy metals and other pollutants into our air, water, and soil. Therefore, to meet the demands of the environmental protection laws, they will need to modify current processes and propose some new ones to replace their traditional methods. This chapter described several ways to evaluate and treat harmful contaminants in air and water (especially those from the surface finishing industry).

References

1. https://en.wikipedia.org/wiki/Air_pollution
2. The Vindictive. File: Venturimistelim.gif. Date: December 2006. This work is in the public domain (It is a result of the U.S. Environmental Protection Agency in collaboration with North Carolina State University). https://commons.wikimedia.org/wiki/File:Venturimistelim.gif, https://en.wikipedia.org/wiki/Wet_scrubber
3. The Vindictive. File: Packedtowerex.gif. Date: December 2006. This work is in the public domain (It is work of the U.S. Environmental Protection Agency). https://commons.wikimedia.org/wiki/File:Packedtowerex.gif, https://en.wikipedia.org/wiki/Wet_scrubber
4. https://en.wikipedia.org/wiki/industrial_wastewater_treatment
5. Lenny CZ. File: Activated Sludge 1.svg. Date: November 2008. This work is in the public domain. https://commons.wikimedia.org/wiki/File:Activated_Sludge_1.svg, https://en.wikipedia.org/wiki/industrial_wastewater_treatment
6. Malyszkz. File: Trickle Filter.svg. Date: April 2011. This work is in the public domain. https://commons.wikimedia.org/wiki/File:Trickle_Filter.svg, https://en.wikipedia.org/wiki/industrial_wastewater_treatment
7. Ivan A. File: Water Filtration System Systems.png. Date: May 28, 2006. License: Creative Commons Attribution-Share Alike, 3.0. https://commons.wikimedia.org/wiki/File:Water_Filtration_Systems.png, https://en.wikipedia.org/wiki/Activated_carbon
8. http://www.pfonline.com/articles/treating-plating-wastewater
9. http://www.pfonline.com/articles/pollution-control-and-the-plating-industry(2)

Frequently Used Evaluations for Effluents

Miho Tanaka and Junichi Hirata

Abstract Heavy elements inevitably contaminate the effluents from surface finishing industries. Determination of the elements is important to estimate the level of contamination and thus proceeds for treatment of the effluents. We offer two types of instruments; ICP-AES, and ICP-MS for determination of the elements. The atomic absorption spectrometer requires an individual lamp for each element to be determined, whereas the ICP-MS offers multi-elemental analysis. For both, ICP-AES and ICP-MS, the apparatus required is to be cleaned in 5 % sodium hydroxide, 5 % nitric acid, and ultrapure water sequentially. Samples of effluents are taken in clean bottles, previously washed three times. All reagents and water are purchased at the ultrapure level. Standard solutions are also bought commercially, and appropriate dilutions are prepared for ICP-AES or ICP-MS. Measurements of the concentration of elements by using the ICP-AES and ICP-MS are repeated at least three times and a blank is checked simultaneously. A calibration line is calculated using 3 or 4 standard solutions and r (the slope) is ascertained, which should be near 1 to achieve accurate determinations. The concentration of elements is calculated from the calibration line and errors are estimated.

From this procedure, the concentration of elements is obtained for effluents. The concentration levels are compared to wastewater quality standards in the respective designated area (Japan or states in the United States). From the results obtained, the concentration of heavy elements can be controlled or monitored.

1 Introduction

During and after the process of surface finishing, effluents are contaminated with heavy metals. The wastewater quality standards are set by the government. In Japan, unique standards are set by the Government. In Table 1 [1], permitted concentrations in effluents are shown.

The concentration is expressed as mg/L = ppm because the density of water is the same as 1 L = 1 kg. In addition, many metal concentrations are required to be

M. Tanaka (✉) • J. Hirata
Tokyo University of Marine Science and Technology, Tokyo, Japan
e-mail: mihotnk@kaiyodai.ac.jp; junichi-hirata@mail.goo.ne.jp

H. Kanematsu, D.M. Barry (eds.), *Corrosion Control and Surface Finishing*,
DOI 10.1007/978-4-431-55957-3_13

Table 1 Waste water quality standards in Japan [1]

	Permitted conc. (mg/L)
Cd and its compound	0.1
CN compound	1
Lead and its compound	0.1
Cr(IV) compound	0.5
As compound	0.1
Hg and its compound	0.005
Se and its compound	0.1

below the wastewater quality standard values. The atomic absorption spectrometer enables us to determine such low concentrations of some metals, but the measurement of each element (e.g. Cd, Pb, Cr, As, Hg and Se) is required to be carried out individually [2]. Recently, these elements were determined simultaneously with instruments, such as an Inductively Coupled Plasma Atomic Emission Spectrometer (ICP-AES) [3–6], a similar instrument known as an Inductively Coupled Plasma Optical Emission Spectrometer (ICP-OES), and an Inductively Coupled Plasma Mass Spectrometer (ICP-MS), [7–11]. ICP-AES and ICP-MS have the same ionization process, but the detection system is different. In the system of ICP-AES, the ionized light is separated by a spectrometer due to the unique wavelength (nm) of the element. The *m*/*z* is detected in the mass spectrometer in the ICP-MS system. It is possible to detect 0.5–20 ppm concentration of metals by using the ICP-AES, and the ICP-MS can detect 1 ppt concentration of metals. The following section covers the details of the ICP system.

2 Principle of ICP System [4]

Optical emission spectra are observed by dispersing the radiation according to wavelength so that photons ($h\nu$) of different frequency (ν) appear in the focal plane of a spectroscopic apparatus as an array of monochromatic images at the entrance of the slit. These images are characterized by their wavelengths (λ) and called "spectral lines" as shown in Fig. 1. Atomic spectra are composed of discrete spectral lines. The wavelength of each line is related to the frequency and to the energies at atomic levels (q,p) between which the transition takes place by the equation

$$h\nu_{qp} = hc/\lambda_{\mathrm{qp}} = \mathrm{E_q} - \mathrm{E_p}$$

where c is the speed of light and h the Plank constant. Generally, the wavelength of spectral lines is expressed as nanometers (nm). For example in Fig. 1, 'a' line corresponds to 285.212 nm, and 'e' line 279.553 nm both of which represent Mg^{2+}.

An atomic absorption spectrometer requires a lamp of unique wavelength (e.g. **b** or **d** nm) for each exciting element [2]. The wavelengths of elements are

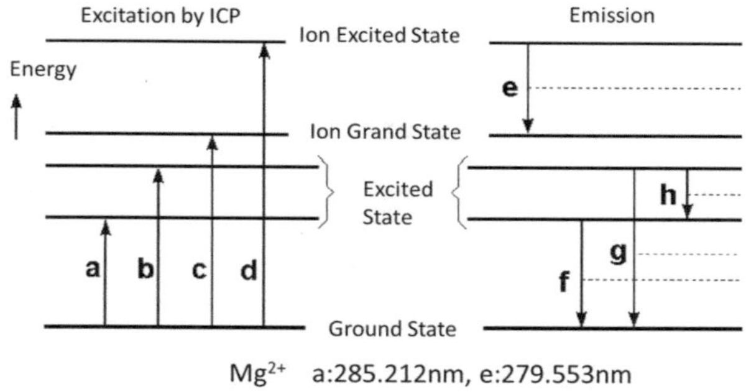

Fig. 1 Atom excitation by ICP and emission

listed in Table 2. As these wavelengths are unique for each element, it is easily identified and determined, that is with the ICP-AES. However, in the case of ICP, once the element is ionized by the plasma source, the spectrometer detects the unique wavelength which is radiated from the excited element (e.g. **e** or **g** in Fig. 1). Therefore, the wavelength is measured by the spectrometer and determination of the element is attained. However, in order to improve high sensitivity, wavelengths of elements are identified by a mass-to-charge ratio (mass number/charge) and their peak intensity. Figure 2 shows basic structures of the ICP-AES and ICP-MS [3].

The torch is where the main ionization process takes place at temperatures of 6000–10,000 K. This is induced by an RF power unit and a plasma RF matching unit.

In case of failure in matching, the torch cannot be maintained and is easily destroyed. The nebulizer is a kind of spray chamber where the sample is converted into a fine mist and transferred to the skimmer of the spectrometer. High salts present in the sample effluents could easily block the tip of the nebulizer due to adiabatic expansion occurring at the tip. The above two mentioned sections of the instrument are to be carefully maintained and treated to obtain optimum results.

3 Instrument

The detection system of ICP-AES is of two types: simultaneous and sequential spectrometers with radial viewed torches in Fig. 3. A simultaneous type, where many elements can be determined in a single scan, is displayed in Fig. 3a, and a sequential type where one element in a scan is determined is in Fig. 3b.

The detection system of ICP-MS is also of two types: Q-mass in Fig. 4a and Sector-Field (SF) types in Fig. 4b, c as shown. In a Q-mass type, the detection limit is lower than ICP-AES, and in the Sector-Field type the detection is the lowest among ICP-AES, Q-mass type ICP-MS, and SF type ICP-MS. However, a SF type ICP-MS is expensive.

Table 2 Recommended wavelengths and estimated instrumental detection limits (IDLs) for radial view ICP/AES [2]

Element	Wavelength (nm)	Estimated IDL (mgL^{-1})
Aluminum	308.215	30
Antimony	206.833	21
Arsenic	193.696	35
Barium	455.403	0.87
Beryllium	313.042	0.18
Boron	249.678 ×2	3.8
Cadmium	226.502	2.3
Calcium	317.933	6.7
Chromium	267.716	4.7
Cobalt	228.616	4.7
Copper	324.754	3.6
Iron	259.94	4.1
Lead	220.353	2.8
Lithium	670.784	2.8
Magnesium	279.079	20
Manganese	257.61	0.93
Mercury	194.227 ×2	17
Molybdenum	202.03	5.3
Nickel	231.604 ×2	10
Phosphorus	213.618	51
Potassium	766.491	–[a]
Selenium	196.026	50
Silicon	251.611	17
Silver	328.068	4.7
Sodium	588.995	19
Strontium	407.771	0.28
Thallium	190.864	27
Tin	189.98 ×2	17
Titanium	334.941	5
Vanadium	292.402	5
Zinc	213.856 ×2	1.2

[a]Detection limit subject to plasma conditions

4 Preparation

4.1 Apparatus

First, the apparatus is to be cleaned to avoid contamination. One hundred millilitre (polyethylene; PE) volumetric flasks, 250 mL (PE) sampling bottles, and tips for pipette filler (polypropylene; PP) etc. are all placed in a 5–10 % sodium hydrate solution with ultrapure water for a day, then after cleaning with ultrapure water,

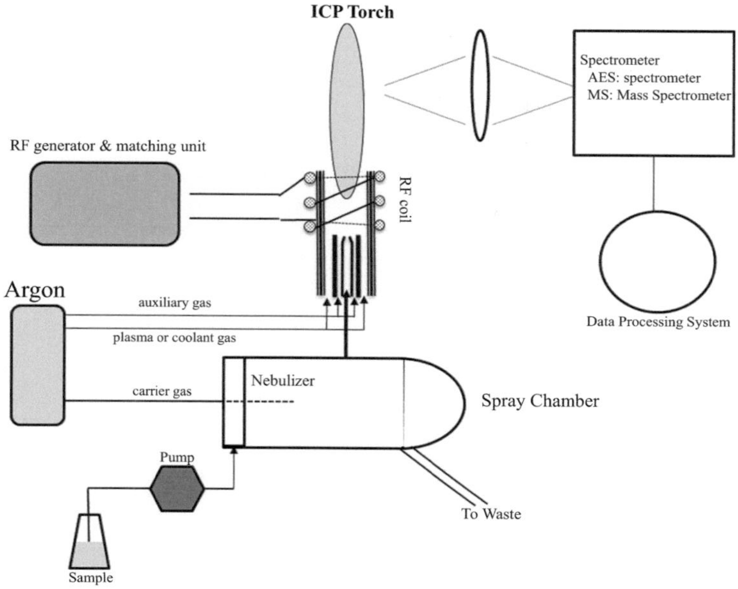

Fig. 2 System of ICP-AES and ICP-MS

they are immersed in 5–10 % nitric acid solution with ultrapure water for a day. They are then cleaned with ultrapure water and soaked in ultrapure water for a day, twice. When the element concentration in the effluent sample (ppm order) is high compared with a blank, determination is sufficient to measure by ICP-AES. For determination by ICP-AES, glass apparatus, PE and PP are permitted, but for ICP-MS, PE apparatus is recommended as glass volumetric flasks contain high concentration of sodium, calcium, silicon, etc. PP volumetric flasks are the best choice when using ICP-MS as the Thermo Scientific concentrations of these elements are low, but the material becomes discolored with long use.

4.2 Sampling

Sampling is important for the determination of elements in effluents. The sampling bottle is to be well cleaned, and it is rinsed with the effluent sample several times before collecting.

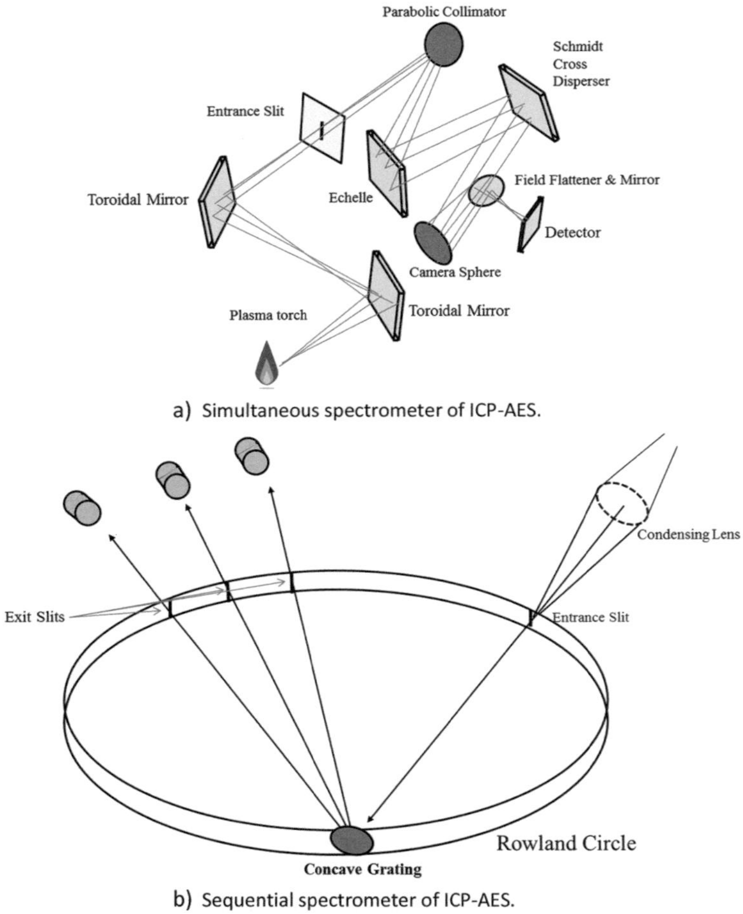

Fig. 3 (**a**) Simultaneous spectrometer of ICP-AES. (**b**) Sequential spectrometer of ICP-AES

4.3 Reagents

For the determination of elements using ICP-AES or ICP-MS, the reagent concentration is to be set lower than the metal concentration in the sample. Reagents may contain elemental impurities which can affect the quality of analytical results. Thus minimum analyte contamination in reagents is sought. Kinds and amount of reagents used should also be minimized. Use of high purity or trace element grade reagents is usually required; ultrapure grade is preferable.

Safety Note: *Reagents should be regarded as potential health hazards and exposure to these compounds should be limited. Safety data sheets for these chemicals (SDS) are to be available to the user.*

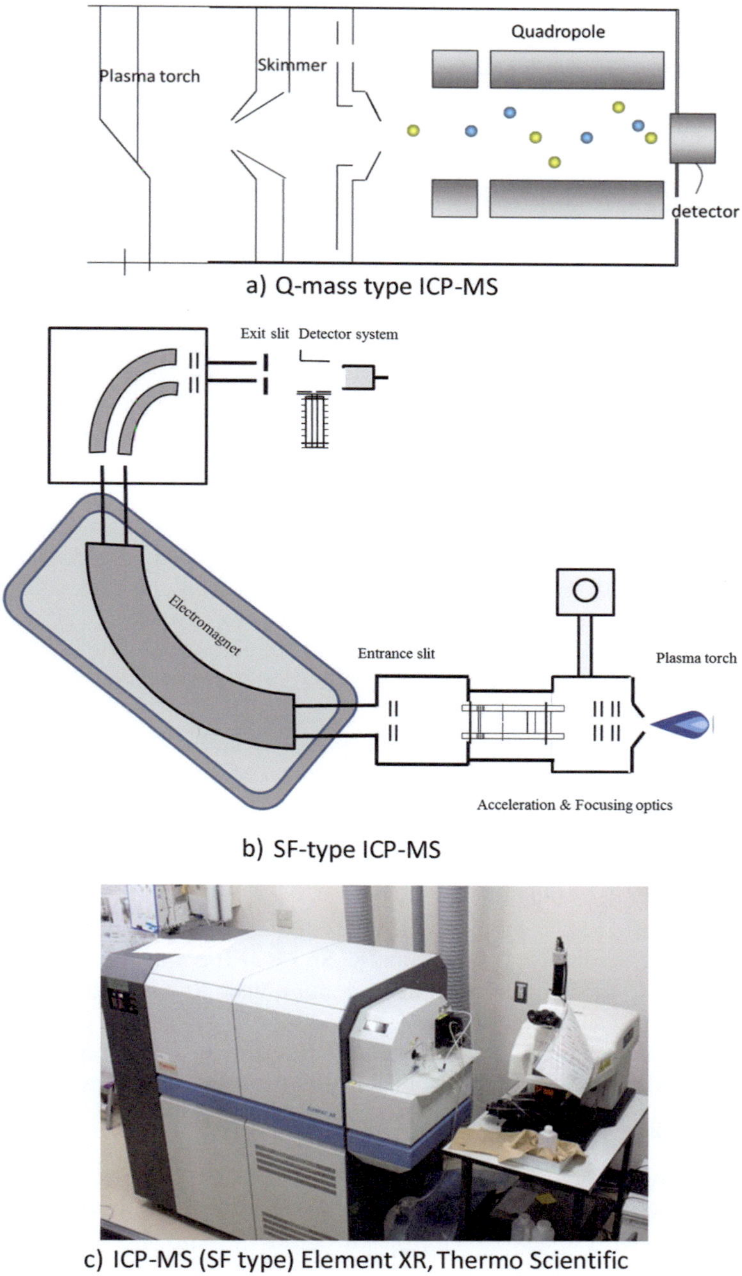

Fig. 4 (**a**) Q-mass type ICP-MS. (**b**) SF-type ICP-MS. (**c**) ICP-MS (SF type) element XR, thermo scientific

1. Reagent water – Water that meets specifications for resistivity 18.2 M Ω and ultrapure water.
2. High purity nitric acid – Concentrated (sp. gr. 1.41), trace element grade or double distilled.
3. Hydrochloric acid – Concentrated (sp. gr. 1.18), trace element grade or double distilled.

Standard solution for ICP-AES

4. Stock standard solutions – Commercially prepared single element solutions prepared specifically for spectrometric analysis (usually 1000 mg/L). Stock standard solutions may also be prepared in the laboratory from high purity ($\geq$99.99 %) metals or salts. Alternatively, commercial multi-element solutions prepared specifically for spectrometric analysis can be used. These multi-element solutions will be much lower in concentration (typically 10–500 mg/L) than single element solutions to avoid compatibility problems.
5. Intermediate standard solution – Prepared to contain appropriate concentration of analytes for preparation of standard solutions. Pipet an appropriate volume of stock standard solution(s) into an acid rinsed volumetric flask and dilute to volume with 1 % nitric acid. Alternatively, intermediate standard solutions may be prepared gravimetrically by measuring stock standard solution and 1 % nitric acid masses multiplied by solution density in a 125 or 250 mL plastic bottle. Store prepared intermediate standard solutions in plastic bottles. Alternatively, commercial multi-element solutions prepared specifically for spectrometric analysis can be used.
6. Standard solutions – Prepare at least three standard solutions by combining appropriate volumes of stock standard solutions or intermediate standard solutions in volumetric flasks. Dilute to volume with 1 % nitric acid. Chemical compatibility (i.e., of analytes, acids, etc.) must be considered to avoid the formation of analyte precipitates when mixing single element stock solutions to prepare standard solutions.

   Calibration range: group-2 elements 0.01–10 ppm, transition metals 0.05–10 ppm, Li and Na 0.1–50 ppm, K, Sr, Cs 1–100 ppm.

   Do not use standard solutions that are more than 30 days old since element concentrations can change with age.
7. Standard blank – 1 % nitric acid. Prepare sufficient amount for use in standardization, determination and for nebulizer rinse between each measurement.

Standard solution for ICP-MS

8. High quality custom-made multi-element solutions are commercially available and are recommended. We recommend "SPEX CertiPrep". Transfer prepared standard solutions to acid cleaned plastic bottles (Teflon® FEP is preferred) for storage. Standard solutions may also be prepared by gravimetric dilution. Gravimetric dilution can be performed by measuring mass of stock or intermediate standard solution and 1 % nitric acid masses in a 125 or 250 mL plastic bottle. Volumes are calculated from solution densities.

Table 3 Internal standard for ICP-MS [11]

Measure elements	Internal standard 1	Internal standard 2
Cr	45 Sc	59 Co
Mn	45 Sc	59 Co
Ni	45 Sc	59 Co
Cu	59 Co	89 Y
Zn	59 Co	89 Y
As	89 Y	115 In, 103 Rh
Se	89 Y	115 In, 103 Rh
Mo	89 Y	115 In, 103 Rh
Cd	115 In	89 Y, 103 Rh
Sb	115 In	89 Y, 103 Rh
Pb	205 Tl	209 Bi

Calibration range: elements except Al, Ge, Sb and As 1 ppt–1 ppm (Q-mass type), 1 ppt–10 ppm (Sector type), Al, Ge, Sb and As 0.1 ppb–1 ppm.

Do not use standard solutions that are more than 30 days old since element concentrations can change with age.

9. Internal Standard

For determination, external standard method or internal standard method is used. In ICP-MS, an internal standard method is applied [9]. For internal standards, Thallium and Indium are generally utilized for heavy metals. In Table 3, element solutions are listed as internal standard solutions. The objective element is sandwiched by two internal standards as shown in Table 3. Usually, these solutions are purchased as reagents. After ascertaining that these elements are very low in an original sample solution, these internal standard solutions are put into a sample solution: concentration of an internal standard should be almost 100 times higher than that of elements in the sample solution.

Standard blank: 1 % nitric acid. Prepare sufficient amount for use in standardization, determination, and for nebulizer rinse between each measurement.

4.4 Sample

Sample solution is prepared in several concentrations, such as 10 times, 10 times and 1000 times diluted with 1 % nitric acid. If element concentration in sample solution is very high, then a 10,000 times diluted solution is required.

Table 4 Elements and measured mass numbers [11]

Element	Mass number	Element	Mass number	Element	Mass number
Na	23	Zn	66	Ta	181
K	39	Se	82	Pb	208
Ca	43	Rb	85	Th	232
Sc	45	Mo	95	In (internal standard)	115
Ti	47	Cd	111		
V	51	Sb	121		
Cr	52	Cs	133		
Mn	55	Ba	137		
Fe	57	La	139		
Co	59	Ce	140		
Ni	60	Sm	147		
Cu	63	Hf	178		

5 Measurement

In Table 2 the wavelengths and instrumental detection limits (IDLs) of elements for ICP-AES are listed. And in Table 4, elements and measured mass numbers are listed for ICP-MS. For example, Cu has a mass number of 63 and 65. This list shows that the peak for mass number 63 is better than that of 65 to detect the intensity of Cu, because the existence of ^{63}Cu is three times higher than that of ^{65}Cu.

Measurement and reproducibility should be at least repeated and checked three times. The blank should be measured and each measurement should be "very low" or negligible.

6 After Measurements

6.1 ICP-AES

As output of ICP-AES is a Lorenz curve represented by wavenumber vs. Intensity count (cps), the concentration of an element is calculated by height of cps or area. The calibration line which is obtained with three or four standard solutions is calculated by a least squares method, and $r = 1$ is favorable. The concentration of the elements is calculated from the calibration line and errors are to be estimated.

From this procedure, detection values are obtained for the elements in the effluents. After determination of the elements, the results are compared with wastewater quality standards in a designated area (Japan or states in the United States). If the determined concentrations are high, they must be decreased by suggesting a method of treatment.

Table 5 Molecule ions disturbed by acids [11]

m/z	Interfared elements	HNO_3	HCl	H_2SO_4
20	Ne(90.5 %)	OH_2		
21	Ne(0.27 %)	OH_3		
22	Ne(9.2 %)			
23	Na(100 %)			
24	Mg(79.0 %)			
25	Mg(10.0 %)			
26	Mg(11.0 %)			
27	Al(100 %)			
28	Si(92.2 %)	CO, N_2		
29	Si(4.7 %)	N_2H, COH		
30	Si(3.1 %)	NO		
31	P(100 %)	NOH		
32	S(95.0 %)	O_2		S
33	S(0.75 %)	O_2H		SH, S
34	S(4.2 %)	O_2		S, SH
35	Cl(75.8 %)	O_2H	Cl	SH
36	S(0.02 %), Ar(0.34 %)	Ar		ClH
37	Cl(24.2 %)	ArH		Cl
38	Ar(0.06 %)	Ar		ClH
39	K(93.2 %)	ArH		S
40	Ar(99.6 %), K(0.01 %), Ca(96.9 %)	Ar		SH
41	K(6.7 %)	ArH		
42	Ca(0.65 %)	ArH_2		
43	Ca(0.14 %)			
44	Ca(2.1 %)	CO_2		
45	Sc(100 %)	CO_2H		

6.2 *ICP-MS*

In the case of Q-mass type ICP-MS, as cps is plotted against mass number, an element and a molecule of a same mass number could emerge simultaneously, such as ^{28}Si(92.2 %), N_2, and CO as shown in Table 5. These interfering molecules are contributed by air or acid, which overlap the mass number of the objective element and should be corrected for by various methods. This can be improved with a SF type of mass spectrometer, where high resolution is attained and peaks derived from ^{28}Si(92.2 %), N_2, and CO are separated from each other.

An example of output of ^{208}Pb observed by a SF type ICP-MS (Element XR, Thermofisher Sci.) is shown in Fig. 5. As you can see, spectrum shape is clearly shown. Spectrum should be a Lorenz curve in Q-mass type, and that should be a trapezoid in SF type ideally.

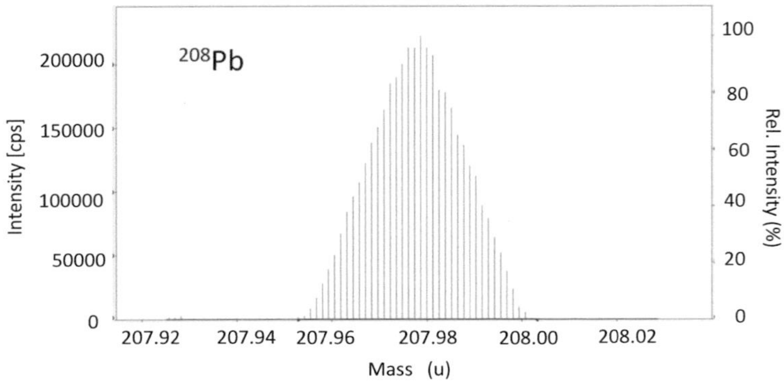

Fig. 5 Output of mass spectra

6.3 Calibration Curve

The calibration curve which is obtained with three or four standard solutions is calculated by a least squares method, and $r = 1$ is favorable as in the case of ICP-AES and ICP-MS. As output of ICP-MS is a clear peak shape with mass number *ver.* Intensity count (cps), concentration of an element is calculated by peak area in Fig. 6. Determination values are calculated from a calibration curve and errors are to be estimated.

7 Conclusion

The elements are inevitably present in effluents produced by surface finishing industries. Determination of these elements is important to estimate the contamination level for monitoring purposes. We offer two types of instruments; ICP-AES, and ICP-MS. The atomic absorption spectrometer requires a single lamp for each element, whereas in ICP-MS a multi-element determination is possible. The apparatus utilized for ICP-AES and ICP-MS, is to be cleaned with 5 % sodium hydroxide, 5 % nitric acid, and ultrapure water sequentially. Sampling of effluent requires the bottles to be cleaned three times. All reagents and water are purchased as ultrapure levels. Standard solutions are also purchased commercially as ICP-AES or ICP-MS solutions. Measurements by ICP-AES and ICP-MS are repeated at least three times, and a blank must be checked. A calibration line is calculated with three or four standard solutions and r should be ascertained almost 1. Concentration of the elements is calculated from the calibration line and errors are to be estimated.

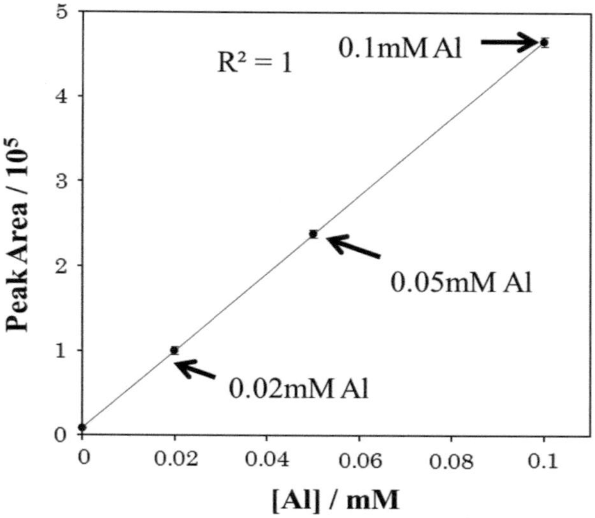

Fig. 6 Calibration curve of aluminum by ICP-MS

References

1. Homepage from Ministry of Environment, Japan. http://www.env.go.jp/water/impure/haisui.html
2. Atomic absorption spectrometry Leaflet from The Royal Society of Chemistry Fine Chemicals and Medicinals Group. http://faculty.rmu.edu/~short/chem3550/. . ./RSC-AA-Leaflet.pdf
3. Tatro ME (2008) Optical emission inductively coupled plasma in environmental analysis. In: Meyers RA (ed) Encyclopedia of analytical chemistry. Wiley, Chichester, pp 1–7
4. Boss CB, Fredeen KJ (1997) Concepts, instrumentation and techniques in inductively coupled plasma optical emission spectrometry, 2nd edn. Perkin Elmer, Shelton, pp 3–19, 3–31
5. Boumans PWJM (ed) (1987) Inductively coupled plasma emission spectroscopy Part 1. Wiley, New York, pp 1–44
6. Hou X, Jones B (2000) Inductively coupled plasma/optical emission spectrometry. In: Meyers RA (ed) Encyclopedia of analytical chemistry. Wiley, Chichester, pp 9468–9485
7. Thomas R (2007) Practical guide to ICP-MS A tutorial for beginners. In: Method: ICP-MS application environmental, 2nd edn. CRC Press, Boca Raton, pp 115–135, 204–208
8. Adams F, Gijbels R, van Grieken R (eds) (1988) Inorganic mass spectrometry. Wiley, pp 257–300 (Gray AL, Inductively coupled plasma source mass spectrometry, pp 257–300)
9. Vanhaecke F, Vanhoe H, Dams R, Vandecasteele C (1992) The use of internal standards in ICP-MS. Talanta 39(7):737–742
10. Giessmann U, Greb U (1994) High resolution ICP-MS-a new concept for elemental mass spectrometry. Fresinius' J Anal Chem 350:186–193
11. Homepage from Ministry of Environment, Japan. www.env.go.jp/air/osen/pm/ca/120419/manual-3.pdf

Frequently Used Evaluations for Aerial and Solid Pollution

Kazuya Takahashi

Abstract Analytical methods, gas chromatography (GC) and ion chromatography (IC), are explained in this chapter. Both methods have been used conventionally for the analyses of air pollution and in aqueous chemistry. Gas chromatography exhibits extremely high sensitivity for organic and inorganic components in gas samples. GC can be applied to the analyses of gas samples to examine atmospheric contaminants. It is used to monitor the pollution in exhausted gas from transportation and manufacturing industries. The other applications are the monitoring of various reactions in the gas phase or the surface of materials in basic chemical research fields. Ion Chromatography is widely used to determine ionic species in solution, including effluent solutions from surface treatment plants. In order to yield a high sensitivity and throughput, many improved devices have been developed for IC. The basics of chromatography are also described with examples of actual applications in this chapter.

1 Introduction and the Principle of Chromatography

Chemical analyses for environmental samples have been supported by several fundamental techniques, such as atomic spectroscopy, chromatography, and so on. One of the most important techniques is chromatography which is a method of continuous chemical separation. In this method, the separating components distribute between two phases. One phase is fixed in place and called the stationary phase, while the other flows over this stationary phase and is known as the mobile phase. The components to be separated are continuously distributed differentially between these two phases to result in a chemical separation. Separation using chromatography is carried out by the continuous movement of the separating components between the stationary and mobile phases to identify and analyze them.

K. Takahashi (✉)
Nishina Center for Accelerator-Based Science, Riken, Tokyo, Japan
e-mail: kazuyat@riken.jp

H. Kanematsu, D.M. Barry (eds.), *Corrosion Control and Surface Finishing*,
DOI 10.1007/978-4-431-55957-3_14

The principle of chromatography can be simply described as multiple extraction stages, which consist of a series of mobile and stationary phases. In this virtual extraction series, a chromatogram can be simulated according to the following equation.

$$C_{n,r} = \frac{n!}{r!(n-r)!} \bullet \frac{D^{r+1}}{(D+1)^{n+1}} \quad (1)$$

In this equation, "n" is the virtual number of extraction stages (number of theoretical plates), $C_{n,r}$ corresponds to an abundance of components in the virtual mobile phase at rth extraction, and D is the coefficient of the component between the mobile phase and the stationary phase. When D of components "A" and "B" is 1 and 4, respectively, the chromatogram of the components "A" and "B" for 30 extraction stages and 100 extraction stages can be shown as Fig. 1a, b) respectively. As shown in this figure, the chromatogram with many stages of extraction exhibits better resolution to distinguish components. Based on this fact, it is better to select a separating system with a "large number of extraction stages". In real chromatography, r in the above equation is the "retention time".

Several types of chromatography have been developed and they are characterized by the nature of the stationary and mobile phases. Gas chromatography (GC) is a system where the mobile phase is a gas and is used for separating volatile components to be detected. In a system where the mobile phase is a liquid, it is called Liquid Chromatography (LC), especially, the system in which the stationary phase is an ion-exchange resin and the separation is an ion exchange process. In this chapter, GC and IC, which are used to examine air and land pollution, will be explained.

2 Gas Chromatography (GC)

For the analyses of gas samples, including mist or aerosol, Gas Chromatography (GC) has been employed as a conventional method. The GC system exhibits the following merits.

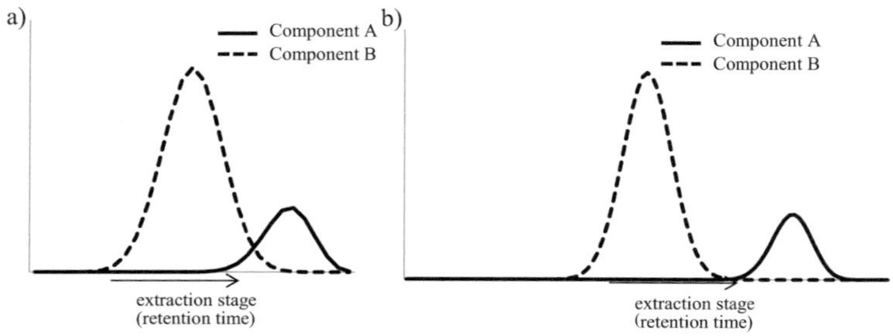

Fig. 1 A chromatogram simulated as a multi stage extraction

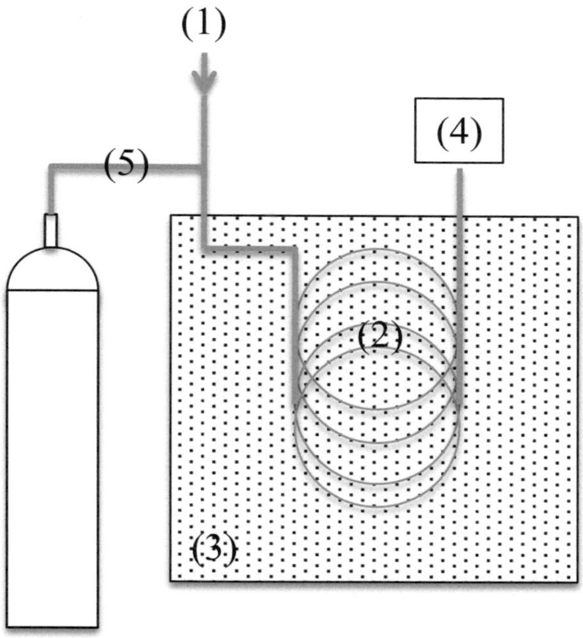

Fig. 2 Schematic diagram of a gas chromatograph

1. GC exhibits such large theoretical extraction stages (more than hundred thousands) that analysis with a high resolution power is achieved.
2. A high abundance sensitivity nearly 10^{-15} g can be achieved by an appropriate combination of carrier gas and detector system.
3. Equilibrium is quickly achieved between the stationary phase and mobile phase. Also the measurement time is very short.

On the other hand, the limitations of GC are as follows; the sample should be a gas or the vapor pressure of the sample should be more than a few hundred Pa under the temperature set by the control unit of that system. Figures 2 and 3 show a schematic GC system and a photograph of a GC instrument, respectively. As shown in Fig. 2, a GC system consists of (1) sample inlet, (2) separation column, (3) a temperature control unit, (4) a detector with a data handling system and (5) a carrier gas with a flow control device. In this GC system, a carrier gas (in many cases, an inert gas such as nitrogen or helium) flows continuously from (a container) through the sample injection port, the column, and then the detector. The sample is injected into an inlet where it is vaporized and carried into the column.

The column is the essential part of GC and there are two types. One is a packed column with various materials, such as porous polymer beads, molecular sieve, silica gel, activated carbon and aluminum oxide. They have been used for a stationary phase packed in the column. These packed columns are usually 1/4- or 1/8-in. outside diameter tubes of stainless steel or glass. The other is the open tubular capillary column where the stationary phase, such as silicon-organic

Fig. 3 A gas chromatograph

polymer, is coated in a narrow-bore fused silica or stainless steel. The latter columns exhibit higher resolution power than the former column. Recently, the open tubular capillary column has been commonly employed on the GC system. After the sample passes through the column, the carrier gas with separated components enters the detector.

There are three types of detectors; Thermal Conductivity Detector (TCD), Flame Ionization Detector (FID) and Electron Capture Detector (ECD). In these detector systems, FID exhibits the highest sensitivity for the detection of hydrocarbon materials and TCD has been used commonly for the detection of inorganic components and organic materials except for hydrocarbons. The ECD is used specifically for the detection of materials with large electronegativity, such as alkyl halides. Recently, GC has been equipped with a mass spectrometry system (GC-MS).

In the process of measurements by GC, the column has a thermostat in order to separate the components at a reproducible temperature. The partition coefficients depend critically on the temperature of column and the temperature control is the most effective method to influence the separation. The system is supplied with an isothermal and programmed temperature control. Isothermal means the chromatographic column is maintained at a constant column temperature and programmed temperature means there is a linear increase in temperature with time to improve the separation. Generally, an isothermal operation, limits the analysis to narrow the boiling range of materials. The programmed temperature control operation is very useful for analysis of a wide range of boiling mixture samples. It is possible to control the temperature according to the materials to be analyzed.

For the evaluation of atmospheric environments, GC is very useful. It is defined as a standard method to analyze the air for pollution. In many nations, some national organizations are permitted to monitor the environment. In Japan, the

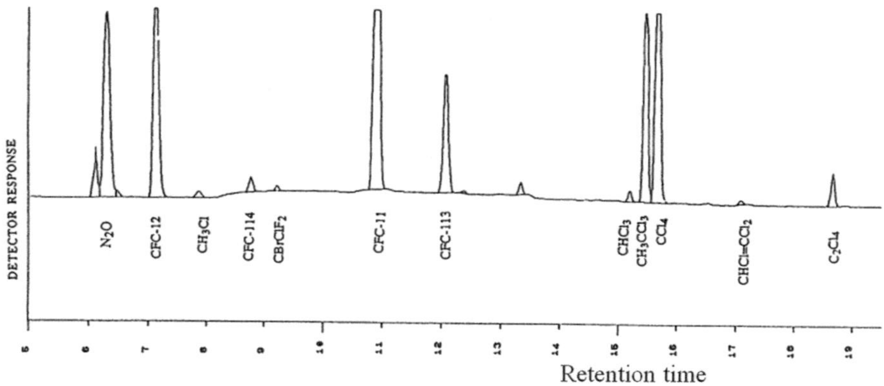

Fig. 4 A GC chromatogram for air sample taken at 1315 m altitude

Ministry of the Environment defines the GC equipped with several devices, such as a mass spectrometer, as one of the essential methods to analyze polluted materials in exhaust gas from automobiles, industries, and so on. For the monitoring of air pollution, toxic substances such as nitrogen oxides (NOx), sulfur oxides (SOx), carbon oxide (CO) and hydrocarbons are the targets to be analyzed. For example, GC has been applied to the monitoring of hazardous materials in atmosphere. Figure 4 shows an example of a chromatogram of an air sample taken at 1315 m altitude and several kinds of halocarbons are observed with nitrogen oxides [1]. Similarly, GC has also been used for the analyses of several organic materials, such as aldehydes or PAH (polycyclic aromatic hydrocarbon), which are produced from building materials which cause the "sick- house syndrome". GC has also been employed widely in scientific research fields to examine exhaust gases from surface-treated materials or catalytic reaction processes. Figure 5 shows a chromatogram of the reactants contained in the air from the surface of heated materials covered with some polymer [2]. As shown in this figure, several components produced by pyrolysis from a polymer covering (material) were identified to characterize the covering polymer.

To identify the components and perform the quantitative analysis, several standard materials have been commercialized. For example, the standard reagents including several aldehydes and aromatic hydrocarbons are provided for the working environmental measurements.

3 Ion Chromatography (IC)

As mentioned before, ion chromatography (IC) is the method to analyze ionic species in solution. Basically, in IC systems, the stationary phase is an ion-exchange resin and the mobile phase (eluent) contains an electrolyte providing

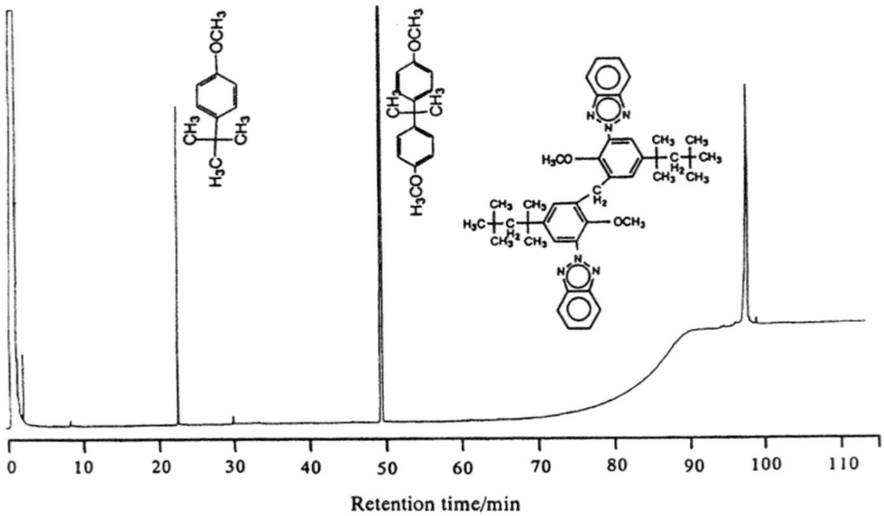

Fig. 5 A GC chromatogram of the reactants from a polymer covering materials

OH^- or H^+ to the eluent. For the analysis of cations, a cation-exchange resin is used as the stationary phase and, for the analysis of anions, an anion-exchange resin is used. Ion chromatography was originally developed to perform process monitoring or process control applications. The capabilities of IC have been improved; faster sample throughput (around 50 samples/day), higher sensitivity (ppb level of species can be detected), improved detection, and a greater ability to determine ionic species which were previously difficult or impossible to be detected. At present, various kinds of ion-exchange resin, eluent (and the control system of eluent) and a detection system with post-separation treatments have been developed. Also a routine analysis of cations and anions is available for many types of samples. Figures 6 and 7 show a schematic diagram for a standard IC system and a photograph of an IC instrument, respectively.

Ion chromatography consists of the following parts; (1) eluent generation system, (2) pump system, (3) sample injection system, (4) column system, (5) thermostat chamber, (6) suppressor, (7) detector and data handling system.

All samples are to be prepared as a solution and each sample is injected through an inlet system with a loop. The volume of injected sample is fixed according to the volume of the loop. An operator may select the appropriate size loop according to the application. First, the sample solution is injected into the loop with an injection syringe. A larger volume than that of the loop is injected so that the loop is filled with the sample solution. Then, the injection pump squeezes out the sample solution into the eluent flow.

The pump system maintains the eluent containing the sample solution with a stable flow speed, such as a 1 ml/min. To keep a stable background in a

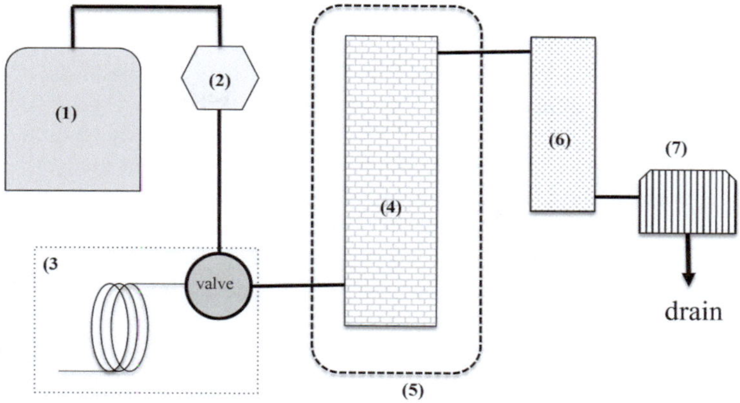

Fig. 6 Schematic diagram of an ion chromatograph

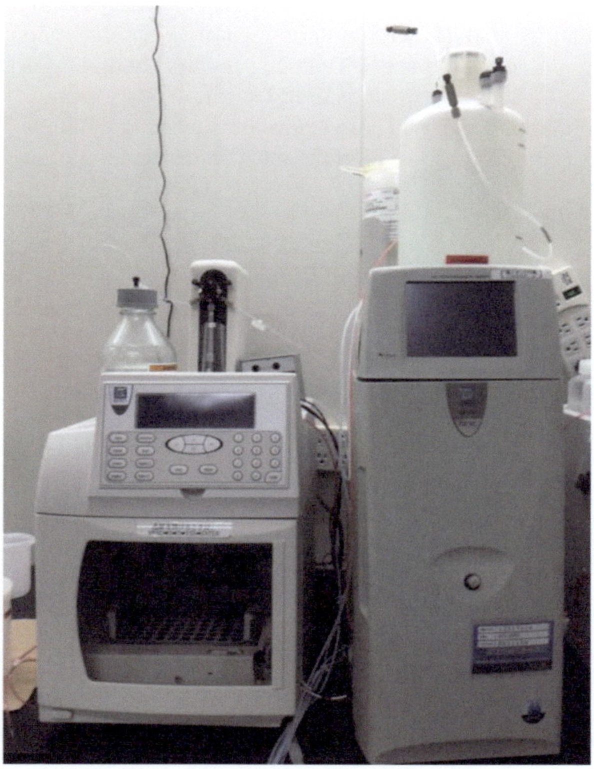

Fig. 7 An ion chromatograph

chromatogram and low detection limits, the pump is designed to be pulsation-free with low drift.

The eluent system is very important in IC and various combinations of eluents and columns have been used for the separations. For cation analysis, acids such as hydrochloric acid or nitric acid are used as eluents. Recently, methane-sulfonic acid has also been employed as the electrolyte in eluent. For anion analysis, sodium hydroxide, potassium hydroxide, sodium bicarbonate, and sodium carbonate have been used as the electrolyte in eluent. When hydroxide eluents are used, the background is effectively reduced by a suppressor compared with the case of the use of carbonate eluents. On the other hand, a significant carbonate peak appears on the chromatogram due to the use of hydroxide eluents and care is taken to locate positions of nitrate and sulfate on the chromatogram. Conventionally, the eluent is manually prepared with an appropriate concentration, but recently, an eluent generator system has been developed. It uses a computer to control varying concentrations of eluent. The condition to separate and detect ionic species under constant concentration of eluent is called isocratic. Sometimes, in order to yield high separation efficiency in a short time, the concentration of electrolyte in eluent is changed (increased). This method is called gradient analysis. Figure 8 shows an example of an anion analysis with programmed gradient eluent concentrations. In this chromatogram, ten anion species can be separated and identified in 17 min.

Basically, cation-exchange resin is used for cation analysis and anion-exchange resin is used for anion analysis. These resins are packed in columns. Various resins are developed for applications to separate a wide variety of cationic species.

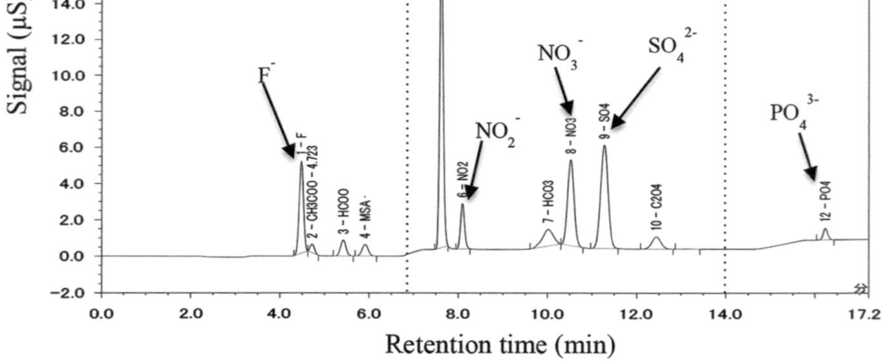

Fig. 8 An example of a chromatogram for anion analysis

In many cases, the column is set in a thermostat chamber and equipped with a guard column device to protect the main column from impurities present in the sample.

Usually a conductivity device is used for the detection of ionic species with a data accumulation system. The conductivity relates the intensity of the signal peak of each ionic species to its concentration and is noted as S (siemens).

The conductivity detector device monitors the change in conductivity of the solution, depending on the concentration of ionic species. In order to yield high sensitivity, suppression of the background signal is essential for the IC system. As mentioned above, the IC system usually has the conductivity detection device as a detector. The eluent itself contains electrolytes rendering a high background which results in very poor signal-noise characteristics. The suppressor device has a role to reduce the background signal. For anion analysis, the suppressor provides H^+ to the sample solution with eluent after the column and the following reactions occur.

$$H^+ + OH^- \rightarrow H_2O \qquad (2)(\text{when alkaline hydroxides are used as eluent})$$

$$H^+ + HCO_3{}^- \rightarrow H_2CO_3 \qquad (3)(\text{when carbonates are used as eluent})$$

As it is well-known, the dissociation constant of water (H_2O) is 10^{-14} and that of carbonate is 4.45×10^{-7} (the first step of dissociation). Thus the reverse reactions hardly occur and OH^- or $HCO_3{}^-$ is removed to reduce the total background conductivity. The anion species in the sample remain and they are enhanced to be analyzed with high sensitivity. On the other hand, for cation analysis, the suppressor device provides OH^- and the same reaction (1) occurs to reduce background problems. The suppressor devices are of several types. Chemical and electrolytic ones have been developed. As a representative device of a chemical suppressor, a cation or anion-exchange column has been used and these suppressor columns exhibit high capacity to provide H^+ or OH^- ions. However, a regeneration process is required after the suppressor column exhausts the H^+ or OH^- ions. Recently, an electrolytic suppressor device has been developed and commercialized, which provides suppressor ions (H^+ or OH^-) to the eluent flow after the separation column. These suppressor ions are electrolytically generated from water through an ion-exchange membrane. The delivery rate of suppressor ions can be controlled according to the concentration of eluent.

For anion analysis, the operator may prepare standard solutions for qualitative and quantitative analyses using several reagents. Figure 8 shows an example of a chromatogram of the standard solution containing 20 ppb of anion species. For cation analysis, a standard solution can be prepared from reagents used in other analytical methods, such as ICP-AES (Inductively coupled plasma atomic emission spectroscopy) or AAS (Atomic absorption spectroscopy). Figure 9 shows a chromatogram of a standard solution containing 50 ppb of Na^+, K^+, $NH_4{}^+$, Mg^{2+} and Ca^{2+} ions prepared from standard reagents available for AAS.

Ion chromatography is a powerful tool for the analysis of ionic species in solution and has been widely applied to the analysis of samples from the hydrosphere or for monitoring the pollution of effluents. In Japan, the Ministry of the

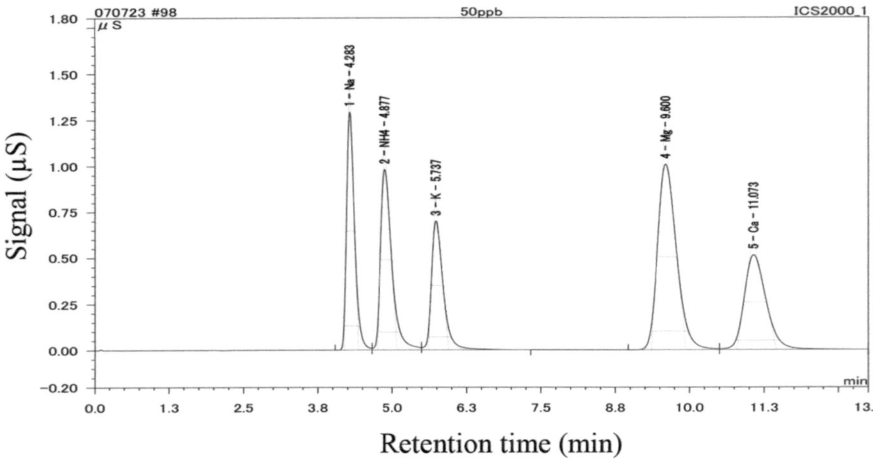

Fig. 9 A chromatogram of the standard solution for cations. All concentrations of ions are 50 ppb

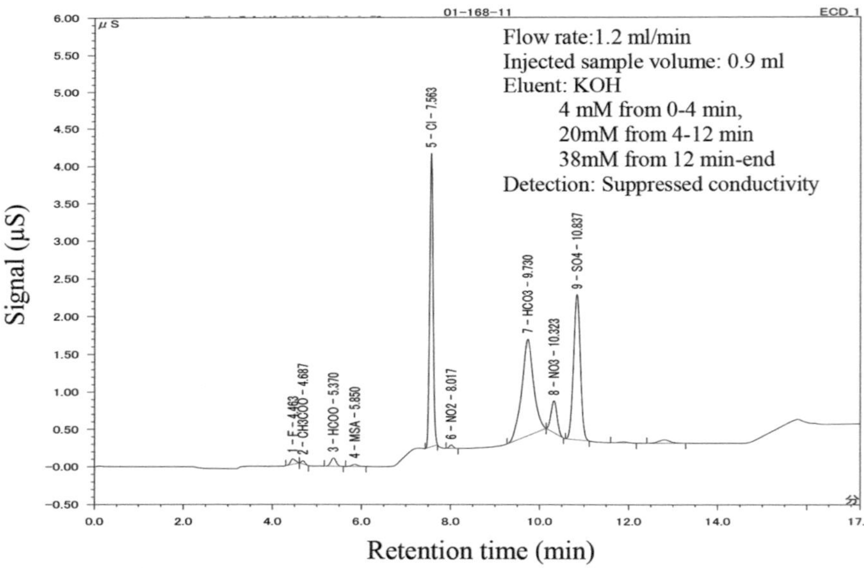

Fig. 10 A chromatogram for the anion analysis of a sample from a glacier in Antarctica

Environment issued the Environmental Quality Standards for ground water pollution. Many hazardous species are required to be monitored and maintained to protect our health. For hazardous substances, ionic species are analyzed with IC, which has also been used to monitor cyan in effluents from metal plating industries. IC has been commonly used not only for monitoring pollution but also for environmental analysis. Figure 10 shows an anion chromatogram of a sample from an

Antarctic glacier. The ice core of the glacier has been considered to contain a variety of information about the history of Earth, including climate changes in the past. Therefore, many researchers pay attention to the chemistry of glaciers in Antarctica and analyze the ions in glacier samples. Due to the advantages of IC, such as high throughput and high sensitivity, many scientists employ IC as a main tool of chemical analysis.

References

1. Blake DR Archives of symposium, Japan Society of Analytical Chemistry http://www.jsac.or.jp/~gc/GCforum_takeda/pdf/GC212.pdf
2. Ishida Y, Ohtani H, Tsuge S (1998) Characterization of synthetic high-polymers and natural organic compounds by reactive pyrolysis-gas chromatography. Bunseki Kagaku 47:673–688

Dissolution Assay

Seiji Yokoyama

Abstract Metallic products rarely have been dumped in final disposal sites because metals are recyclable materials. Slag and dust are generated when metallic scraps are melted and refined. The dust includes environmentally regulated substances, so it must be treated appropriately. The slags have been reused as construction materials, roadbed materials, and others. However, dissolution of regulated substances such as arsenic, cadmium, mercury, lead and others, from the slags becomes the problem for these uses. The leaching test is provided to check the dissolution quantities of the regulated substances. If the eluted concentrations of them are higher than the environmental quality standards, the slags cannot be used and dumped. The various leaching tests are summarized in this chapter. In addition the tramp elements, which cannot be removed from molten steel in a refining process for scraps, are described in this chapter.

1 Dissolution of Metals in Environmental Issues

The dissolution of material into an aqueous solution and so on, has a merit as well as a demerit to the environment. Consider coated materials for an example. One of the objectives of a surface treatment such as plating is to improve corrosion resistance. In this case, it is preferable that the materials do not dissolve in water. This improvement of corrosion resistance enables the long-term use of materials, and will contribute to the reduction of resources and energy required for materials production. In Japan, JIS (Japanese Industrial Standards) H8502 (1999) regulates the methods of corrosion resistance tests for metallic coatings so as to evaluate the corrosion of a material. This standard includes the ISO 4540 (1980), ISO 4541 (1978) and others.

Next, we will think about the recycling of coated materials. Recycling is the key technology to establish a sustainable society. If a base material or a coating material

S. Yokoyama (✉)
Toyohashi University of Technology, Toyohashi, Japan
e-mail: yokoyama@me.tut.ac.jp

H. Kanematsu, D.M. Barry (eds.), *Corrosion Control and Surface Finishing*,
DOI 10.1007/978-4-431-55957-3_15

individually dissolves in a solution such as acid, we can collect it separately. In general, selective extraction is preferable to recycling. However, in the present state, we cannot perform such a separation economically. If a coating material such as nickel, tin, zinc, and others, is as expensive as gold and platinum, separation of these coating metals from base metals with the dissolution method may be profitable.

One of the issues related to dissolution from materials is the dissolutions of environmentally regulated substances. This problem mainly occurs on damping and recycling products. Whether materials dissolve in various solutions or not is of great concern in regards to environmental issues.

2 Dissolution of Slag

Typical environmentally regulated substances are arsenic, cadmium, mercury, and lead. Presently, there are few products including mercury. Some electronic devices include arsenic or cadmium, and are used in a computer and so on. Lead is used in the battery of a car, and its alloys are used as solders. In addition, it is included in polyvinylchloride (PVC) cables as a stabilizer. The battery has been treated appropriately. The solder was replaced by a lead-free solder after the RoHS directive. However, a lead-free solder includes lead below 1000 (mass) ppm. The quantity of lead in PVC cables has been decreased. The consumption of the environmentally regulated material seems to decrease, but the discarded products may be the products produced before the enactment of the regulations.

These environmentally regulated substances cannot dissolve in water, etc. as long as products such as a computer are in use. These substances may dissolve in water after the products that include them are discarded. Therefore, leaching tests are provided by laws to determine whether these products may be discarded or not. If the results of the leaching tests for these products are beyond the environmental quality standards (EQSs), these products must be discarded in a final disposal site where effluent cannot leak out. Disposal of wastes to a final disposal site is very expensive. To begin with, it is very difficult to secure the last disposal ground.

The processes for producing metals from metallic wastes are less expensive and less energy-intensive than those for producing metals from ores of metals [1–3]. The metal industry has been actively recycling metallic wastes for a long time. Metal is an honor student of recycling. Figure 1 shows the schematic of a steelmaking process using iron and steel scraps.

Environmentally regulated substances transfer from scrap to newly produced steel, slag, and dust. Usually, we do not need to think about dissolution of environmentally regulated substances from the steel because the steel is used on the premise that it does not dissolve in water. However, substances incorporated into the steel influence the mechanical properties, its workability, etc. This is described at the end of the chapter. Dissolution of the regulated substances from slag and dust is an important problem. The dust exhaust, from the steel-making process using

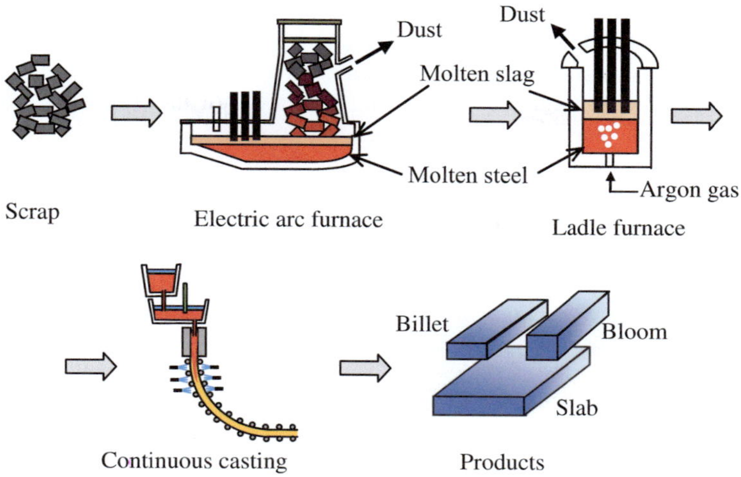

Fig. 1 Schematic of steelmaking process using iron and steel scrap

scrap as a raw material, includes the regulated substances such as zinc, lead, and others. Almost all the recycled zinc depends on collections from this dust with the pyro-metallurgical method, because it is difficult for zinc to be collected with a hydrometallurgical method. This dust cannot be dumped because it still contains the regulated substances. Therefore, further processing is needed for reuse and discarding of the dust.

In comparison with the dust, slag has less environmentally regulated substances because the slag is held at a high temperature. The slag components are similar to Portland cement and rocks such as andesite. There is another slag, which is called the waste molten slag or the molten slag, except slag exhausted from a metallurgical process. This slag is produced by burning ashes of wastes and sludge with SiO_2 at high temperatures of more than 1300 °C to prevent the elution of the regulated substances from the ashes and to reduce the waste volume. Presently, these slags have been reused as construction materials, roadbed materials and for other materials. This does not lead to the destruction of nature. However, elution of the regulated substances from these slags becomes a problem when the slag is reused for construction materials, etc. The leaching tests are provided to check whether the waste such as slag may be reused or not. The results of the leaching tests are collated with the EQSs.

3 Leaching Test

There are two ideas for leaching tests of wastes. One leaching test is performed to find out the maximum dissolvable quantity from wastes under the severe conditions for the dissolution of wastes. The dissolution quantity is less than or equal to the contents of waste. The maximum dissolvable quantity is called availability. Therefore, this leaching test is performed by using a buffer solution or a solution of which the pH is kept constant with a pH controller. This type of leaching test is a good method from a viewpoint of guaranteeing the safety of the materials. USA, the Netherlands, and other countries, provide this type of leaching test. The other leaching test is performed to determine the dissolvable quantity from wastes under conditions similar to those where wastes are used or dumped. A solution with relatively high pH and a relatively large sample is used in the leaching tests. Japan and others adopted these types of leaching tests.

Leaching tests for different countries are summarized as follows. If we perform the leaching test, then we must refer to each regulation for more information about a specific leaching test.

3.1 Japan (Public Announcement No. 19 and No. 14 of the Environment Agency. JIS K0058-1 (2005))

In the case that materials and wasted material are used on land, dissolved concentrations must be below the environmental quality standards (EQSs) for soil. Because these materials may pollute soil, the use and the disposal of them on an area of land is judged from the EQSs for soil. These EQSs for soil are the same as for groundwater in Japan. In the case that materials are disposed of at sea, dissolved concentrations must be below the EQSs for seawater by the marine pollution prevention law. These EQSs are the criteria to determine whether waste is to be disposed of at sea or not.

There are two leaching tests for soil in Japan. One is a method based on the public announcement no. 19 of the Environment Agency. The other is based on JIS K0058-1 (2005) (Table 1). The leaching test for the marine/sea area is based on the public announcement no. 14 of the Environment Agency.

3.1.1 Leaching Test for Soil

The outline of the leaching test based on the public announcement no. 19 is as follows.

1. Preparation of test sample

   After crushing a sample roughly, all crushed particles are passed through a nonmetallic sieve with an opening of 2 mm. The passed sample is mixed sufficiently.

2. Preparation of solvent

   A solvent is prepared by adding hydrochloric acid to pure water so that the concentration of HCl is 1 mol/l.

Table 1 Leaching condition for each leaching test

	Public announcement no. 19	JIS K0058-1	
Sample weight	≥6 g	Depends on sample size	100 g Size: <2 mm
Solvent	HCl + pure water	HCl + pure water	HCl + pure water
	Concentration of HCl: 1 mol/l	pH of solution: 5.8–6.3	pH of solution: 5.8–6.3
Ratio of sample weight [g] to solvent [ml]	0.03	0.1	0.1
Temperature	20 °C	–	–
Atmospheric pressure	100 kPa (1 atm)	Ambient pressure	Ambient pressure
Mixer	Shaker	Rotary stirrer	Shaker
Amplitude	4–5 cm	–	4–5 cm
Frequency	200 rpm	200 rmp	200 rpm
Leaching time	6 h	6 h	6 h

Fig. 2 Schematic of leaching test based on public announcement no. 19 of the Environment Agency in Japan

3. The crushed sample of which weight is more than 6 g is added to the solvent so that the ratio of sample weight [g] to solvent [ml] is 0.03. Furthermore, the solution with a volume more than 500 ml must be used for leaching. The following container can be used for this test. One is a container made of polyethylene (PE). Another is a container for which the measuring substances do not adsorb and from which they do not dissolve.
4. Leaching

   Figure 2 shows a schematic of the leaching test. The leaching test is performed at approximately 20 °C under approximately 100 kPa (1 atm) using a shaker of which the horizontal amplitude is 4–5 cm and of which the frequency is approximately 200 rpm. The leaching test is performed in succession for 6 h.
5. Preparation of the solution for analyzing

   The shaken solution is left silently for 10–30 min. Subsequently, the solution is centrifuged at approximately 3000 rpm for 20 min. A supernatant liquid is passed through a membrane filter with a pore of 0.45 μm in diameter.
6. Analysis

   We analyze this filtrate with an official analytical method.

The JIS K0058-1 regulates the two leaching tests. One leaching test is for a sample with a practical form and size. Another leaching test is for a sample with a size below 2 mm. This leaching test is mainly applicable to the former sample. The leaching test for the latter sample will be described later.

A molded sample with a weight less than 5 kg is used for a leaching test without crushing. For example, a cylindrically molded piece of concrete that is prepared for the compression test can be used in this test. Even if the weight of a sample is more than 5 kg, the sample is crushed roughly. The relatively large lump is used for the leaching test in the JIS K0058-1, as listed in Table 2. This table also shows the minimum weight of sample required for this leaching test.

An aqueous solution is prepared by adding hydrochloric acid to pure water so that the pH range is from 5.8 to 6.3. The leaching test is performed on the condition that the ratio of sample weight [g] to solvent [ml] is 0.1. Therefore, the necessary volume of the solvent is determined according to the weight of the sample, as listed in Table 2.

The leaching test is performed in a tank with a mixer, as shown in Fig. 3. The aqueous solution is mixed by an impeller at a rotary speed of approximately 200 rpm. The tank is cylindrical and is made of plastic such as PE. The container for the leaching test becomes large according to sample weight as listed in Table 2. The diameter of the tank determines the length of an impeller blade. It is regulated that the ratio of the tank diameter to the length of the stirring bade ranges from two

Table 2 Size, minimum weight, solvent volume and diameter of cylindrical tank for leaching test

Crushed lump size [mm]	Minimum weight of sample [g]	Volume of solvent [ml]	Diameter of tank [mm]
37.5–53.0	3,000	30,000	400
31.5–37.5	2,000	20,000	350
26.5–31.5	1,000	10,000	300
16–26.5	500	5,000	230
9.5–16	200	2,000	160
<9.5	100	1,000	130

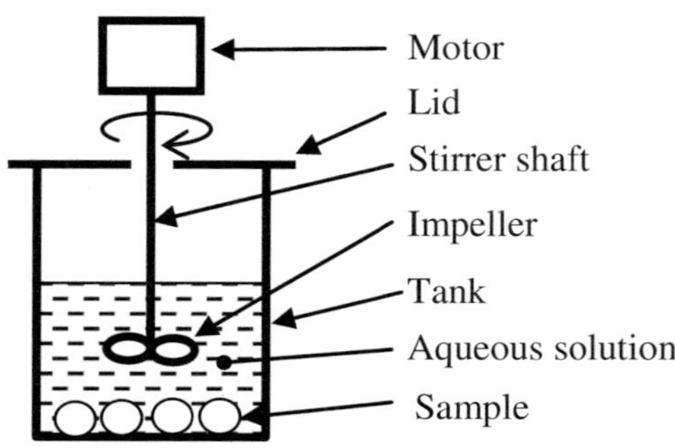

Fig. 3 Schematic of apparatus for leaching test

Table 3 Environmental quality standards (EQSs) for soil and results of leaching test of slags [mg/l]

Element	As	B	Cd	Cr	Pb	Hg	Se	F
EQS	0.01	1	0.01	0.05[a]	0.01	0.0005	0.01	0.8
Slag A	ND	0.28	ND	ND[b]	ND	ND	0.003	0.5
Slag B	ND	0.16	ND	0.024[b]	ND	ND	0.012	ND

ND not detected
[a]Hexavalent chromium
[b]Total chromium

to four. An impeller and a stirrer shaft shall be made of resin such as the fluoric resin without the elution and adsorption of the measurement constituents.

The results of the leaching test of slags are listed in Table 3 [4]. The test is regulated by JIS K0058-1 (Sample size <2 mm). Here, EQS for soil regulates the dissolution concentration of hexavalent chromium. The chromium concentration listed in this table is total chromium concentration that is the sum of trivalent and hexavalent chromium concentrations. The dissolution concentrations of Slag A were below the EQSs. This slag can be used at an area of land. The dissolution concentration of selenium for Slag B exceeded the EQS of selenium. This slag cannot be used at a land area. Also a serial batch elution of the slag was investigated [5].

Japan has the other leaching test. It is the method regulated by JIS K 0058-2 (2005), (Test method for acid extractable contents of chemicals). This test is almost the same as the public announcement no. 19 except for sample size and weight. For this test, a sample is dissolved in a hydrochloric acid solution of 1 mol/l. The dissolved concentration given by this test is estimated to be equivalent to the maximum extractable concentration.

3.1.2 Leaching Test for Marine Area

The leaching test based on the public announcement no. 14 is almost the same as that on the public announcement no. 19. The difference points are the pH of solvent and the ratio of sample weight to solvent volume. The solvent with a pH of 7.8–8.3 is used in this leaching test. The ratio of the weight of sample [g] to solvent volume [ml] is 0.03 for inorganic sludge or inorganic sediment. In addition, the mixed solution volume is more than 500 ml. The results of this test should be compared with EQSs determined by the marine pollution prevention law.

Now, the average pH at the surface layer of seawater is 8.1 [6]. Acidification of the ocean is progressing by an increase in the concentration of carbon dioxide in the air. It may be necessary to revise the pH of a solvent used in the elution test if the pH continues to fall in the future.

3.2 USA (EPA SW−846 Method 1311, Method 1312)

US-EPA provides two leaching test methods. One is the Toxicity Characteristic Leaching Procedure (EPA SW−846 Method 1311), and is called the TCLP for short. The other is the Synthetic Precipitation Leaching procedure (EPA SW−846 Method 1312), and is called the SPLP. The TCLP is intended to simulate dissolution of waste materials inside a landfill, and is the only leaching test specified by the regulation for characterization of the hazardous waste (40 CFR 261.24−Toxicity Characteristic, waste code D004-D043) under the Resource Conservation and Recovery Act (RCRA).

The SPLP is designed to simulate waste materials in the ground surface or on the top of the surface. The substances there are exposed to rainfall. Leachate forms by dissolution of them. Data given from the SPLP are utilized to develop site-specific soil remediation criteria that protect ground water. Federal and/or State specific guidance utilize the data for a conceptual site model.

3.2.1 TCLP

One of the characteristics of this TCLP is to use the buffer solution which is composed of acetic acid and sodium hydroxide. The pH of the solution is 4.93 ± 0.05. Generally, the pH becomes large as the waste such as slag dissolves in an aqueous solution. Dissolution from wastes becomes hard with an increase in pH. The buffering action prevents the pH of the solution from becoming large. Because the pH will be kept constant during the test due to the buffering action, the dissolved concentration will be large. When the pH of the solution at the end of the test is more than 5, then the pH exceeds the buffering capacity. Thus we must perform the leaching test again using acetic acid solution with a pH of 2.88 ± 0.05.

A sample of more than 100 g is used in the TCLP. The TCLP employs the ratio of liquid volume [ml] to sample weight [g] of 20. The solution is stirred for 18 ± 2 h with a rotary agitation apparatus as shown in Fig. 4. Filtration of the solution is performed with a filter press to prevent volatile substances from evaporating. The filter is a glass fiber filter with a pore size of 0.6–0.8 μm.

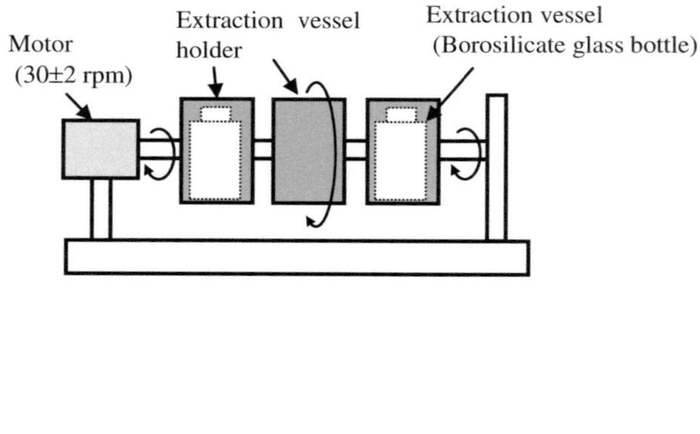

Fig. 4 Schematic of rotary agitation (in end-over-end fashion) apparatus

3.2.2 SPLP

The SPLP is almost the same as the TCLP. However, an extraction solution used in the SPLP is different from that in the TCLP. The extraction solution for the SPLP is prepared by adding the 60/40 wt.% mixture of sulfuric and nitric acids to reagent water until the pH is 4.20 ± 0.05. This solution is used to determine the leachability of soil from a site that is east of the Mississippi River. For the area that is west of the Mississippi River, the extraction solution is made by adding the 60/40 wt.% mixture of sulfuric and nitric acids to reagent water until the pH is 5.00 ± 0.05.

3.3 Germany (DIN 38414 S4)

There is a regulation on using burned ash of municipal wastes as construction materials for roads. The standard value is determined about Hg, Cd, Pb, Cr, Cu, Zn, Ni, Cl, SO_4, CN, conductivity and pH. The elution test method is regulated by the DIN 38414. This method is applied to solid- and paste-like wastes or slime. We add 100 g of samples of which size is adjusted to less than 10 mm into distilled water of 1 l. The solution is mixed for 24 h using a rotary agitation device. After filtrating it by a membrane filter with a pore size of 0.45 μm, the concentrations of metals, conductivity, and pH will be measured.

3.4 French (AFNOR X31-210)

There is a regulation for using burned ash of wastes as construction materials for roads. The standard value is determined about Hg, Cd, Pb, Cr (VI), As, SO_4, TOC, etc. The elution test method is regulated by the AFNOR X31-210. As a general rule, a sample with an existence figure is used for this leaching test. For the molded sample, it will be immersed in distilled water for 4 h. If its strength is less than 1 N/mm^2, the molded sample is crushed to become the size of less than 4 mm. The sample of 100 g is added to pure water of 1 l. After shaking it for 16 h, the solution is filtered by a membrane filter with a 0.45 μm pore size. Or, the solution is separated from the residue by a centrifuge with more than 2,000 G. In addition, a constant temperature tank is used to suppress the solubility change by the temperature of the solvent.

3.5 Switzerland (TVA)

This test method regulated by the TVA takes into account the influence of carbonate formation with atmospheric carbon dioxide in a reclaimed land. Solid waste with the strength of more than 1 N/mm^2 is used for the test as an existence figure.

A sample of 100–200 g is added to distilled water. Water volume is determined so that the ratio of water volume [ml] to sample weight [g] becomes ten. Carbon dioxide gas is introduced into the solution at the flow rate of 100 ml/min, and the solution is stirred in order to prepare a solution saturated with carbon dioxide gas. The solution is filtered 24 h later. Similarly, the residue in the container and on the filter is tested again.

3.6 Netherlands (NEN 7341, 7343, 7345)

Several kinds of leaching test methods are provided in the Netherlands from the viewpoint that the single batch leaching test shows one side of the elution behavior of the waste. Because the behavior of elution for powdery samples is different from those for a molded body, the leaching tests are different depending on the form of each sample. In addition, three kinds of test methods for the powdery waste are prepared to investigate the time dependency of dissolution behavior, as listed in Table 4.

3.6.1 Leaching Test Through a Column (NEN 7343)

The NEN 7343 is a test method to understand the short-term elution behavior of the powdery waste. A column made of plastic is 5 cm in diameter and more than 20 cm in height. Sample is filled in the column so that the height of the sample in the column becomes more than four times the diameter. A nitric acid solution with pH of 4 is poured from the bottom to the top of the column. It continues until the ratio of liquid volume [ml] to solid weight [g] becomes ten. We will sample the solution at the ratio of 0.1–0.2, 0.2–0.5, 0.5–1.0, 1.0–2.0, 2.0–5.0, and 5.0–10.0, totally six times, and analyze each solution.

3.6.2 Serial Batch Leaching Test (NEN 7343)

The NEN 7343 is a test method to understand the medium-term elution behavior of the powdery waste. The sample size is less than 3 mm. The leaching test is performed using a nitric acid solution with pH of 4 under the condition that the ratio of liquid volume [ml] to sample weight [mg] is 20. The solution is stirred for 24 h.

Table 4 Dissolution period targeted by each regulation

Regulation	Manners	Period
NEN 7344	Column test	Short-term (~50 years)
NEN 7344	Serial batch test	Medium-term (50–100 years)
NEN 7341	Availability test	Maximum dissolvable quantity (ca. 500 years)

After the leaching, the solution is filtered by a membrane filter with a pore size of 0.45 μm. The leaching test for the residue is performed with the same conditions again. We repeat this operation five times so that final accumulated ratio of liquid volume to solid weight is 100.

3.6.3 Leaching Test for Availability (NEN 7341)

NEN 7341 is the test method to check the maximum dissolvable quantity of wastes. It is supposed that waste dumped in a final disposal place releases pollutants to the environment by elution. In this test, a sample of 8 g is added to distilled water of 400 ml. The pH of the solution will be kept 7 for 3 h by adding a nitric acid of 1 N (mol/l) with a pH monitoring automatic titration device. The solution is filtered by a membrane filter of which the pore size is 0.45 μm. The residue is added to distilled water of 400 ml. Subsequently, the pH of the solution will be kept 4 for 3 h by adding a nitric acid of 1 N (mol/l) with an automatic titration device.

3.6.4 Leaching Test in a Tank (NEN 7345)

NEN 7345 is the leaching test for wastes or reused substances. A molded body which is larger than a cube 4 by 4 by 4 cm in size is used in this test. The sample is tested in a nitric acid solution with a pH of 4. A sample is immersed in a nitric acid solution on the condition that the ratio of the solution volume [ml] to sample weight [g] is 5. We leave the solution at rest. The solvent will be exchanged with a new solvent at the time of 6 h, 1 d (day), 2.25 d, 4 d, 9 d, 16 d, 36 d and 64 d. Each time the solvent is collected before the exchange is analyzed.

The building material decree regulates the quantities dissolved from a construction material to the neighboring soil and the surface water. The standard values are established about Hg, Cd, Pb, Cr, As, Se, Cu, Zn, Ni, Ba, Sn, Sb, Co, Mo, V, Cl, SO_4, Br, F, CN(free), CN^-, SCN^-. Dissolution quantities per 100 years are calculated from the results given by the tank leaching test (NEN 7345) and availability leaching test (NEN 7341).

4 Dissolution of Non-environmentally Regulated Substances

In general, most plants grow by absorbing nutrients from the soil. There are essential elements for plant growth. The quantity that is necessary for each plant is different. Nonetheless, the following Table 5 gives the general requirements of plants [7]. Silicon is not considered an essential element for a plant, but it is an essential element for poaceae such as rice, wheat, and others. If the soil contains

Table 5 Typical elements and their concentrations required for plant growth

Element	mg/kg	Percent	Relative number of atoms	
Nitrogen	15,000	1.5	1,000,000	Macronutrients
Potassium	10,000	1.0	250,000	
Calcium	5,000	0.5	125,000	
Magnesium	2,000	0.2	80,000	
Phosphorus	2,000	0.2	60,000	
Sulfur	1,000	0.1	30,000	
Chlorine	100	–	3,000	Micronutrents
Iron	100	–	2,000	
Boron	20	–	2,000	
Manganese	50	–	1,000	
Zinc	20	–	300	
Copper	6	–	100	
Molybdenum	0.1	–	1	
Nickel	0.1	–	1	

smaller amounts of these essential elements than the values shown in this table, then the plants may be nutrient deficient. Contrarily, excess intake of nutrients by the plants may cause disorders, especially if there are too many essential elements, particularly micronutrients, in the soil.

Calcium and magnesium are dissolved from the slag that is exhausted from the steel-making process. This is preferable to plants. Copper, nickel and zinc are environmentally regulated substances in many countries, but they are included with the micronutrients. Aluminum is not an environmentally regulated substance for soil in many countries. It has been known that aluminum dissolved in water inhibits the growth of the plant root. This is the so-called aluminum toxicity [8]. Aluminum ranks third in abundance among the elements included in the Earth's crust, and is mostly incorporated into soil as aluminosilicates. Aluminum does not dissolve in a neutral solution, but begins to dissolve in an acid solution with a pH below approximately 5. The harvest of cereal grain, etc. is small in an acidic soil.

Acidification of the soil is not necessarily caused by acid rain. The rain is acidic all over the world. The pH of rain in Japan was 4.72 (average, 2008–2012). The acid rain can dissolve aluminum from wastes. We should be careful about the influence of aluminum dissolution on agriculture that directly relates to food.

5 Tramp Element

The recycling process for steel was described in Sect. 2. Some elements, which are called the tramp elements, cannot be removed from molten iron with present pyrometallurgical refining methods. The tramp elements and the influence of them on the mechanical properties and workability of steel are listed in Table 6. Copper content in scrap is 0.1–0.5 mass%. Measures to prevent copper from mixing

Table 6 Typical tramp elements in steel

Element	Influence
Copper	Hot workability (Copper >0.2 mass%)
Tin	Hot workability (Tin >0.04 % mass%)
	Cold workability and temper brittleness (Tin >0.2 mass%)
Nickel	Hardening (steel sheet)
Chromium	Hardening (steel sheet)

in a scrap are necessary. Tin content in scrap is 0.005–0.016 mass%. Presently, utilization of tin plated canning as a raw scrap has been limited. The high content of chromium in steel can be eliminated from steel, but the low content of chromium is difficult to remove. Nickel and chromium harden steel, and cause the problem for a steel sheet. Nickel and chromium contents in scrap are 0.06–0.1 mass% and 0.03–0.3 mass% respectively. In addition to them, molybdenum hardens steel, and causes the problem for a steel sheet. Furthermore, lead, arsenic, antimony and bismuth influence the hot workability, respectively, even if their contents in steels are very low.

The tramp elements listed in this table are also typical coating materials. The reason that they are included in scraps is not always dependent on their plating materials. The nickel-plated steel scrap seems to become the raw material for stainless steel production. However, the nickel- phosphorus plated steel scrap cannot be used as the raw material because phosphorus in the molten stainless steel cannot be removed with the present refining method.

Metallic copper is produced from copper ore. Crude copper has been refined by the electrolytic refining process throughout the world. Usually, nickel is considered not to be a tramp element for copper because nickel is easily eliminated from the crude copper. However, the reason that electrolytic refining is adopted as a refining process is to recover gold and platinum. Copper products have not already contained such noble metals. The electrolytic refining process takes an enormous amount of energy. Accordingly, electrolytic refining is not used for waste such as nickel plated copper except in a country where electric power is inexpensive. One must know the refining processes of plating and base materials, when selecting plating materials in consideration of their recycling methods.

References

1. System Analysis Group (RITR) (2010) nenjitenno enerugi-genntannino suikei (Tekkoubumon-tennrokou). https://www.rite.or.jp/Japanese/labo/sysken/about-global-warming/download-data/Comparison_EnergyEfficiency2010steel.pdf (in Japanese)
2. System Analysis Group (RITR) (2010) nenjitenno enerugi-genntannino suikei (Tekkoubumon-sukurappudenkkirokou). https://www.rite.or.jp/Japanese/labo/sysken/about-global-warming/download-data/Comparison_EnergyEfficiency2010steelEAF.pdf (in Japanese)

3. Green JAS (2007) Aluminum recycling and processing for energy conservation and sustainability. ASM International, p 147
4. Yokoyama S, Suzuki A, Izaki M, Umemoto M (2009) Elution behavior of electric arc furnace oxidizing slag into fresh water. Tetsu-to-Hagane 95(5):434–443
5. Yokoyama S, Suzuki A, Nor NHBM, Kanematsu H, Ogawa A, Takahashi T, Izaki M, Umemoto M (2010) Serial batch elution of electronic arc furnace oxidizing slag discharged from normal steelmaking process into fresh water. ISIJ Int 50(4):630–638
6. Rhein M, Rintoul SR, Aoki S, Campos E, Chambers D, Feely RA, Gulev S, Johnson GC, Josey SA, Kostianoy A, Mauritzen C, Roemmich D, Talley LD, Wang F (2013) Observations: ocean. In: Stocker TF, Qin D, Plattner G-K, Tignor M, Allen SK, Boschung J, Nauels A, Xia Y, Bex V, Midgley PM (eds) Climate change 2013: the physical science basis. Contribution of Working Group I to the Fifth Assessment Report of the Intergovernmental Panel on Climate Change. Cambridge University Press, Cambridge, UK, p 295
7. Epstein E (1965) Mineral metabolism. In: Bonner J, Varner JE (eds) Plant biochemistry. Academic, London, pp 438–466
8. Rout GR, Samantaray S, Das P (2001) Aluminium toxicity in plants: a review. Agronomie EDP Sci 21:3–21

Part IV
Novel Biological Environmental Evaluation Processes

Gene Analysis for the Evaluation of the Effect of Environmental Factors

Satoshi Wakai

Abstract A variety of pollutants are present in natural and artificial environments. These pollutants are divided into two groups, chemical and biological pollutants. They cause various problems such as metal corrosion, environmental damage, and human health issues. Chemical pollutants contain heavy metals, hydrocarbons, exhaust gases, and endocrine disruptors. Such pollutants may possibly be detected by using a biosensor. On the other hand, the main biological pollutants are microorganisms. Microorganisms in environments can be detected by using culture-dependent and molecular biology-dependent methods. In particular, this chapter focuses on the molecular biology-dependent evaluation for biological pollution. First I introduce some methods of gene analysis as molecular biology-dependent evaluation techniques. Then I discuss gene analysis for endocrine disruptor detection and the application of a biosensor. Materials may be exposed to various pollutions in their respective environments so a variety of evaluation techniques for different pollutants exist. Thus, the adoption of a proper evaluation technique would be useful for maintaining the quality of materials.

1 Introduction

The surfaces of materials are exposed to a variety of environmental factors. Therefore, when considering the contamination of a surface, we must also consider the combination of environmental factors to which the surface is exposed (e.g., liquid, gas, or solid as physical features of the environment) and the type of material making up the surface (e.g., metal, resin, glass, or porcelain). Furthermore, materials may be exposed to various pollutants in their respective environments. Thus, different surfaces may be defined by combinations of environmental factors, materials, and pollutants. It is challenging to establish an appropriate evaluation technique from the currently available options. The adoption of a proper evaluation

S. Wakai (✉)
Organization of Advanced Science and Technology, Kobe University, Kobe, Japan
e-mail: wakaists@pegasus.kobe-u.ac.jp

H. Kanematsu, D.M. Barry (eds.), *Corrosion Control and Surface Finishing*,
DOI 10.1007/978-4-431-55957-3_16

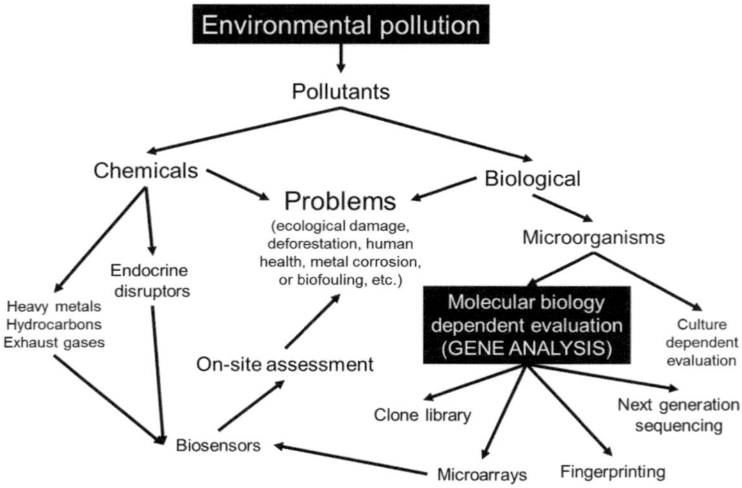

Fig. 1 Overview of the correlation between environmental pollution and gene analysis

technique would be useful for maintaining the quality of materials. This chapter discusses environmental pollutants, the problems, and evaluation techniques for materials based on gene analysis (Fig. 1).

2 Environments and Materials

A substance may be found in three states: liquid, gas, and solid. This is also true for the environmental factors with which the substance interacts. For example, the bottom of a ship, a bridge, a heat exchanger, and a pipeline are exposed to seawater and fresh water in the liquid form; a chimney is exposed to exhaust gas in the gas form; and an underground pipeline and steel tower are exposed to soil in the solid form. Moreover, a material will be exposed to pollutants in its respective environment, e.g., a petroleum pipeline will be exposed to various hydrocarbons, while a chimney will be exposed to corrosive gases, such as hydrogen sulphide.

When considering the quality of a material, we also must consider the type of material (e.g., metal, resin, glass, or ceramic, etc.). In particular, metal materials exposed to different weather events may be oxidized and corroded. Here, I will focus on corrosion. Carbon steel, stainless steel, and alloys are considered metal materials, and these different types of steel may be chosen according to the purpose of the material and the environment in which the material will be used. However, changes in the environment will shorten the lifetime of the material, and the presence and activity of microorganisms can accelerate corrosion; this process is called microbiologically influenced corrosion (MIC) [1–5]. Thus, when investigating the environmental factors associated with MIC, biological molecules, in addition to chemicals, would have to be considered. This will be described later as a type of biological pollution.

3 Environmental Pollutants

Environments contain various substances, including organic solvents, heavy metals, endocrine disrupters, chlorine compounds, sulphur compounds (SOx), ammonia, nitrogen oxides (NOx), and phosphate compounds. Although industrial development during the nineteenth and twentieth centuries has improved the quality of life for many individuals, these advancements have also led to the release of an abundance of pollutants into various environments. For example, with the development of automobiles and industrial manufacturing, SOx and NOx in exhaust gas can cause acid rain, which then leads to deforestation. Pollution by organic solvents and heavy metals is a serious health issue [6–13]. Indeed, leaking of organic solvents and heavy metals can destroy ecosystems and cause toxicities in humans and animals. Ground pollution by organic solvents has been observed at sites of chemical factories, and various processing strategies have been used to remove such ground contamination (e.g., passive treatment and bioremediation) [14–16]. Pollution by heavy metals can also cause various sicknesses. They may be able to be removed from the contaminated material by chemical and biological processing. In recent years, endocrine disrupters have also become an issue in environmental pollution [17–19]. Indeed, while hormone analogues were thought to be highly specific to the corresponding hormone receptor, studies have shown that some hormone analogues may function as endocrine disrupters and negatively influence disease progression in patients.

Biological pollution has also become an issue. For example, biofilm production on various surfaces can result in poor functioning of the affected system (e.g., biofilm formation on heat exchangers or pipes may lead to reduced heat recovery efficiency and corrosion, respectively) [20–22]. Moreover, the formation of biofilms containing shellfish, protists, and algae on the bottom of ships reduces fuel efficiency by increasing resistance while moving through the water. In the medical field, biofilm formation on the surfaces of catheters and implants may become sources of hospital infection by opportunistic bacteria.

4 Evaluation of Environmental Pollution by Gene Analysis

4.1 Microbial Community Analysis

This chapter discusses the application of gene analysis for evaluation of pollution by bio-based substances. It describes endocrine disrupters and the development of biosensors for analysis of gene responses to various chemicals.

In terms of pollution by bio-based substances, it is necessary to identify pollutants based on the amount of biomass and the type of microorganism. Analysis of pollution by biological substances is commonly carried out by culture-dependent methods using either agar-plate cultivation, in which microorganisms are cultivated on an agar plate and counted, or the sequential dilution method, in which microorganisms are inoculated into and cultivated in culture medium after sequential dilution. However, culture-dependent methods are difficult to use for analyzing unknown microorganisms because many microorganisms cannot be cultured; appropriate cultivation conditions may not have been established, or microorganisms may not grow under artificial cultivation, taking on a so-called viable but nonculturable (VBNC) state [23]. For example, only about 0.1 % of microorganisms in seawater are thought to be cultivatable [24]. Therefore, molecular biology-based methods, which have become more mainstream in recent years, may facilitate the identification of microorganisms from natural environments such as seawater, lakes, rivers, and underground water and from artificial environments such as cooling water and processed water. However, despite the applicability and utility of molecular biology methods, identification and quantification of specified microorganisms in some contexts, such as bacteria responsible for food poisoning and faecal coliform bacteria, are carried out using culture-dependent methods.

Various molecular biology-based methods have been reported for identification of biological pollutants. In many cases, methods aim to target the 16S rRNA gene (16S rDNA). The 16S rRNA is one of the components of the ribosome, which functions to synthesise proteins within the cell. Therefore, all living organisms have ribosomal RNA (rRNA); all prokaryotes (Bacteria and Archaea) have 16S rRNA and all eukaryotes have 18S rRNA. The rRNA gene is often selected as a housekeeping gene in a variety of molecular biology methods. First, this gene is conserved and moderately rich in sequence variations, allowing for discrimination among species and strains [25, 26]. However, phylogenetic identification based on sequence analysis requires many reference sequences. Fortunately, many 16S rRNA gene sequences for prokaryotes have been deposited in public nucleotide databases, facilitating a variety of applications.

In addition to 16S rDNA, the sequences of some housekeeping genes are also available. When the sequences of 16S rDNA between different species are identical, multilocus sequence analysis (MLSA), which targets other genes such as RNA polymerase and gyrase, is recommended [27]. However, this method requires further work to create databases with a sufficient number of sequences; indeed, while information on 16S rRNA gene sequences is abundant, genetic information for other genes is not as easily available. Therefore, it is necessary to determine reference sequences of related species. Analysis of eukaryotic microorganisms using the 18S rRNA sequences is hampered by similar problems. Thus, phylogenetic identification of fungi and yeasts is generally performed based on morphology and physiology rather than molecular genetics.

The primary methods used, ranging from conventional methods to the latest molecular biology techniques, are explained below.

4.1.1 Clone Library

The 16S rDNA sequence consists of about 1,500 base pairs. For high-resolution analysis, the full-length sequence will be targeted. In contrast, for rapid analysis, only the partial fragment will be targeted. For both cases, microorganism cells in an environmental sample are first collected by centrifugation or filtration, and genomic DNA (gDNA) is then extracted from the cells. The gDNA recovered from environmental samples is sometimes referred to as environmental DNA (eDNA). Next, 16S rDNA fragments are amplified from the eDNA by a polymerase chain reaction (PCR) using universal primers that are designed to match the conserved region of the 16S rDNA.

The amplified 16S rDNA fragments in the reaction mixture are derived from more than one microorganism, and it is therefore necessary to store the fragments separately. Thus, each fragment is introduced separately into a plasmid (a circular nucleotide molecule), and the plasmids containing the amplified fragments are then introduced to and are maintained in *Escherichia coli* cells. Next, the recombinant *E. coli* cells are spread on agar plates and form colonies. Each colony has the plasmid harbouring a single fragment introduced from the amplified gene fragments. A set of *E. coli* colonies is stored as a library of genetic clones, and the sequences of the plasmids can be determined when necessary. In order to determine the sequences, *E. coli* cells are separately cultivated in liquid medium, and plasmid DNAs are extracted from the cells. The nucleotide sequences are then analysed using a cycle sequencing reaction and nucleotide sequencer. Next, the determined sequences should be compared to known sequences in a database in order to identify the microorganism species. To be identified as a known species, sequence homology should be at least 97 % [26]; therefore, if there are no sequences with over 97 % homology, the microorganism is considered to be of an unknown species. Overview of the process is depicted in Fig. 2.

The resolution of microbial community analysis by the clone library method depends on the number of analysed clones. For example, if one microorganism is identified out of 100 clones, the resolution is 1 % of the community, whereas if one microorganism is identified out of 1000 clones, the resolution is 0.1 %. Therefore, the clone library method can be used to analyse a microbial community with high resolution by increasing the number of clones as much as possible; however, it is unsuitable for the analysis of more than one environmental sample. For example, if one wishes to compare 10 samples, analysing a total of 1000 clones (100 from each

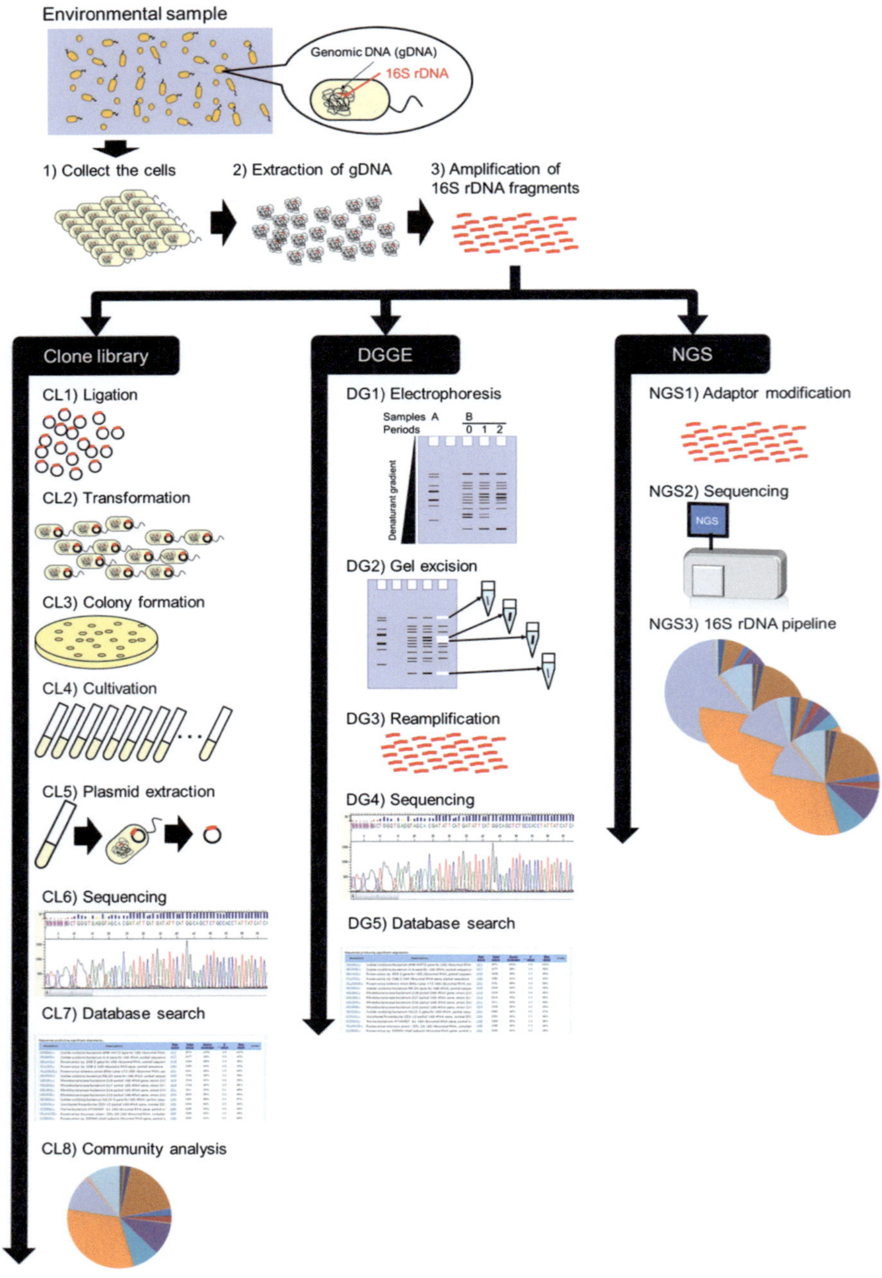

Fig. 2 Representative methods of microbial community analysis. The first four steps are almost the same as each other. The clone library method is the most complicated. The DGGE method is not suitable for quantitative analysis. The NGS method is the most simplified and suitable for comparison and quantitative analysis

sample), rare microorganisms representing less than 1 % of the community cannot be detected.

4.1.2 Fingerprinting

When comparing the microbial community structures of different samples, the comparisons of gene fragment patterns (fingerprints) are useful because they reduce the time and cost of the analysis. Common DNA fingerprinting methods include restriction fragment length polymorphism (RFLP) [28, 29], terminal-RFLP (T-RFLP) [30], amplified fragment length polymorphism (AFLP) [31, 32], denaturing gradient gel electrophoresis (DGGE) [33], and temperature gradient gel electrophoresis (TGGE) [34, 35].

In RFLP, 16S rDNA fragments are digested using restriction enzymes that recognise an arbitrary nucleotide sequence, and DNA band patterns are evaluated by electrophoresis. RFLP uses a targeted 16S rRNA gene sequence derived from an independent microorganism; therefore, this method is not directly applicable to environmental samples containing unknown microorganisms. Instead, the above-mentioned clone library method can be combined with RFLP. Namely, RFLP analysis of each clone can be performed rather than determining each sequence, and microorganism types can then be grouped. I and co-workers have reported the microbial community analysis in acid-rock drainage from pyrite mines by such a method [36]. This method reduces the cost and time requirements compared with direct determination of nucleotide sequences. However, a representative sequence must be known in order to identify the microorganism species. In contrast, T-RFLP is thought to be suitable for analysing environmental samples. In T-RFLP, fluorescence-labelled primers are used when amplifying gene fragments, which are then digested by a restriction enzyme. After digestion, the fragments are applied to a capillary type sequencer, and the intensity and retention time of the detection peaks are analysed. The intensity and time indicate the amount and type of each microorganism. Therefore, T-RFLP is suitable for the comparative analysis of more than one sample.

In AFLP, the whole genome, rather than a specific gene or 16S rDNA, is targeted. First, genomic DNA is digested with one or more restriction enzymes. After digestion, the DNA fragments are ligated with a nucleotide adaptor and then amplified with primers that anneal to the adaptor regions. Therefore, the AFLP method may be appropriate for comparison between two species or among a comparatively small numbers of samples.

DGGE analysis is quite effective for the comparative analysis of multiple samples. In contrast to RFLP analysis, partial 16S rDNA fragments are used and resolved by electrophoresis. When amplifying the gene fragment, a GC clamp primer, which has 30 nucleotides with high G+C content, is used. Then, 16S rDNA fragments containing GC clamps from a variety of microorganisms (such that the fragments are almost the same size but differ in nucleotide composition) are loaded on a denaturing gradient polyacrylamide gel prepared by mixing and applying an acrylamide solution with no or low concentrations of denaturants (urea and formamide) and an acrylamide solution with a high concentration of denaturants using a gradient mixer or a special system (e.g., the gradient delivery system released from Bio-Rad Laboratories, Inc.). During electrophoresis, 16S rDNA fragments with different nucleotide compositions can be separated on the gel, based on the nucleotide compositions rather than the fragment sizes. The migration of DNA fragments containing the GC-clamp would exhibit a distinct pattern owing to its molecular structure, because the DNA fragments would migrate towards the part of the gel with a high concentration of denaturants and then become partially expanded by the disruption of intramolecular hydrogen bonds. However, the GC-clamps would not be melted by the change in denaturant concentration. Therefore, the DNA fragments are separated depending on the number of intramolecular hydrogen bonds, which is determined by nucleotide composition (G+C contents and G or C distribution). In TGGE, a temperature gradient rather than a gradient of denaturants is adopted during electrophoresis; the intramolecular hydrogen bonds are disrupted by heating.

In DGGE and TGGE, respective DNA bands are derived from different microorganisms (species or strains). Therefore, the number of DNA bands may indicate the diversity in the microbial community. DGGE and TGGE analyses are both appropriate for comparative analysis of multiple samples because 10–20 samples can be analysed simultaneously by electrophoresis. For example, changes in microbial communities by environmental pollution, e.g., heavy metals, organic solvents, and petroleum, can be analysed by DGGE analysis. This type of analysis may identify increased and/or decreased species among samples. Furthermore, the extraction of DNA from the gel after electrophoresis allows DNA re-amplification and determination of nucleotide sequences, which may facilitate identification of the specific microbial species through a database search. Overview of DGGE analysis is depicted in Fig. 2.

4.1.3 Quantitative PCR (qPCR)

Although the fingerprinting methods mentioned above are appropriate for comparative analyses, qPCR is an excellent method for quantification. In particular, qPCR is appropriate for detecting and monitoring several microorganisms in environments or during a time-course investigation.

There are two primary types of qPCR; the first involves the use of fluorescence reagents that are intercalated into the double-strand structure of the amplified PCR product (i.e., the intercalation method), and the other involves the use of a probe molecule that yields dual fluorescent signals (i.e., the probe method) [37]. The former is less costly than the probe method because an expensive fluorescence probe is not required. However, use of the intercalation method can also result in nonspecific amplification by intercalation of the fluorescence reagents into an undesired PCR product. Therefore, qPCR assays using the intercalation method should include a dissociation curve to examine nonspecific amplification. If double peaks or a split peak is observed in the dissociation curve, nonspecific PCR products have been amplified. In such cases, primers have to be redesigned. However, lack of an abnormal dissociation curve indicates an effective reaction. In contrast, the probe method is thought to be more specific than the intercalation method.

The probe method can be further divided into two additional approaches: (1) use of a DNA probe, and (2) use of a chimeric probe consisting of DNA and RNA, plus addition of RNase H [38, 39]. Both methods utilise fluorescence emission from the hydrolysed probe during the PCR. The probes first hybridise to the target region on the template DNA and are then hydrolysed by the 5′–3′ exonuclease activity of DNA polymerase during the extension step in PCR. An intact probe has two fluorescence labels, and labels on the same molecule interfere with emission. Therefore, only free labels released from the probe can be detected. This type of system shows high specificity. Furthermore, the cycling probe method using a DNA-RNA chimeric probe shows the highest specificity of all of these methods. The chimeric probe hybridises to template DNA, and then only the RNA portion of the hybridised probe is digested by RNase H. Therefore, this method can detect differences in single nucleotides for sequence targeted by the probe.

4.1.4 Microarray

Although qPCR can be used to detect and evaluate from one to five types of microorganisms (limits of available fluorescence probes and detector), microarray analysis can be used to evaluate a microbial community consisting of multiple species. Microarrays are conventionally used to monitor intracellular gene expression under different conditions. Recently, microarray analyses have been developed to evaluate the microbial community by targeting 16S rDNA and some functional genes [40, 41].

Microarrays are used for detection and quantification of the target gene using a DNA chip on which known DNA molecules are fixed. Complementary DNAs (cDNA) are reverse transcribed from RNAs isolated from culture samples, and the DNA extracted from environmental samples is hybridised to the nucleotide probes on the DNA chip. The signal intensities of these DNA-probe hybridised molecules are then detected using a fluorescence scanner.

To date, four types of microarrays have been developed for environmental experiments [40]. Phylogenetic oligonucleotide arrays (POAs) target 16S rDNA and use short oligonucleotide probes (typically 18–25 nucleotides). This method is appropriate for analysis of the microbial community structure in natural environments. Moreover, this method is easy to carry out because the 16S rDNA fragments have been routinely amplified and used in other analytical methods, as described above. Moreover, there are two methods for targeting of functional genes: (1) PCR product-based functional gene arrays (FGAs), and (2) oligonucleotide-based FGAs. These FGAs are useful for estimating the ecological function of the microbial community compared with the POAs. In addition to these, community genome arrays (CGAs) have also been used for identification of species or strains because these arrays apply whole genome sequences from known microorganisms. However, it is difficult to prepare DNA chips for this method, and CGAs cannot be applied to analyse the genomes of uncultivable microorganisms.

In all of these cases, microarray methods are restricted to detection and evaluation of known microorganisms and cannot be used to evaluate unknown microorganisms in microbial communities. Therefore, microarray analysis would be useful for the evaluation of functional microorganism contents and distribution.

4.1.5 Next-Generation Sequencing (NGS)

NGS or pyrosequencing has become possible based on technological advancements in the development of novel sequencers [42]. Although clone libraries and DGGE analyses use sequencers for a determination of nucleotide sequences, these methods still apply the terminator method, which is based on the Sanger method. However, NGS can be used to detect chemical reactions that occur through an extension reaction of the DNA chain as a signal [43, 44], and the total volume of analysable nucleotide sequence per assay is markedly improved. Conventional sequencing can analyse approximately 800 bases per capillary, and more advanced equipment, which can have up to 96 capillaries, can analyse approximately 8×10^4 bases per assay. In contrast, NGS yields over 10^{11} bases per assay. Notably, while NGS was originally very costly, reduced costs per assay in recent years have allowed for broader applications of NGS technology in the microbial community.

For gene analysis of environmental samples using NGS, two approaches namely metagenome analysis and meta-16S rDNA analysis, can be used. The former targets the whole genome of all microorganisms in the environmental sample. Researchers may obtain several circular sequences by metagenome analysis. However, in most cases, whole genome sequences from all microorganisms cannot be obtained because the total amount of genetic information is too great. Therefore, many researchers have used metagenome analysis as a gene resource in order to explore novel enzymes [45, 46]. In contrast, meta-16S rDNA analysis is often used to investigate microbial community structures in samples from natural environments [47, 48]. To date, this method may be the most useful method for microbial community analysis. Meta-16S rDNA analysis allows for identification of more microorganisms than the clone library and microarray methods; it also has the capacity for analysis of multiple samples and for targeting unknown microorganisms.

In meta-16S rDNA analysis, partial fragments of 16S rDNA are amplified by PCR, and these PCR products are used as amplicons. Utilisation of primers with different tag sequences allows amplicons derived from different samples to be differentially labelled. These tags then allow researchers to assign read sequences to each sample, thereby permitting analysis of multiple samples in a single run. After running the operation in NGS, an abundance of read sequences is obtained. It is not possible to manually conduct the assignments and homology searches for all these reads. Therefore, 16S rDNA pipeline analysis software, including QIIME and Mothur programs, have been developed [49, 50]. Overview of meta-16S rDNA analysis by NGS is depicted in Fig. 2.

4.2 Gene Analysis for the Detection of Endocrine Disruptors

In recent years, environmental pollutants called endocrine disruptors have had a major effect on human health [18]. Endocrine disruptors are defined as exogenous chemicals that adversely affect the endocrine system. Dioxins, polychlorinated biphenyls (PCB), agricultural chemicals, bisphenol A (a feedstock of plastic), nonylphenol (a feedstock of non-ion surface-active agents), and phthalic acid ester are thought to act as endocrine disruptors and can cause chronic toxicity, even at very low levels (approximately 10^{-9} mol) [51]. Endocrine disruptors have been shown to exert their effects through disruption of gene expression, interfering with the effects of endogenous hormones and preventing adequate transcriptions of target genes, thereby causing a variety of metabolic issues.

Because endocrine disruptors affect gene expression, even in trace amounts, recent studies have reported attempts to develop detection systems for endocrine disruptors by measuring changes in gene expression [51]. Endocrine disruptors initially bind to receptors and are then internalized and can affect the gDNA. Therefore, monitoring the behaviours of specific genes that are affected by particular ligand-receptor complexes may allow for the indirect detection of endocrine disruptors. Furthermore, the identification of correlations between the concentration of an endocrine disruptor and the expression level of the target gene in model systems may allow for the estimation of environmental concentrations of endocrine disruptors through analysis of changes in transcription. Indeed, studies have identified a cluster of genes whose expression is significantly altered following treatment with several endocrine disruptors in reference cells. Therefore, these genes may be used as reference genes for analysis of the concentrations of endocrine receptors in environmental samples.

Similarly, analysis of the indirect relationships of gene expression and environmental pollutants may be applied to detect other types of pollutants. For example, this method may be used for detection of heavy metals. Viable cells have the ability to respond to various substances in the environment; therefore, detection of this cellular response by gene analysis may allow for detection of a variety of environmental pollutants.

4.3 Biosensor

A biosensor is a device that converts a biological signal (from enzymes, antibodies, genes, cells, and tissues) into a physicochemical signal (e.g. electrons and light).

For example, the commercial blood glucose sensor consists of a glucose oxidase, a small electrode, and a small electrochemical detector, and converts the oxidation signal of glucose into the electron signal, and then detects the electron using the electrode and the detector. In another case, a DNA biosensor consists of a sequence-specific hybridization module and an optical or electrochemical detector [52]. The target DNA as an environmental factor is trapped on the single strand DNA immobilized on the surface of the device, and then the emission or current taking place concomitantly during the hybridization event is detected by the appropriate detector. The microarray described above is one of the DNA biosensors.

In recent years, the microarrays based on killifish complementary DNA (cDNA) and algae *Synechocystis* cDNA have been constructed in order to evaluate environmental stresses at the levels of genes [53–55]. The application and development of such microarrays (against a variety of environmental pollutants) may result in rapid and easy detection of pollutants. However, such microarray technology requires a lab-based specific device. In order to use it as the DNA biosensor, downsizing of the whole system is required. Success of such downsizing may make it possible to do on-site assessment.

5 Gene Analysis of MIC

Gene analysis can also be used for evaluation of MIC. A microorganism is collected from the location where corrosion is observed or from an environment promoting corrosion, and the microbial community structure is analysed using the isolated DNA [56–58]. Alternatively, the DNA can be used to determine whether MIC has occurred and to assess the risk of MIC. When analysing the microbial community structure in order to assess the corrosion, only those in the corrosion products were analysed in many cases. Such analysis may not be appropriate. Indeed, the microbial community structure fluctuates with changes in natural environmental conditions. Ideally, microorganisms in the corrosion products and substances in surrounding environments should be analysed and chased in a time series (certain amount of elapsed time). For example, the microbial community structure may vary during the corrosion process, and variations in the community structure may similarly influence the corrosion process. Therefore, it is necessary to compare the corrosion product, fouling substances on the material where corrosion is not occurring, and samples of environmental water in order to determine whether the microbial community structure in the corrosion product truly differs from those of the other samples. Ideally, it is better to extract a sample at every specific amount of elapsed time and monitor the progress of corrosion to determine changes in the microbial community structure. However, this is difficult to carry out after corrosion has already occurred. Therefore, it may be possible to combine microbial community analysis with corrosion tests on the laboratory scale and compare environmental samples as described above. I and co-workers have revealed that iodide-oxidizing bacteria correlate to the corrosion in the iodide-producing facility

[59]. At that time, the DGGE analysis of the corrosion product, environmental water samples, and lab-scale corrosion test samples showed the correlation of iodide-oxidizing bacteria to the corrosion. Then the corrosion test, using the iodide-oxidizing bacteria isolated from them, directly showed iron-corrosive ability of these isolates. Therefore, analytical methods dependent on cultivation are still important. The details of this are not within the scope of this chapter. The most important aspect of the corrosion test on the laboratory scale is achieving reproduction of the actual environment in which corrosion occurs.

6 Conclusion

In this chapter, I described the gene analysis of environmental pollutants, with a focus on the effects of microorganisms within the environment. However, various substances, including those derived from organisms, are found as contaminants to the environment. Gene analysis against unknown pollutants may become possible by technical developments in the field of technology (e.g. biosensors) and accumulation of knowledge about correlations between pollutants and biological responses. Appropriate assessment of environmental pollution would be essential to maintain the quality of materials.

References

1. Beech IB, Sunner JA, Hiraoka K (2005) Microbe-surface interactions in biofouling and biocorrosion processes. Int Microbiol 8:157–168
2. Coetser SE, Cloete TE (2005) Biofouling and biocorrosion in industrial water systems. Crit Rev Microbiol 31:213–232
3. Kjellerup BV, Thomsen TR, Nielsen JL, Olesen BH, Frolund B, Nielsen PH (2005) Microbial diversity in biofilms from corroding heating systems. Biofouling 21:19–29
4. Muthukumar N, Rajasekar A, Ponmariappan S, Mohanan S, Maruthamuthu S, Muralidharan S, Subramanian P, Palaniswamy N, Raghavan M (2003) Microbiologically influenced corrosion in petroleum product pipelines – a review. Indian J Exp Biol 41:1012–1022
5. Videla HA, Herrera LK (2005) Microbiologically influenced corrosion: looking to the future. Int Microbiol 8:169–180
6. Samanta SK, Singh OV, Jain RK (2002) Polycyclic aromatic hydrocarbons: environmental pollution and bioremediation. Trends Biotechnol 20:243–248
7. Järup L (2003) Hazards of heavy metal contamination. Br Med Bull 68:167–182
8. Mastrangelo G, Fadda E, Marzia V (1997) Polycyclic aromatic hydrocarbons and cancer in man. Environ Health Perspect 104:1166–1170
9. Goldman R, Enewold L, Pellizzari E, Beach JB, Bowman ED, Krishnan SS, Shields PG (2001) Smoking increase carcinogenic polycyclic aromatic hydrocarbons in human lung tissue. Cancer Res 61:6367–6371
10. Seidal K, Jorgensen N, Elinder CG, Sjogren B, Vahter M (1993) Fatal cadmium-induced pneumonitis. Scand J Work Environ Health 19:429–431

11. Barbee JY Jr, Prince TS (1999) Acute respiratory distress syndrome in a welder exposed to metal fumes. South Med J 92:510–512
12. Weiss B, Clarkson TW, Simon W (2002) Silent latency periods in methylmercury poisoning and in neurodegenerative disease. Environ Health Perspect 110:851–854
13. Lidsky TI, Schneider JS (2003) Lead neurotoxicity in children: basic mechanisms and clinical correlates. Brain 126:15–19
14. Wilson SC, Jones KC (1993) Bioremediation of soil contaminated with polynuclear aromatic hydrocarbons (PAHs): a review. Environ Pollut 81:229–249
15. Haritash AK, Kaushik CP (2009) Biodegradation aspects of polycyclic aromatic hydrocarbons (PAHs): a review. J Hazard Mater 169:1–15
16. Liang SH, Kao CM, Kuo YC, Chen KF, Yang BM (2011) In situ oxidation of petroleum-hydrocarbon contaminated groundwater using passive ISCO system. Water Res 45:2496–2506
17. Depledge MH, Billinghurst Z (1999) Ecological significance of endocrine disruption in marine invertebrates. Mar Pollut Bull 39:32–38
18. Lutz I, Kloas W (1999) Amphibians as a model to study endocrine disruptors: I. Environmental pollution and estrogen receptor binding. Sci Total Environ 225:49–57
19. Sun Y, Huang H, Sun Y, Wang C, Shi XL, Hu HY, Kameya T, Fujie K (2013) Ecological risk of estrogenic endocrine disrupting chemicals in sewage plant effluent and reclaimed water. Environ Pollut 180:339–344
20. Flemming HC (2002) Biofouling in water systems–cases, causes and countermeasures. Appl Microbiol Biotechnol 59:629–640
21. Huttunen-Saarivirta E, Honkanen M, Lepistö T, Kuokkala VT, Koivisto L, Berg CG (2012) Microbiologically influenced corrosion (MIC) in stainless steel heat exchanger. Appl Surf Sci 258:6512–6526
22. Almeida MAN, De Franca FP (1999) Thermophilic and mesophilic bacteria in biofilms associated with corrosion in a heat exchanger. World J Microbiol Biotechnol 15:439–442
23. Xu HS, Roberts N, Singleton FL, Attwell RW, Grimes DJ, Colwell RR (1982) Survival and viability of nonculturable Escherichia coli and Vibrio cholerae in the estuarine and marine environment. Microb Ecol 8:313–323
24. Amann RI, Ludwig W, Schleifer KH (1995) Phylogenetic identification and in situ detection of individual microbial cells without cultivation. Microbiol Rev 59:143–169
25. Neefs JM, Van de Peer Y, De Rijk P, Chapelle S, De Wachter R (1993) Compilation of small ribosomal subunit RNA structures. Nucleic Acids Res 21:3025–3049
26. Stackebrandt E, Goebel BM (1994) Taxonomic note: a place for DNA-DNA reassociation and 16S rRNA sequence analysis in the present species definition in bacteriology. Int J Syst Bacteriol 44:846–849
27. Stackebrandt E, Frederiksen W, Garrity GM, Grimont PA, Kämpfer P, Maiden MC, Nesme X, Rosselló-Mora R, Swings J, Trüper HG, Vauterin L, Ward AC, Whitman WB (2002) Report of the ad hoc committee for the re-evaluation of the species definition in bacteriology. Int J Syst Evol Microbiol 52:1043–1047
28. Tanksley SD, Young ND, Paterson AH, Bonierbale MW (1989) RFLP mapping in plant breeding: new tools for an old science. Nat Biotechnol 7:257–264
29. Poly F, Monrozier LJ, Bally R (2001) Improvement in the RFLP procedure for studying the diversity of nifH genes in communities of nitrogen fixers in soil. Res Microbiol 152:95–103
30. Osborn AM, Moore ER, Timmis KN (2000) An evaluation of terminal-restriction fragment length polymorphism (T-RFLP) analysis for the study of microbial community structure and dynamics. Environ Microbiol 2:39–50
31. Franklin RB, Taylor DR, Mills AL (1999) Characterization of microbial communities using randomly amplified polymorphic DNA (RAPD). J Microbiol Methods 35:225–235
32. Franklin RB, Garland JL, Bolster CH, Mills AL (2001) Impact of dilution on microbial community structure and functional potential: comparison of numerical simulations and batch culture experiments. Appl Environ Microbiol 67:702–712

33. Muyzer G, de Waal EC, Uitterlinden AG (1993) Profiling of complex microbial populations by denaturing gradient gel electrophoresis analysis of polymerase chain reaction-amplified genes coding for 16S rRNA. Appl Environ Microbiol 59:695–700
34. Heuer H, Krsek M, Baker P, Smalla K, Wellington EM (1997) Analysis of actinomycete communities by specific amplification of genes encoding 16S rRNA and gel-electrophoretic separation in denaturing gradients. Appl Environ Microbiol 63:3233–3241
35. Felske A, Rheims H, Wolterink A, Stackebrandt E, Akkermans AD (1997) Ribosome analysis reveals prominent activity of an uncultured member of the class Actinobacteria in grassland soils. Microbiology 143:2983–2989
36. Okabayashi A, Wakai S, Kanao T, Sugio T, Kamimura K (2005) Diversity of 16S ribosomal DNA-defined bacterial population in acid rock drainage from Japanese pyrite mine. J Biosci Bioeng 100:644–652
37. Heid CA, Stevens J, Livak KJ, Williams PM (1996) Real time quantitative PCR. Genome Res 6:986–994
38. Duck P, Alvarado-Urbina G, Burdick B, Collier B (1990) Probe amplifier system based on chimeric cycling oligonucleotides. Biotechniques 9:142–148
39. Harvey JJ, Lee SP, Chan EK, Kim JH, Hwang ES, Cha CY, Knutson JR, Han MK (2004) Characterization and applications of CataCleave probe in real-time detection assays. Anal Biochem 333:246–255
40. Zhou J (2003) Microarrays for bacterial detection and microbial community analysis. Curr Opin Microbiol 6:288–294
41. Loy A, Lehner A, Lee N, Adamczyk J, Meier H, Ernst J, Schleifer KH, Wagner M (2002) Oligonucleotide microarray for 16S rRNA gene-based detection of all recognized lineages of sulfate-reducing prokaryotes in the environment. Appl Environ Microbiol 68:5064–5081
42. Voelkerding KV, Dames SA, Durtschi JD (2009) Next-generation sequencing: from basic research to diagnostics. Clin Chem 55:641–658
43. Nyrén P (1987) Enzymatic method for continuous monitoring of DNA polymerase activity. Anal Biochem 167:235–238
44. Harrington CT, Lin EI, Olson MT, Eshleman JR (2013) Fundamentals of pyrosequencing. Arch Pathol Lab Med 137:1296–1303
45. Ferrer M, Martínez-Abarca F, Golyshin PN (2005) Mining genomes and 'metagenomes' for novel catalysts. Curr Opin Biotechnol 16:588–593
46. Ferrer M, Golyshina O, Beloqui A, Golyshin PN (2007) Mining enzymes from extreme environments. Curr Opin Microbiol 10:207–214
47. Ducey TF, Hunt PG (2013) Microbial community analysis of swine wastewater anaerobic lagoons by next-generation DNA sequencing. Anaerobe 21:50–57
48. Fierer N, Lauber CL, Ramirez KS, Zaneveld J, Bradford MA, Knight R (2012) Comparative metagenomic, phylogenetic and physiological analyses of soil microbial communities across nitrogen gradients. ISME J 6:1007–1017
49. Caporaso JG, Kuczynski J, Stombaugh J, Bittinger K, Bushman FD, Costello EK, Fierer N, Peña AG, Goodrich JK, Gordon JI, Huttley GA, Kelley ST, Knights D, Koenig JE, Ley RE, Lozupone CA, McDonald D, Muegge BD, Pirrung M, Reeder J, Sevinsky JR, Turnbaugh PJ, Walters WA, Widmann J, Yatsunenko T, Zaneveld J, Knight R (2010) QIIME allows analysis of high-throughput community sequencing data. Nat Methods 7:335–336
50. Schloss PD, Westcott SL, Ryabin T, Hall JR, Hartmann M, Hollister EB, Lesniewski RA, Oakley BB, Parks DH, Robinson CJ, Sahl JW, Stres B, Thallinger GG, Van Horn DJ, Weber CF (2009) Introducing mothur: open-source, platform-independent, community-supported software for describing and comparing microbial communities. Appl Environ Microbiol 75:7537–7341
51. Kiyama R, Wada-Kiyama Y (2015) Estrogenic endocrine disruptors: molecular mechanisms of action. Environ Int 83:11–40
52. Mikkelsen SR (1996) Electrochecmical biosensors for DNA sequence detection. Electroanalysis 8:15–19

53. Gracey AY, Cossins AR (2003) Application of microarray technology in environmental and comparative physiology. Annu Rev Physiol 65:231–259
54. Oleksiak MF, Kolell KJ, Crawford DL (2001) Utility of natural populations for microarray analyses: isolation of genes necessary for functional genomic studies. Mar Biotechnol (NY) 3 (Supplement 1):S203–S211
55. Kanesaki Y, Suzuki I, Allakhverdiev SI, Mikami K, Murata N (2002) Salt stress and hyperosmotic stress regulate the expression of different sets of genes in *Synechocystis* sp. PCC 6803. Biochem Biophys Res Commun 290:339–348
56. AlAbbas FM, Williamson C, Bhola SM, Spear JR, Olson DL, Mishra B, Kakpovbia AE (2013) Influence of sulfate reducing bacterial biofilm on corrosion behavior of low-alloy, high-strength steel (API-5L X80). Int Biodeterior Biodegrad 78:34–42
57. Duncan KE, Perez-Ibarra BM, Jenneman G, Harris JB, Webb R, Sublette K (2014) The effect of corrosion inhibitors on microbial communities associated with corrosion in a model flow cell system. Appl Microbiol Biotechnol 98:907–918
58. Murali Mohan A, Hartsock A, Hammack RW, Vidic RD, Gregory KB (2013) Microbial communities in flowback water impoundments from hydraulic fracturing for recovery of shale gas. FEMS Microbiol Ecol 86:567–580
59. Wakai S, Ito K, Iino T, Tomoe Y, Mori K, Harayama S (2014) Corrosion of iron by iodide-oxidizing bacteria isolated from brine in an iodine production facility. Microb Ecol 68:519–527

The Application of Mammalian Cells for Assessment of Metallic Products (*In Vitro* Process)

Akiko Ogawa

Abstract This chapter describes the use of mammalian cells to evaluate environmental friendliness in regards to biological responses, metallic toxicity, etc. It discusses several *in vitro* assays using mammalian cell cultures to measure some features of metallic products. Information is provided about both the colony formation assay and the viable cell number assay for detecting changes in surface conditions of several metallic materials. It should be noted that this type of detection is difficult using traditional analyzing techniques including conventional chemical analyses, scientific analyses of materials with various microscopes, etc. Being compared with conventional methods, the biological methods with a mammalian cell culture system (described in this chapter) could make it possible to quantify the effect of environmental factors such as the components of alloys and heavy metals and to easily realize the relationship between them and organisms.

1 Background

1.1 *Concept: Why We Use Mammalian Cells for Analysis of Metallic Materials and Products?*

Mammals including human beings consist of several types of cells that cooperate with each other in order to maintain the functions such as growing, digestion and metabolism. These cells can survive *in vitro* with the satisfied conditions of nutrition, temperature, and pH. Mammalian cell cultures (*in vitro*) are usually used for research in molecular biology, embryology, etc. because we can simplify the complicated system of the body (*in vivo*) by revealing the mechanism of several functions. Mammalian cell culture is also used for biomedical production and several chemical safety tests that are free from using animals because of several

A. Ogawa (✉)
National Institute of Technology, Suzuka College, Suzuka, Japan
e-mail: ogawa@chem.suzuka-ct.ac.jp

H. Kanematsu, D.M. Barry (eds.), *Corrosion Control and Surface Finishing*,
DOI 10.1007/978-4-431-55957-3_17

reasons: it costs much lower than animal testing, it has no moral problems, and caring for mammalian cells is easier than caring for animals. Also many samples can be examined at one time and you can get reproductive experimental results.

Mammals use several metals in the body to maintain their physiological functions such as metabolism, differentiation, and survival. These metals are known as essential elements: chromium, manganese, iron, cobalt, nickel, copper, zinc, molybdenum, vanadium, and selenium [1]. They exist as ions or bind to specific proteins in the body. Each essential element has a different appropriate amount in the body. For instance, iron, zinc, and manganese exist at 85.7 mg/kg, 28.5 mg/kg and 1.43 mg/kg respectively in the human body [2]. If the amount of an essential element is higher than the optimal concentration, physiological activity will be decreased and the organism will have excess symptoms or die. On the other hand, the loss of an essential element will cause deficiency symptoms or stop growth (Fig. 1). That is, the physiological changes relate to the amount of an essential element. What happens in each cell at that time? Cells change not only protein and enzyme activities but also several gene expressions, resulting in survival or death, inducing differentiation, and maintaining an undifferentiated state. We can detect these changes by counting the number of colonies, morphological change (Fig. 2) or cell-based assay methods such as measuring viability and the MTT assay. MTT is an abbreviation for 3-(4, 5-dimethythiazol-2-yl)-2,5-diphenyl tetrazolium bromine, which enters cells and passes into the mitochondria where MTT is reduced to a colored formazan compound. The reduction activity of MTT depends on the activity of cellular aerobic metabolism. In a recent study, I tested how zinc ion (Zn^{2+}) and nickel ion (Ni^{2+}) influenced the cell proliferation of PC-12, rat pheochromocytoma cell line. Figure 3 shows the viable cell number of each culture condition where nickel ions or zinc ions were added to the culture medium. The control culture condition was PC-12 cultured without any addition of ions. When PC-12 was cultured with zinc ion concentration at 0–450 μM, the viable cell

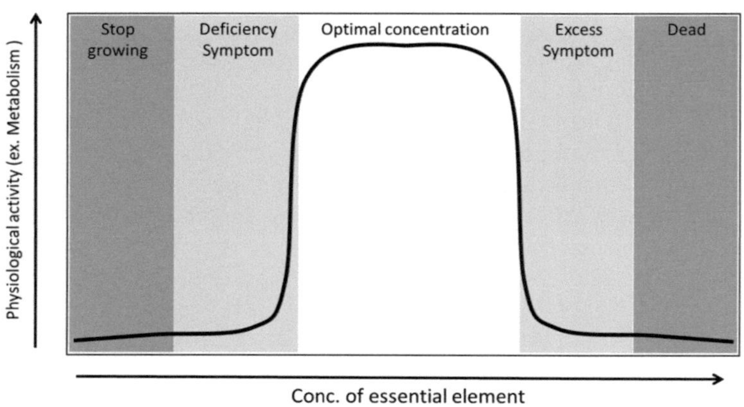

Fig. 1 The interaction between the amount of essential elements and physiological functions (This picture was modified based on figure 3.1 from Ref. [2])

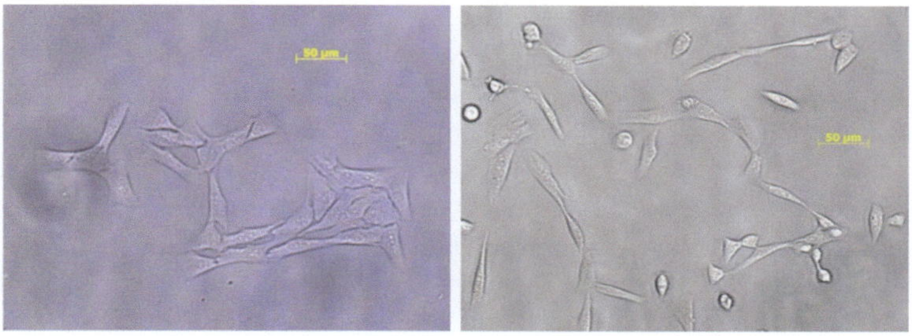

Fig. 2 The morphological change of V79. V79 is derived from a Chinese hamster lung. The morphology of V79 looks like fibroblast. V79 attached on the surface of a culture dish is sticky in normal culture conditions (*left photo*). On the other hand, when V79 is cultured with an excess amount of zinc chloride, then the shape becomes thinner and smaller than that of normal conditions (*right photo*)

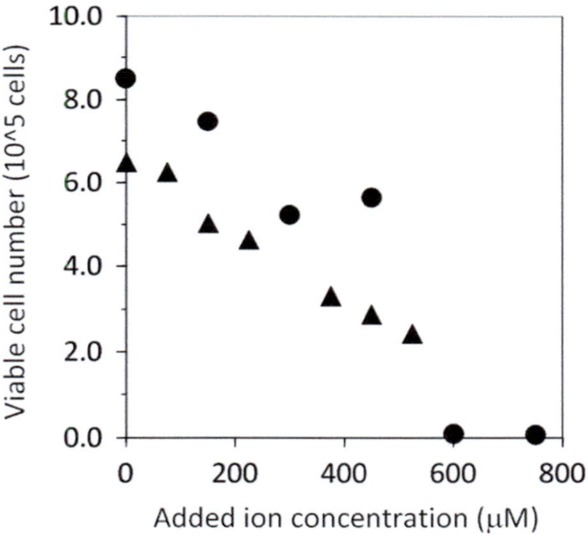

Fig. 3 The effect of Ni^{2+} or Zn^{2+} on PC-12 cell growth. *Circles* and *triangles* indicate the average viable cell numbers cultured with Ni^{2+} and Zn^{2+} respectively (n = 3). Added ion concentration indicates the final concentration of the supernatant. The initial viable cell number was 1.66*10^5 cells. The culture period was 44–45 h after Ni^{2+} or Zn^{2+} addition

number was decreased in a dose-dependent manner and the coefficient of determination was 0.82. Almost all cells were dead at 600 μM. When PC-12 was cultured with nickel ion concentration at 0–525 μM, the viable cell number was also decreased in a dose-dependent manner and the coefficient of determination was 0.98. Compared to the concentration where the viable cell number was reduced by around 40 % of control, zinc ion (350 μM) was smaller than nickel ion (450 μM).

These results indicate that (1) we can use mammalian cells for detecting and quantifying metals by measuring the cellular change, and (2) the range of reactive concentration and degree of cellular change will differ with metals. In addition, using mammalian cells (*in vitro* assay) has some advantages: it can simplify the complicated phenomena of *in vivo*, we can easily realize the relationship between cellular reaction(s) and metal function(s), (which is very important information to make metal products), we can measure many samples very quickly at once and we can detect sensitively the changes of cells by metal because most of the cell based assays are optimized for measuring a small amount of cells.

1.2 Introduction of the Mammalian Cell Culture [3–5]

Mammals consist of multiple tissues and organs and each tissue or organ consists of specific types of cells. For example, the brain consists of neurons and glia. The neurons transmit information such as feelings and thinking, while glia supports neurons. These cells get nutrients (oxygen, glucose and so on) from blood. We can culture these cells and maintain the functions in a culture dish (*ex vivo*) if we prepare a suitable environment for them. For starting a mammalian cell culture, we need to prepare storage space, a work space with low risk of contamination, sterile materials including media, culture dishes, and all kinds of items that are used for experiments. Specific experimental equipment includes a carbon dioxide incubator, a laminar-flow hood or biological safety cabinet, microscopes, autoclaves, centrifuges, refrigerators, and freezers. The carbon dioxide incubator can regulate temperature and the carbon dioxide concentration for sustaining the pH of culture medium neutrality (pH 7) because a lot of mammalian cell culture media use sodium bicarbonate as a buffer for pH control. A laminar-flow hood or biological safety cabinet is a work space where we can reduce the risk of contamination and perform all experiments aseptically. The biological safety cabinet is also suitable for the situation where you treat biohazardous materials, some pathogens, virally infected cells and so on. A microscope is used for cell observation which is very important to check the cellular conditions such as morphology, viability, and occupancy. We usually use a phase contrast (inverted) light microscope to enhance optical contrast because cultured mammalian cells are very thin and transparent in color. An autoclave is a sterilizer that uses high-pressured steam. Some reagents cannot be autoclaved because of denaturing. In these cases, we use filter to remove small particles such as microorganisms. We can also purchase sterile products such as culture dishes, plastic equipment, some media, and some chemicals. A centrifuge is used for collecting suspended cells and for removing the old media in subcultures. Refrigerators and freezers are used for storage of culture media, some nutrients, antibiotics, and cytokines. If you want to store mammalian cells for a long period (over 1 year), a liquid nitrogen freezer or a cryostorage container is needed. We can keep them (the cells) almost permanently there.

Table 1 Worldwide cell banks

Organization/Company	Website	Location (country)
American Type Culture Collection (ATCC)	http://www.atcc.org/en.aspx	10801 University Boulevard Manassas, VA 20110 (USA)
BioResource center of RIKEN	http://en.brc.riken.jp/	3-1-1 Koyadai, Tsukuba, Ibaraki 305-0074 (Japan)
DZMS	https://www.dsmz.de/home.html	Inhoffenstraβe 7B 38124 Braunschweig (Germany)
European Collection of Cell Cultures (ECACC)	https://www.phe-culturecollections.org.uk/collections/ecacc.aspx	Public Health England Porton Down Salisbury SP4 0JG (UK)
JCRB cell bank	http://cellbank.go.jp/english/	7-6-8 Saitoasagi Ibaragi Osaka 567-0085 (Japan)

Many mammalian cells can be purchased from public cell banks or private companies. Table 1 shows the summary of worldwide cell banks (Table 1). You can get information about several mammalian cells including characteristics, culture conditions and references. Depending on the purpose of the experiments, one usually uses cell lines which are immortalized, developed from a single cell, and with uniform features. They are suitable for repeating a lot of tests. Information about them is provided in articles, websites and books. This helps us realize the interaction between metals and cells.

1.3 Mechanism for the Changing Cells by Metals

Mammalian cells need some metals as nutrients to maintain functions such as proliferation, metabolism, and differentiation. Optimal metal concentrations vary with the types of cell. For example, estimates of the free zinc ion intra cellular concentration are 5–10 pM for pheochromocytoma (PC-12) cells, 170 pM for primary human monocytes and 350 pM for lymphocytes [6]. In addition, the regulating mechanism of metals is also diverse in cells. Mammalian cells have two ways for metal uptake and ejection: one is endocytosis, another one is through metal transporters that can transport specified metal(s) selectively. Several kinds of transporters are known. Zinc transporters have two families: ZnT (Slc30a family) effuses zinc ion from cytoplasm and ZIP (Slc39a family) infuses it into cytoplasm [7]. Divalent metal transporter 1 (DMT1) apparently transports divalent metals not only divalent iron but also copper ion, zinc ion, divalent manganese and so on [8]. In mammalian cells, some proteins combine with specific metals which are called metal proteins. For instance, transferrin is an iron carrier, ferritin can store iron [9], metallothionein (MT) can hold divalent metals such as zinc, copper, cadmium and so on [10, 11]. In chromosomes, there are some regions where metal proteins combine and regulate gene expressions. Zinc-finger motif proteins are famous for regulating transcription and translation [12]. These metal-related proteins are expressed in several mammalian cells. However, cells vary in expression level and type. For example, MT has four sub-classes: MT-I, II, III and IV. MT-I and II are expressed in wide tissues but MT-III is mainly in the brain, and MT-4 is unknown. The highest concentration of MT is found in the liver, kidney, intestine and pancreas [13]. These metal-related proteins and the expression level settle cellular reactions to specific metals. That is why we should garner the information about cellular characteristics and we should test whether or not a cell is suitable for detection and measuring specific metal(s). In the next section, I introduce several studies that are related to the measurement of metals and applied to comparing the features of metal alloys.

2 Case Studies

2.1 Application of Colony Formation Assay

In medical devices and apparatus such as a contact lens and metals, cytotoxicity tests must be determined with reference to ISO 10993-5. One of them is the colony formation test that consists of two parts. One part is making an extract from a medical device. The other one is cytotoxic assessment using colonies (by counting the number of colonies) after a cell culture with the extract. Colony means a group formed by a lot of cells derived from single cell. A cell can multiply well to create a colony when the culture condition is good for growing. Otherwise, it cannot (Fig. 4). Some cell lines are known as strong colony-forming cells: V79 is derived from a Chinese hamster lung fibroblast, Balb/3T3 clone A31 is derived from a mouse embryo and L929 is derived from a mouse's connective tissue.

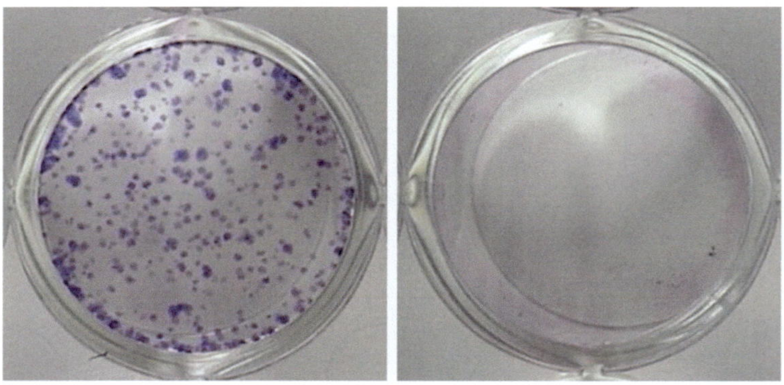

Fig. 4 Colony forming test. *Left photo* shows colonies formed on a culture plate. *Right photo* shows no colony. V79 was seeded on a 12-well culture plate at 500 cells/well. Culture medium was 5 % FBS containing MEM medium. V79 was cultured with or without zinc diethyldithiocarbamate for 4 days. After the culture, V79 was stained using the Giemza dye method

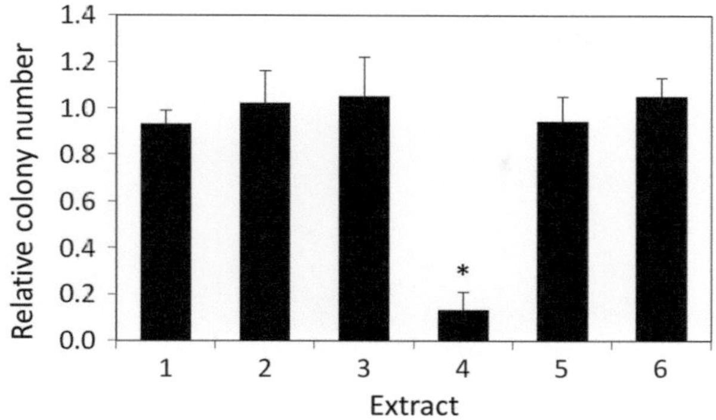

Fig. 5 Results of the colony formation test. Each colony number was visually counted (n = 3). Normal culture (no extract) was used as a control. The colony number of each extract added culture was divided by that of the control. Each *column* and *error bar* indicates means and standard deviation, respectively. *1* SUS304; *2* SS400; *3* nickel coated; *4* chromate conversion and zinc-coated SS400; *5* nickel plate: *6* chromium plate. * $p < 0.01$ (Student t test)

We examined whether this colony formation was changed by metals. We used V79 and prepared several metal plates: stainless steel SUS304, carbon steel SS400, chromate conversion and zinc-coated SS400, nickel-coated SS400, chromium plate and a nickel plate. First these metal plates were sterilized by autoclave treatment (121 °C, 15 min). Then they were immersed in phosphate buffer and incubated at 36.5 °C for 24 h. We collected the extract and added it to the supernatant of V79 culture plate (the ratio of extract to supernatant was 1:9). After a 5-day culture, each culture plate was stained with Giemsa dye and the number of colonies formed in each culture condition was counted. Only when extract of chromate conversion and zinc-coated SS400 was added to the V79 culture, the number of colonies was very low (about 1/10) compared to that of the control (Fig. 5). This result indicates that the extract of chromate conversion and zinc-coated SS400 contain enough amounts of metals to inhibit proliferation of V79. We also analyzed the surface of these samples by laser microscope. We observed that chromate conversion and zinc-coated SS400 had rough surfaces before and after extraction but the tendency was the same between them. Meanwhile, other coupons had flat surfaces before extraction and only nickel-coated SS400 had some holes on the surface after extraction (Fig. 6). Rough surface tends to take local cell action. We think that chromate conversion and zinc-coated SS400 would have a reaction and the zinc ion and hexavalent chromium would be eluted from the surface because of an ionization tendency (Zn > Cr > Fe). On the other hand, nickel-coated SS400 flowed out much more iron ions than nickel ions. The rest of the coupons were not corroded. Next we estimated the amount of zinc ion and hexavalent chromium in the culture supernatant by using colony formation test. We used zinc sulfate and potassium chromate as a zinc ion supplier and hexavalent chromium supplier, respectively. In Fig. 7, the colony number becomes less than 1/10 of the control when the zinc ion concentration is over 100 μM and the hexavalent chromium is 1.9–2.2 μM. From this result,

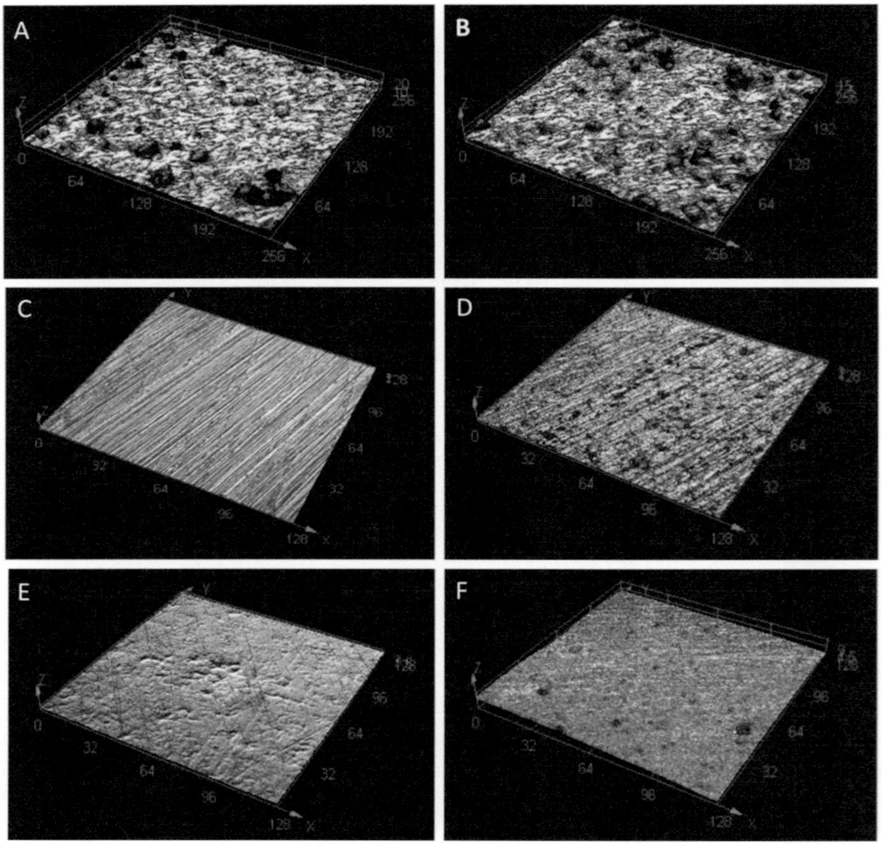

Fig. 6 The surface condition of metal coupons. We observed the surface of chromate conversion and zinc-coated SS400 (*upper panel*), SS400 (*middle panel*) and nickel-coated SS400 (*lower panel*) before extraction (**a**, **c** and **e**) and after extraction (**b**, **d** and **f**) by LEXT OLS4000 (Olympus)

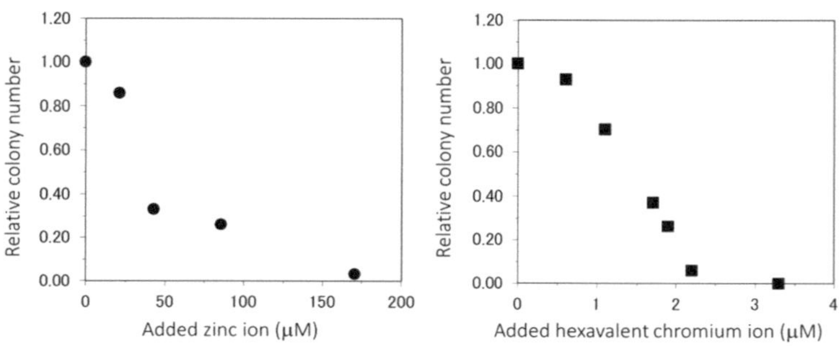

Fig. 7 Colony formation assay of V79 in the culture condition of zinc sulfate (*left*) or potassium chromate (*right*). Each colony number was visually counted ($ZnSO_4$ was n = 3, K_2CrO_4 was n = 2). Normal culture (no addition of ion) was used as a control. Colony number of each culture was divided by that of the control

the extract of chromium conversion zinc-coated SS400 would have over 1 mM of zinc ion and (or) 19–22 μM of hexavalent chromium ion. These results show that colony formation assay will reflect the differences of metal products in the material features [14].

We applied this assay to compare the property of several tin-zinc alloy filmed carbon steel (SS400) coupons that had different histories of making process [15, 16]. We used V79 for the colony formation assay because V79 was sensitive to the amount of zinc, [17]. We prepared three tin-zinc alloy filmed SS400 as follows. First we purchased zinc-coated SS400 that was coated with tin by a radio frequency magnetron sputtering machine. Next they (the coupons) were heated for 1 h at three different temperatures: 250 °C treated coupon was SnZn-250, 350 °C treated one was SnZn-350 and 450 °C treated one was 450 °C. Next they were cut into 1 × 1 cm and their edges were covered with white marker pen. These coupons were wiped with 100 % ethanol then soaked in 70 % ethanol for 18 h to sterilize them. Sterile coupons were transferred to a phosphate buffer containing tube and incubated at 36.5 °C for 24 h to make the extract. Finally, soaked coupons were taken out from the tube and analyzed by an X-ray diffractometer. Extract was used for the colony formation assay. Zinc, iron, and tin concentrations were measured by an inductively coupled plasma atomic emission spectrometer (ICP-AES). Figure 8 shows the results of the colony formation assay. The order of colony numbers was SnZn-450 > SnZn-250 > SnZn-350. We also measured the metal concentration of extract (Table 2). The order of zinc concentration was SnZn-350 > SnZn-250 > SnZn-450, which was opposite from the order of the colony number. This result

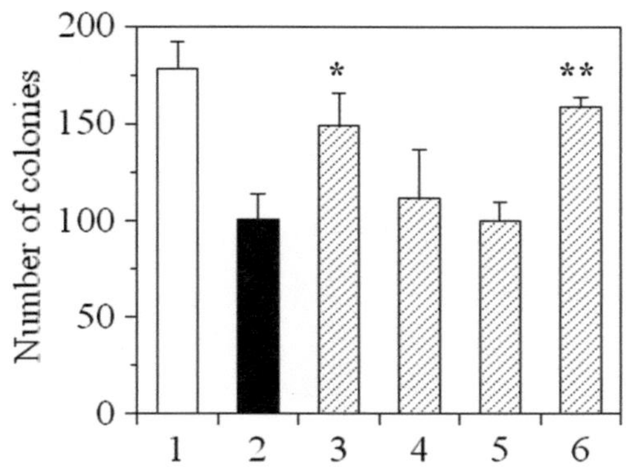

Fig. 8 Colony formation assay of tin-zinc alloy film SS400 [18]. Each extract was added to the supernatant of V79 culture plate (the ratio of extract to supernatant was 2:5). After a 5-day culture, each culture plate was stained with Giemsa dye and we counted the number of colonies formed in each culture condition (n = 3). *1* No extract (phosphate buffer was added in spite of extract); *2* Added extract of zinc-plating SS400; *3* Added extract of SnZn layered SS400 (no heat treatment); *4* Added extract of SnZn-250; *5* Added extract of SnZn-350; *6* Added extract of SnZn-450. * and ** show $p < 0.05$ v.s. SnZn-350 and $p < 0.01$ v.s. SnZn-350 of the result of Student's t test respectively

Table 2 Metal concentrations of extract (modified Table 3 of reference 18)

Coupon	Zn (ppm)	Sn (ppm)	Fe (ppm)
Zinc coated SS400	13.0	ND	0.05
SnZn-layered	5.1	6.9	0.02
SnZn-250	4.6	2.5	0.03
SnZn-350	6.8	1.2	0.04
SnZn-450	2.0	0.08	0.3

ND means not detected

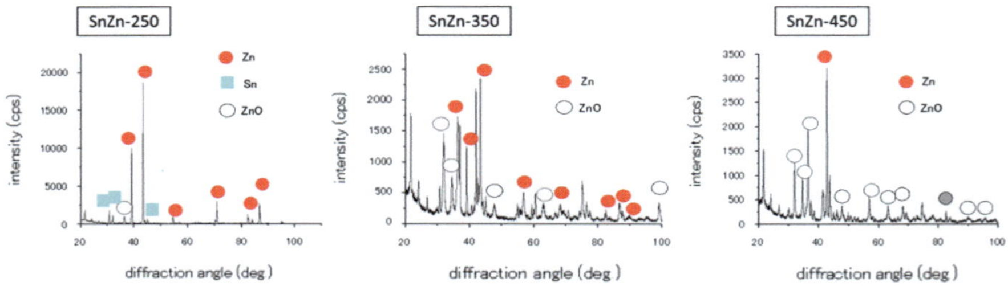

Fig. 9 The surface of tin-zinc alloy film SS400. *Red circles*, *open circles*, and *blue squares* mean zinc, zinc oxide, and tin, respectively

indicates that zinc concentration will be the key factor that determines the environmental risk (especially against mammals) of the coupon. We analyzed the surface of these coupons using an X-ray diffractometer. SnZn-250 had both tin and zinc but SnZn-350 had mainly zinc and no tin. SnZn-450 had zinc oxide and no tin (Fig. 9). From these results, we suppose that SnZn-350 accelerated the elution of zinc ion from the surface. Meanwhile SnZn-450 was protected because the surface was almost covered with zinc oxide that is an insoluble compound. SnZn-250 would be formed (tin-zinc intermolecular compound) which reduced the elution of zinc ion.

2.2 *Application of Viable Cell Number*

Cell lines usually proliferate once or twice a day in normal culture conditions. When they arrest the cell cycle, this brakes DNA synthesis or reduces metabolism, and the proliferation rate will change. Consequently, each culture condition is different in a grown (proliferated) cell number. We tried to detect zinc ion and measure the amount of zinc ion and nickel ion by comparing viable cell number.

For detecting zinc ion, we used three different cell lines: V79, PC-12 is derived from rat pheochromocytoma and Hep G2 is human hepatocyte carcinoma. These cells were cultured in fetal bovine serum (FBS) containing medium to supply not only an energy source and amino acids, but also some hormones and cytokines. Each cell was seeded on a culture plate by appropriate cell number and cultured for 1 day with a CO_2 incubator (CO_2 5 %, 36.5 °C). On the next day, sterile zinc sulfate solution was diluted

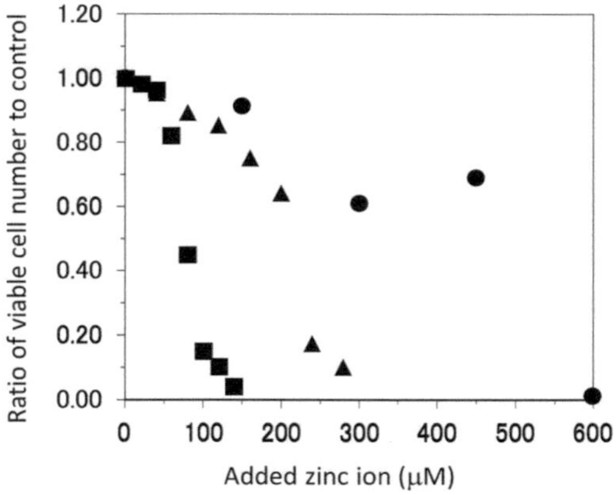

Fig. 10 Comparison of the effect of zinc ion concentration on cell proliferation among three cell lines. Zinc sulfate was used for the zinc ion supplier. *Circles* indicate the standardized viable cell number of PC-12. *Triangles* indicate the cultured standardized viable cell number of Hep G2. *Squares* indicate a standardized viable cell number of V79 ($n = 3$). No addition of zinc sulfate was used for the control culture condition

with each FBS containing medium to a desired concentration and then added to the culture supernatant. Each cell was cultured more than 1 day. On the second day, cells were removed from each culture plate by trypsin treatment. They were collected by centrifugation and then re-suspended by a new FBS containing medium. The re-suspended cells were counted with trypan blue-exclusion test and a hemocytometer. We compared the viable cell number of each zinc concentration among three cell lines (Fig. 10). V79 was the smallest zinc ion concentration in which the viable cell number decreased by 20 % (at 60 μM). At 80 μM, the ratio of viable cell number to control was almost half ($p < 0.01$). The decrease was in a dose dependent manner from 40 to 100 μM ($R^2 = 0.98$). Hep G2 decreased the ratio of a viable cell number to control by 15 % at 120 μM, and the ratio of the viable cell number to control was half between 200 and 240 μM. The decrease was in a dose dependent manner from 80 to 200 μM ($R^2 = 0.97$). On the other hand, PC-12 was over 0.90 at 150 μM and the ratio of the viable cell number to the control maintained over 0.60 until 450 μM. These results indicate that V79 can detect the smallest amount of zinc ion and V79 and Hep G2 can measure the zinc concentration in different ranges of concentration.

For detecting nickel ion, we used four different cell lines: PC-12, V79, Hep G2, MOLT-3 and U937. MOLT-3 is derived from human T cell leukemia and U937 is derived from human lymphoma and can be cultured in a suspension. We performed with the same culture conditions and procedures as for zinc detection about PC-12, V79 and Hep G2. While U937 and MOLT-3 were cultured in a FBS containing medium, seeded on a culture plate with or without nickel chloride solution, and cultured for 2 days with a CO_2 incubator (CO_2 5 %, 36.5 °C). On the second day, U937 was re-suspended by pipetting, collected from each well of the culture plate, and viable cell concentrations were counted with the trypan blue exclusion method. Figure 11 shows the effect of nickel ion on cell proliferation about five cell lines.

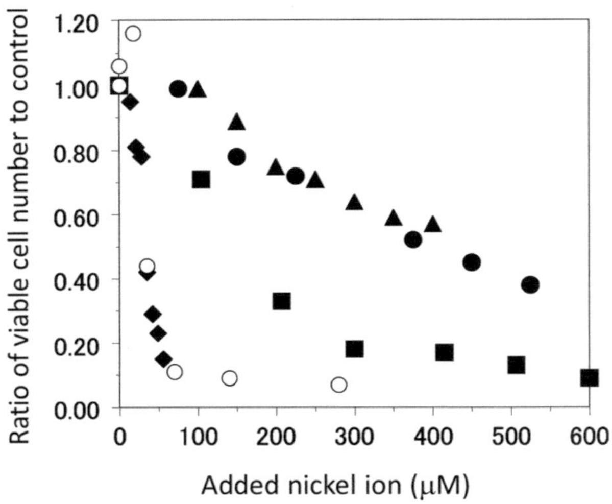

Fig. 11 Comparison of the effect of nickel ion concentration on cell proliferation among five cell lines. Nickel chloride was used for the nickel ion supplier. *Closed circles* indicate the standardized viable cell number of PC-12. *Open circles* indicate the standardized viable cell number of MOLT-3. *Triangles* indicate the cultured standardized viable cell number of Hep G2. *Squares* indicate the standardized viable cell number of V79. *Diamonds* indicate the standardized viable cell number of U937 (n = 3). No addition of nickel chloride was used for the control culture condition

U937 was the smallest nickel ion concentration in which the viable cell number decreased by about 20 % (at 21 μM). The decrease was in a dose-dependent manner from 14 to 56 μM ($R^2 = 0.94$). Hep G2 and PC-12 were very similar in response to nickel. Both of them started to reduce the ratio of the viable cell number at 150 μM and continued the reduction equally until 400 μM in a dose-dependent manner. MOLT-3 had a similar tendency as U937 over 35 μM. The responsibility of V79 was between that of U937 and that of Hep G2. These results show that U937 and MOLT-3 will be suitable cell lines to detect and measure nickel ion under 60 μM and HepG2 and PC-12 will be good to do higher concentrations (over 150–400 μM).

Next we applied this assay to the assessment of a nickel-titanium alloy [18]. We selected MOLT-3 that was the most sensitive against nickel ion (see Fig. 11). Nickel-titanium (NiTi) alloys were made by an arc melting method. Those alloys varied greatly in composition (Table 3) and phase (Fig. 12). First these alloys were sterilized by soaking them in 70 % ethanol solution. Then they were immersed in phosphate buffer and incubated at 36.5 °C for 25 h. We collected the extract and added it to a MOLT-3 culture plate (the ratio of extract to supernatant was 1:1). After a 4-day culture, we measured the viable cell number of each culture condition. The order of the viable cell numbers was coupon 5> coupon 2> coupon 1, coupon 3> coupon 4 (Fig. 13). We also measured the nickel concentration of the extract by atomic absorption spectrophotometry. The order of the nickel concentration was coupon 4> coupon 1, coupon 3> coupon 2> coupon 5 (Table 4), which was

Table 3 The element composition of NiTi alloys

Coupon	Element		
	Ni (wt%)	Ti (wt%)	Others
1.	98.7	0	Y, Al, Si
2.	94.4	4.8	Al, Fe
3.	85.2	14.5	Br
4.	74.5	25.1	Br, Si
5.	49.2	50.4	Si, Br, Y

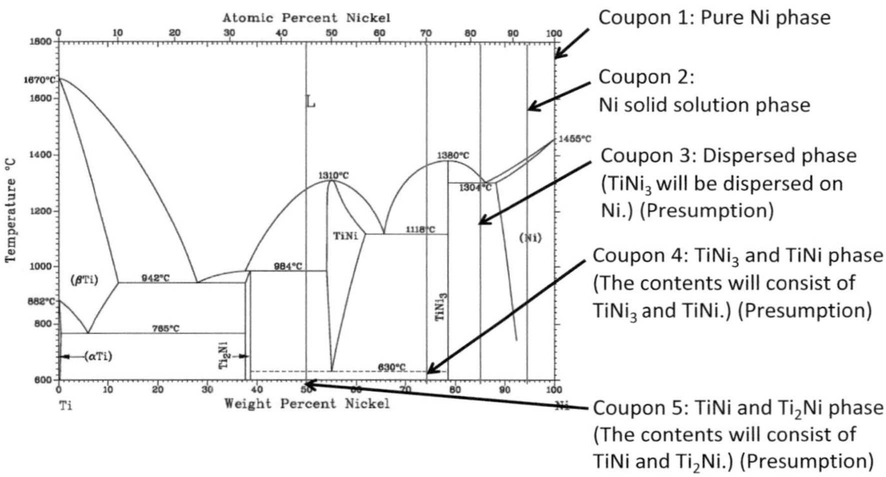

Fig. 12 Nickel Titanium constitution diagram and presumed phase of nickel-titanium alloys

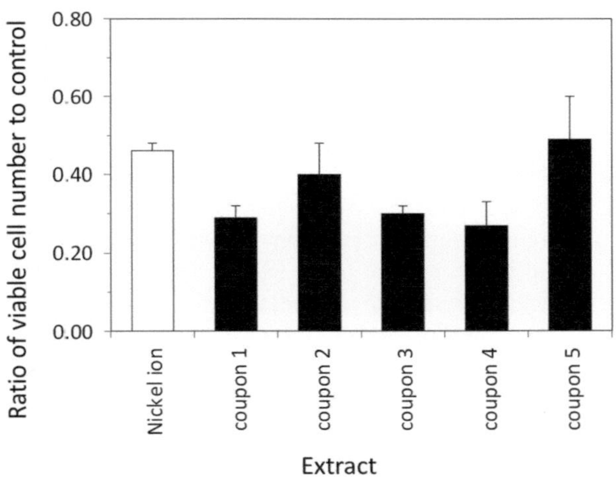

Fig. 13 Relative viable cell number of each culture condition added to an extract of nickel-titanium alloy. No addition of extract was used for the control culture condition (n = 3). Nickel ion indicates 84 μM of nickel chloride solution was added to the supernatant of MOLT-3

Table 4 The nickel concentration of extract of nickel-titanium alloys

Coupon	Ni (ppm)
1.	1.23
2.	0.53
3.	1.17
4.	1.60
5.	0.30

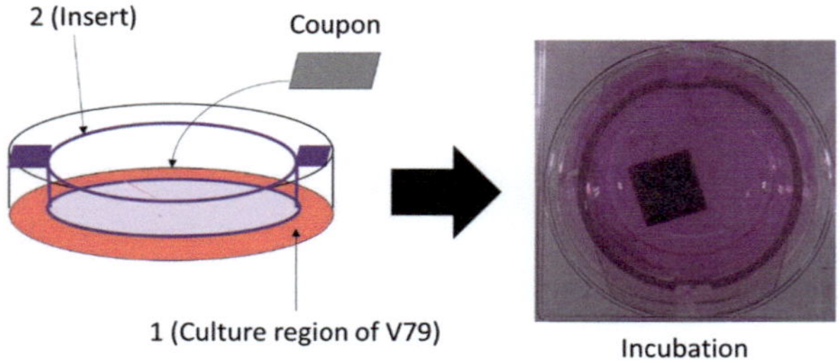

Fig. 14 Outline of insert culture. V79 was seeded on the surface of a culture plate (*1*). Insert was put on it (*2*) and each coupon was plated on the surface of the insert. They were cultured in a CO_2 incubator for 3 days. These steels were sterilized by soaking in 70 % ethanol solution. Then they were washed by sterile ultra-pure water and rinsed with culture medium

completely correlated with that of viable cell number. These results show that nickel ion will be the key factor that influences proliferation of MOLT-3 and that nickel and titanium constitution will affect the tendency of nickel leakage from a nickel-titanium alloy.

We also applied this assay to the assessment of steels: SUS304, SK4 and SPCC. We selected V79 and cultured it with a steel coupon using the insert (Fig. 14). After a 3-day culture, we counted the viable cell number of each culture condition and measured the metal concentration of the culture supernatant by ICP-AES. The order of viable cell numbers was SUS304 > SPCC > SK4 (Fig. 15). As for the metal concentration of the culture supernatant, SK4 was the most abundant manganese. Meanwhile SUS304 was almost the same as the control. The order was SK4 > SPCC > SUS304 (Table 5) and it was negatively correlated to the viable cell number. We compared the surface conditions of these steels in regards to element content before and after the culture by an X-ray fluorescence spectrometer. The three steels had very little change in regards to elements, by less than 1 % (wt) (Table 6), which indicates that the surface component of any steel did not change. We also observed the surface conditions of them before culture and after culture by using a laser microscope. SK4 and SPCC changed the color of their surfaces and were covered with rust but SUS304 did not change (Fig. 16). From these results, it is assumed that measuring the viable cell number gives us some useful information about features of steels related to corrosion that are detected by instrumental, analytical techniques.

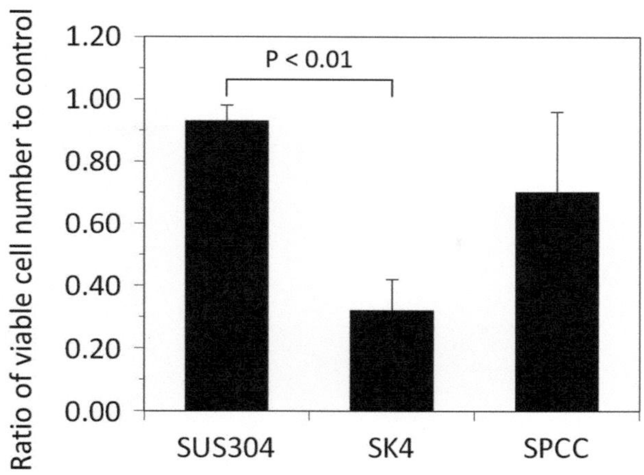

Fig. 15 Comparison of viable cell number among three culture conditions. No addition of steel was used for the control culture condition (n = 3)

Table 5 Metal concentration of culture supernatant-: not detected. Control means the supernatant of V79 cultured without coupon

Coupon	Fe (ppm)	Mn (ppm)	Ni (ppm)	V (ppm)	Cr (ppm)	Al (ppm)
Control	0.10	0.004	0.007	–	–	0.010
SUS304	0.07	0.004	0.011	–	–	0.012
SK4	168	0.312	0.011	–	0.009	0.010
SPCC	164	0.258	0.013	–	–	0.047

Table 6 Metal components of the surface of steels

	SPCC		SK4		SUS304	
	Before (wt%)	After (wt%)	Before (wt%)	After (wt%)	Before (wt%)	After (wt%)
Fe	99.85	99.79	99.39	99.33	73.04	73.28
Ni	–	–	–	–	8.06	7.99
Cr	–	–	0.13	0.16	18.30	18.37
Mn	0.15	0.11	0.37	0.38	–	–
Si	–	–	–	–	0.35	0.31
Al	–	–	–	–	0.17	–
V	–	–	–	–	0.08	0.04
Rb	–	0.10	0.11	0.14	–	–

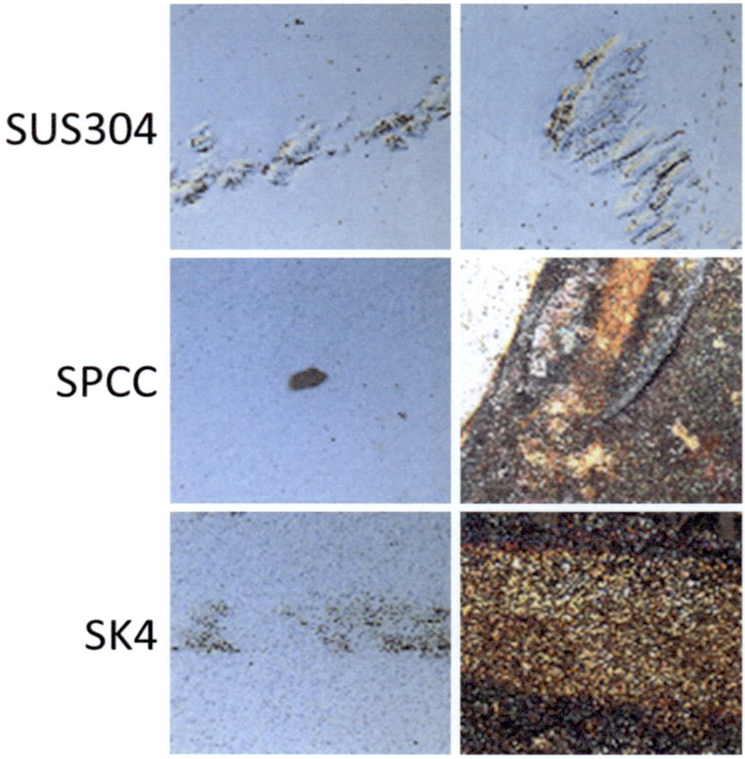

Fig. 16 The surface condition of thee steels before the culture (*left panels*) and after the culture (*right panels*)

2.3 Application of MTT Assay

Colony formation assay and measuring the viable cell number are very useful assays for metallic products. However, they have a limitation on the number of samples, which depends on the person examining them. In order to examine many samples at once, we tried to apply MTT assay for measuring metals and metal ions such as zinc ion and hexavalent chromium. For the zinc ion, absorbance was decreased by 10 % of the control (no addition of zinc ion) at 80 μM and the zinc ion concentration negatively correlated with absorbance from 80 to 150 μM ($R^2 = 0.98$) (Fig. 17). For hexavalent chromium, absorbance was decreased by 10 % of the control at 2.5 μM and hexavalent chromium negatively correlated with absorbance from 2 to 6 μM ($R^2 = 0.98$). Comparing the range of detecting metal ions of MTT assay with that of the viable cell number assay, they were almost the same. As a result, we were able to reduce the time needed for measuring metal concentration without losing the sensitivity of detection. Additionally, we could scale-down and test more samples (all together) than colony formation assay and viable cell number assay.

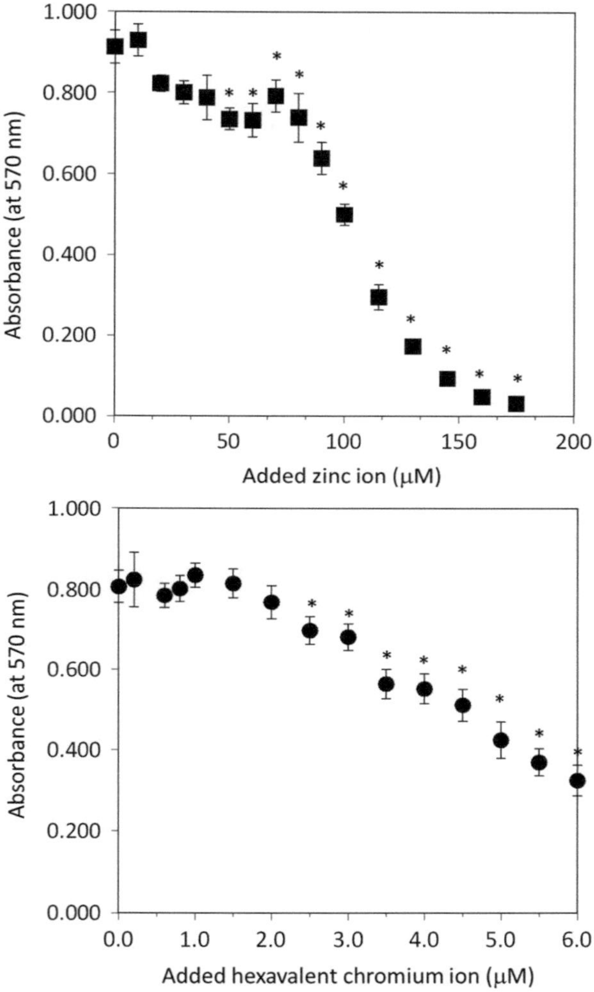

Fig. 17 MTT assay of V79 cultured with zinc ion (*upper panel*) and hexavalent chromium (*lower panel*). V79 was seeded on a 96-well culture plate and incubated for 24 h in a CO_2 incubator. Next, zinc sulfate solution, diluted with a culture medium or potassium dichromate solution diluted with culture medium, was added to each well and cultured 24 additional hours. On day 2, MTT assay was performed according to the recommended protocol (nacalai tesque). *: $p < 0.01$ v.s. control (no addition of metal ion)

3 Conclusion

In this chapter, we proposed several *in vitro* assays using mammalian cell cultures to measure some features of metallic products. Both the colony formation assay and the viable cell number assay can detect changes in the surface conditions of metallic materials, which are difficult to detect using traditional material analyzing techniques such as microscope analysis and X-ray fluorescence spectrometric analysis. The MTT assay is more effective method to quickly analyze multiple samples at one time. These *in vitro* assays are applied in assessment of metallic products and understanding the biological functions of metal(s). This information is very important in order to generate environmental friendly metallic products and biocompatible ones.

Other cellular change of features, such as morphological changes and differentiation stage, will be suitable to assess metallic products. However, we must overcome some problems about data processing techniques.

Acknowledgement Some parts of these works were supported by KAKENHI (21760600: Grant-in-Aid for Young Scientists (B) of 2009–2010; 24710035: Grant-in-Aid for Young Scientists (B) of 2012–2013; 26550084: Grant-in-Aid for Exploratory Research of 2015–2016).

References

1. da Frausto Silva JJR, Williams RJP (2001) The biological chemistry of the elements – the inorganic chemistry of life, 2nd edn. Oxford University Press, Oxford
2. Sakurai H, Kinzoku ha naze jintai ni hitsuyouka (Japanese). Kodansha, 1996
3. Nema R, Khare S (2012) An animal cell culture: advance technology for modern research. Adv Biosci Biotechnol 3:219–226
4. Landecker H (2007) Culturing life: how cells became technologies. Harvard University Press, Cambridge, MA
5. Fundamental techniques in cell culture. Laboratory handbook, 2nd edn
6. Haase H, Maret W (2010) Chapter 8: The regulatory and signaling functions of zinc ions in human cellular physiology. In: Zalups RK, Koropatnick J (eds) Cellular and molecular biology of metals. CRC Press, Boca Raton
7. Kanbe T (2011) An overview of a wide range of functions of ZnT and Zip zinc transporters in the secretory pathway. Biosci Biotechnol Biochem 75:1036–1043
8. Garrick MD, Dolan KG, Horbinski G, Ghio AJ, Higgins D, Porubcin M, Moore EG, Hainsworth LN, Umbreit JN, Conrad M e, Feng L, Lis A, Roth FA, Singleton S, Garrick LM (2003) DMTI: a mammalian transporter for multiple metals. Bio Met 16:41–54
9. http://sickle.bwh.harvard.edu/iron_transport.html
10. Voltrova S, Hidasova D, Genzer J, Srogl J (2011) Metallothionein-inspired prototype of molecular pincer. Chem Cmommun 47:8067–8069
11. Vasak M, Mcloni G (2011) Chemistry and biology of mammalian metallothioneins. J Biol Inorg Chem 16:1067–1078
12. Cousins RJ (1994) Metal elements and gene expression. Ann Rev Nutr 14:449–469
13. Coyle P, Philcox JC, Carey LC, Rofe AM (2002) Metallothionein: the multipurpose protein, CMLS. Cell Mol Life Sci 59:627–647
14. Ogawa A, Okuda N, Kanematsu H (2009) Comparison of biological toxicity of several plating products by mammalian cells, 2009 Sur/Fin Technical Conference Proceeding, Louisville, Kentucky, USA, June 15–17, National Association for Surface Finishing (NASF), Surface Finishing Research Session, CD-ROM
15. Kanematsu H, Murata K, Ogawa A, Miyano Y (2008) Sn-Zn alloy films produced by HSSL and their environmental characteristics, 2008 Sur/Fin Technical Conference Proceeding, NASF, Indianapolis, Indiana, USA, June 16–18, Process and Environmental Session, CD-ROM
16. Kanematsu H, Kobayashi T, Oki T, Method for producing a tin-zinc alloy film. Japan patent JP, 3438–000030, J
17. Ogawa A, Okuda N, Hio K, Kanematsu H, Tamauchi H (2012) The monitoring possibility of some mammalian cells for zinc concentrations on metallic materials. Cytotechnology 64:341–347
18. Ogawa A, Akatsuka R, Tamauchi H, Hio K, Kanematsu H (2011) Influence of nickel-titanium alloy components on biological functions. BMC Proc 5:79

Metal Ion Sensor of Pore-Forming Toxin for Environmental Evaluation

Hajime Ikegai

Abstract Today, one of the materials attracting attention in the nanotechnology field includes proteinaceous toxins (pore-forming toxins, PFTs) that form nanosize pores in cell membranes. They originally attracted attention as bacterial virulence factors, with which research has been conducted on their biological effects on human beings and animals. In the 1980s, it was found that PFTs, when coming into contact with membranes, were transformed from water-soluble proteins into membrane proteins, causing pores (ion channels) to form in the target membrane. They have been studied by protein chemists and X-ray crystallographers. In the twenty-first century, PFTs that have pores about the size of 2 nm could be made available in a large quantity through genetic recombination at any time. Interesting research on nanotechnology started after such availability was noted by nanotechnology researchers. This chapter is a review mainly based on the author's research findings that the PFT becomes a new probe that works as a metal ion sensor.

1 Bacterial Toxins Transformed into Useful Materials

It has been known since long ago which virulent bacteria bring about infectious diseases, that are caused by the production of special substances (mainly proteins) of bacteria or toxins. A variety of studies have been conducted on their structures and effects to prevent or treat the infectious diseases [1]. Bacterial toxins include the types that affect nerves, destroy cells, and inhibit protein synthesis. These effects often lead to the appearance of peculiar clinical symptoms at local sites or in the whole body [1]. These terrifying toxins have the potential to become useful in the sector of metal materials which is different from biology.

H. Ikegai (✉)
National Institute of Technology, Suzuka College, Suzuka, Japan
e-mail: ikigai@chem.suzuka-ct.ac.jp

H. Kanematsu, D.M. Barry (eds.), *Corrosion Control and Surface Finishing*,
DOI 10.1007/978-4-431-55957-3_18

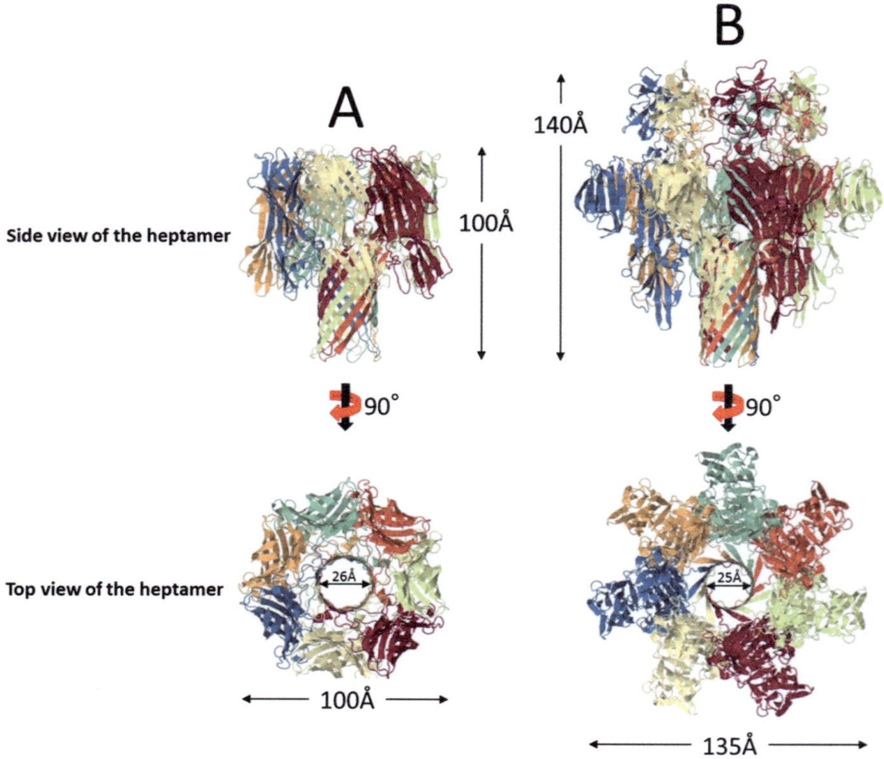

Fig. 1 Structural comparison of aHL and VCH heptamers. Analysis was done using JSmol of 3D viewer. (**a**, **b**) are aHL (PDB No.: 7AHL) and VCH (PDB No.: 3O44), respectively. Display style of these PFTs was selected as a cartoon style of display modes

1.1 Pore-Forming Toxin (PFT)

There is a group referred to as a pore-forming toxin (PFT) among bacterial toxins, which forms a cylindrical structure containing a pore with a diameter of 1–50 nm in its center in the target membrane. Figure 1a shows *Staphylococcus aureus* alpha-hemolysin (aHL, Ref. [2, 3]) produced by S. *aureus* that is known most commonly among PFTs and the tertiary structure of *Vibrio cholerae* hemolysin (VCH) * (Fig. 1b, Ref. [4, 5]), which is discussed in this study. Interestingly, tertiary structures of these two toxins have the relationship of almost the same structural homology [4]. However, aHL has almost no specificity with respect to the permeability of pores to ions, whereas VCH possesses a Cl^- channel property (Fig. 2, Ref. [6]).

A characteristic of PFT lies in the fact that it aggregates instantly and makes holes in the target cell membranes immediately after the binding of water-soluble

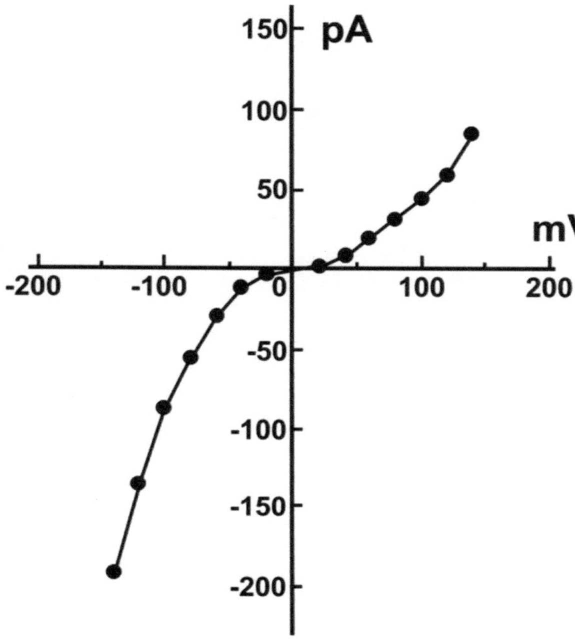

Fig. 2 Relationships between current and voltage in a VCH heptameric pore. VCH monomer (10 ng in 5 μl) was added to 2.5 mM N-(2-hydroxyethyl)-piperazine-N′-2-ethanes sulfonic acid (HEPES, pH 7.2) with 1 M NaCl in *cis*-compartment. When one VCH pore was incorporated into the planar lipid membrane, the membrane current was recorded automatically under fluctuating applied membrane potentials (Figure modified from Ref. [6])

toxin proteins, which are released from the bacterial cells, to the target cells. When PFT monomers that were dissolved in a buffer solution bind to the target cells, multiple PFT monomer molecules spontaneously aggregate on the membrane. Thereafter, they form cylindrical heptamers in the membrane of the target cell (Fig. 3, Ref. [3, 5]). Ions or substances smaller than the diameter of the pore enter the cytoplasm through the pore from outside of the cell. The aHL heptamers are formed in its membrane. Then the osmotic pressure of the cytoplasm is enhanced. To maintain a balance between osmotic pressures in the inner and outer aqueous solution of the cell, water influxes in its cytoplasm, causing swelling of cells. Eventually, it is no longer able to withstand the inner pressure of the cell, resulting in its rupture.

PFTs can be divided into two types by the size of pores. One of them is a PFT that forms large pores with a diameter of 20–50 nm. Streptolysin O produced by *Streptococcus pyogenes* [7] and pneumolysin produced by *Streptococcus pneumonia* [8] have often been studied. The second is the PFT that forms small pores with a diameter of approximately 1 nm as previously introduced. aHL and VCH are included in this category of PFT. Many PFTs in the former type bind to cholesterol (Chol) molecules on the target cell membranes, forming pores through the instant aggregation of dozens of monomer molecules on the target membrane [9]. Receptors of aHL [10] and VCH [4, 11] have been clarified in the latter type.

*VCH is a protein toxin that differs from a cholera toxin which causes cholera. It is also reported, however, that VCHs cause diarrhea in humans [12], and proper

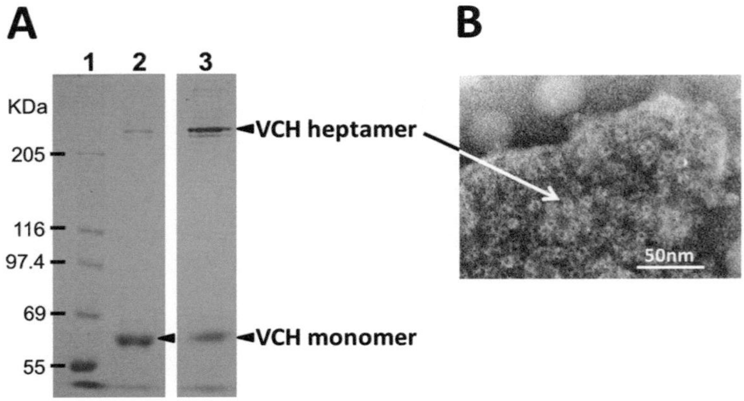

Fig. 3 Electrophoretic and ultrastructural analyses of VCH heptamers formed on target membranes. (**a**) SDS-PAGE profiles of VCH. *1* Markers, *2* VCH monomers, *3* VCH heptamers formed on PC-Chol liposome membranes. (**b**) Ultrastructural analysis of VCH heptamers formed on rabbit erythrocyte ghosts. Sample was stained with 2 % phosphotungstate, pH 7.4 and was observed under a JEOL JEM-1200 EX II transmission electron microscope

measures thus need to be taken to prevent them from harming the cells. Their virulence exists in monomers, and they have no cytotoxic activities unless the monomers bind to the target cell membranes. In contrast, heptamers do not affect the cells because they are not incorporated into biological membranes. It is possible, therefore, to incorporate them into lipid bilayers in advance through applying a special method. It is believed that the virulence of VCHs can be minimized.

1.2 The Ion Permeability of VCH

The material permeability of pores in PFT heptamers depends on the size of pores, amino acid residues located within the pores, and the shape of permeable molecules. Even 5 kDa of insulin can pass through pores of aHL [13], whereas it is difficult to allow materials larger than 1 kDa to permeate through pores of VCH [11].

Additionally, types of permeable ions vary with the properties of the pores, occasionally affecting the extent of cytotoxic activity. For example, VCH has the hemolytic activity about ten times as high as aHL as the property of the former's chloride ion channel is effectively affecting it, as discussed earlier [14]. aHL allows monovalent and divalent ions to permeate through it; however, it has a low ion selectivity [3]. VCH also allows monovalent and divalent ions to permeate through it as aHL, whereas it has a very high selectivity of chloride ions (Fig. 2, Ref. [6]). VCH has higher hemolytic activity and cytotoxic activity than aHL even if their 3D structures are quite similar, and this fact can be attributed to differences in the permeability of ions and types of permeating ions. As discussed earlier, it is believed that the existence of receptors on the target membrane in both of the toxins is associated with the strength of cytotoxic activity.

2 Application of PFT to Metal Ion Sensors

It was introduced in *Nature* in 2008 that Oxford Nanopore Technologies, Inc. [15] was trying to use aHL as a probe of the DNA sequencer in order to read DNA sequences of genes [16]. Many researchers came to know that nanodevices could be created from PFTs. Bayley who developed nanopores had originally been interested in the engineering application of PFTs, developing a Drug Delivery System that was capable of arbitrarily discharging drugs encapsulated in liposome through controlling the opening and closing of pores and a metal ion detection technology with the use of ion channel properties in the early 1990s [17–19]. However, aHL pores have a low selectivity of various types of ion permeability and no research papers have been presented on the development of metal ion sensors since then. Bayley et al. published an article on Ag^{+} and Cd^{2+} ion sensors using aHL pores again in 2013 [20].

3 Metal Ion Sensors Using VCH

There would be a possibility that VCH could be used as a probe for the DNA sequencer because it has a tertiary structure similar to that of aHL. However, another way should be pursued for VCH because aHL already established a dominant position as a probe for the fourth-generation of the DNA sequencer. Thus, we would like to consider the application of PFTs to metallic sensors utilizing the ion permeability known as the properties of many PFT pores. Ions passing through channels are detected by recording the single-channel current of VCH pores using the planar lipid bilayer method. The conductance is defined with a reciprocal of resistance. It is used as an index for ion permeation through pores.

3.1 Preparation of Planar Lipid Bilayers

A planar lipid bilayer was prepared in a Teflon hole by partitioning a Teflon cell with a Teflon film (25 μm in thickness) where a single 0.2–0.3 mm hole was made, filling each compartment with a salt solution and applying phosphatidylcholine (PC) – Chol (molar ratio 1:1) that was made soluble with *n*-decane onto it using a wire brush (Fig. 4).

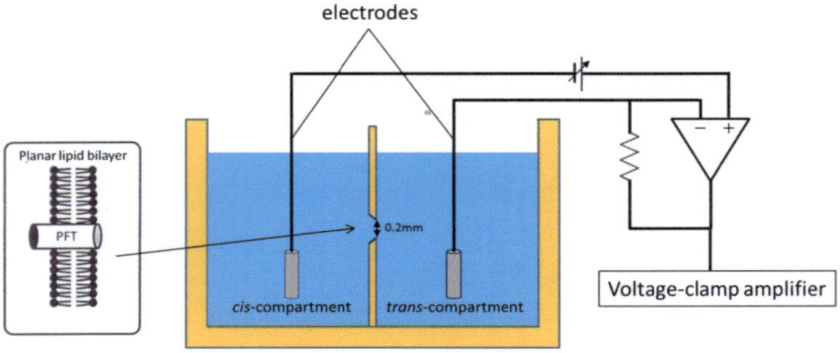

Fig. 4 Equipment for observation of single-channel recordings of PFT currents. *Cis*-compartment is hyperosmotic side

3.2 Measurement of Single-Channel Conductance

VCH proteins were added to a salt solution on either the *trans*-side or the *cis*-side, after making sure that a stable planar lipid bilayer had been formed on the hole in the Teflon film using an oscilloscope (Fig. 4). The single-channel current should fluctuate, if PFT pores are formed on the planar lipid bilayer preventing the movement of the salt solution put in the *cis* and *trans* positions. This measurement was performed using silver-silver chloride electrodes. Since the signal is weak, it may be possible to guess conditions inside pores of ion channels from its wave form if amplified with a patch-clamp amplifier.

3.3 The Single-Channel Conductance of the VCH Channel Varies with Types of Ions

VCH was added to the *cis*-side solution in the Teflon cell where the PC-Chol planar lipid bilayer was formed. At the same time pore-shaped VCH heptamers were inserted into a planar membrane, and continuous stepwise signals, of the single-channel current which showed the ion permeation, were recorded into a recorder (Fig. 5, see Ref. [21]). Moreover, we could clarify that the channel conductance varied largely with types of ions, when examining various types of ion permeabilities via the pores of VCH heptamers (Table 1, Fig. 6). Additionally, while almost no filtering (phenomena of channel opening/closing) occurred in the presence of monovalent cations, it was found that it occurred frequently in the presence of divalent cations (Fig. 6, Ref. [6]). The divalent cations are considered to perform the opening and closing of channels in a disorderly manner, seeing that this filtering occurs frequently as a result of adding divalent cations (Ca^{2+}, Mn^{2+}, Mg^{2+}, Co^{2+}, Ni^{2+}) in either a *cis*- or *trans*-side solution of the Teflon cell where PC-Chol planar lipid bilayers are formed. Filtering occurs due to the frequent opening and closing of

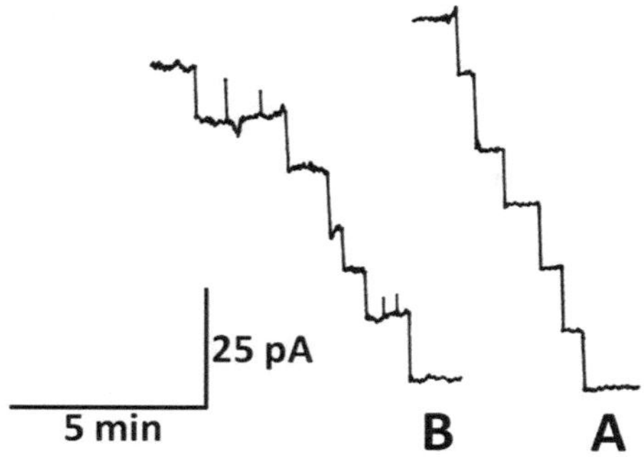

Fig. 5 Single-channel recordings of VCH heptameric pores in planar lipid bilayer membranes. Single-channel currents of the pores were recorded with PC-Chol planar bilayer membranes bathed in 2.5 mM HEPES (pH 7.2) with 1 M salt. (**a**) 65-kDa VCH monomers (300 ng in 3.6 μl of Tris-HCl (pH 8.0) with 1 mM EDTA and 3 mM NaN_3 (TEN). (**b**) 65-kDa VCH monomers (300 ng in 3.6 μl of TEN) and PC-Chol planar lipid bilayer membranes bathed in 2.5 mM HEPES (pH 7.2) with 1 M KCl. The applied membrane potential was +50 mV and then VCH was added to the *trans*-compartment of a Teflon cell

Table 1 Single-channel conductances of VCH heptameric pores in various salts

Salt	Single-channel conductance of VCH heptamers (pS)[a]	Conductance divided by number of ions per molecule (pS)[b]
CH_3COONa (pH7.0)	49	25
KCl (pH7.0)	237	119
LiCl (pH7.0)	213	107
NaBr (pH7.2)	179	90
NaCl (pH7.2)	246	123
NH_4Cl (pH7.2)	317	159
$NiSO_4$ (pH6.9)	113	57
$CaCl_2$ (pH7.2)	715	238
$MgCl_2$ (pH7.2)	413	138
$MnCl_2$ (pH6.9)	607	202

[a]The average of single-channel conductances was calculated from at least 150 channel events, which were measured under the voltage-clamp mode at the membrane potential of +50 mV in 1 M salt solution containing 2.5 mM HEPES (pH7.2)

[b]The conductance per each ion was calculated by the formula which is (measured VCH channel/the number of ions per one molecule) (Table modified from Ref. [6])

pores in PFTs. It suggests, therefore, that the identification of types of metallic ions passing through the pores in the VCH and the calculation of what content of metallic ions are contained in the solution are possible from the value of channel current and the presence or absence of filtering with its frequency (Fig. 7, see Ref. [6]). All of the data shown here represents the results of detection up to the order of mM. It would be possible to measure the channel conductance of metallic ions contained in a solution even if the concentration of the sample solution to be

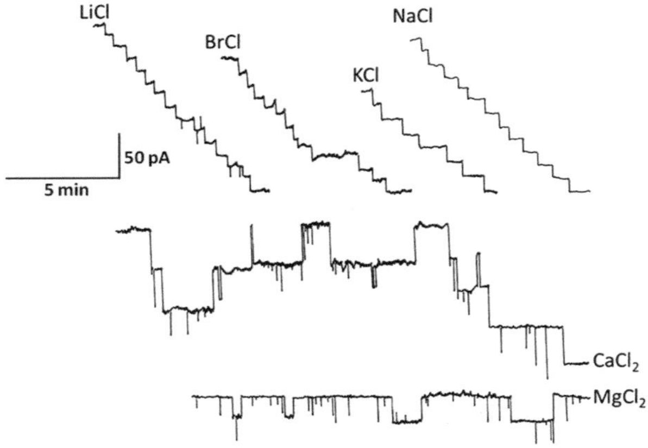

Fig. 6 Single-channel recordings of VCH heptameric pores in various salts. Single-channel currents of the pores were recorded with PC-Chol planar bilayer membranes bathed in 2.5 mM HEPES (pH 7.2) with 1 M salt. The applied membrane potential was +50 mV

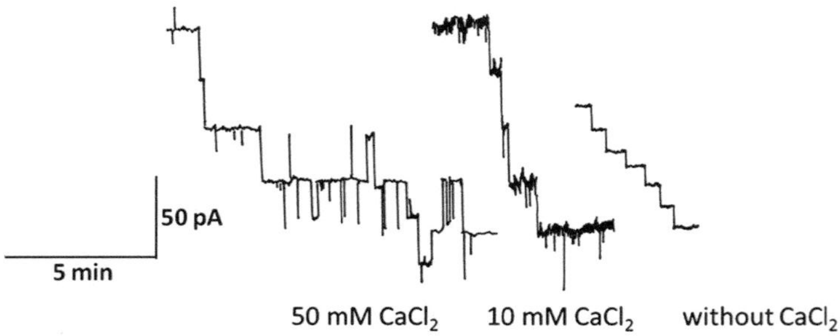

Fig. 7 Effect of Ca^{2+} on fluctuation of single-channel currents of VCH heptameric pores in PC-Chol planar lipid bilayer membranes. After forming VCH heptameric pores in a PC-Chol planar lipid bilayer membrane in 2.5 mM HEPES (pH 7.2) with 1 M NaCl, the indicated concentration of $CaCl_2$ was added to the bath solution. Immediately, 30 ng of VCH in 5 μl of TEN was added to the solution. The applied membrane potential was +50 mV

examined is reduced. However, we think that it would not be difficult to increase the concentration of metallic ions contained in the solution per unit volume until it becomes possible to measure the channel conductance, if we carefully evaporate the water content after removing impurities by filtering the sample solution, since it is to detect metallic ions.

4 Conclusions

It has become clear that it should be possible to effectively utilize PFTs as innovative sensor probes for use in environmental evaluations involving nanotechnology. Currently, aHL that is an exotoxin of *S. aureus* is a pioneer of biosensor probes which are capable of measuring various substances. There are not too many difficult issues that would obstruct the practical application of metallic ion detection using the VCH. Additionally, it would be the results of having conducted interdisciplinary research that links the fields of nanochemistry, X-ray crystallography, biological informatics, electrical engineering, etc. besides the up-to-date biology, that PFTs could be manipulated freely at the researchers' will. We expect that further practical PFT research will be conducted through interactions with researchers in different fields in the future.

References

1. Henkel JS, Baldwin MR, Barbieri JT (2010) Toxins from bacteria. NIH Publ Access Author Manuscr 100:1–29
2. Bhakdi S, Waley I, Husmann M, Valeva A (2005) Staphylococcal alpha-toxin, topics in current genetics. In: Schmitt MJ, Schaffrath R (eds) Microbial protein toxins, vol 11. Springer, Berlin, pp 91–110
3. Song L, Hobaugh MR, Shustak C, Cheley S, Bayley H, Gouaux JE (1996) Structure of staphylococcal α-hemolysin, a heptameric transmembrane pore. Science 274:1859–1866
4. Olson R, Gouaux JE (2005) Crystal structure of the *Vibrio cholerae* cytolysin (VCC) pro-toxin and its assembly into heptameric transmembrane pore. J Mol Biol 350:997–1016
5. De S, Olson R (2011) Crystal structure of the *Vibrio cholerae* cytolysin heptamer reveals common features among disparate pore-forming toxins. Proc Natl Acad Sci U S A 108:7385–7390
6. Ikigai H, Ono T, Iwata M, Nakae T, Shimamura T (1997) El Tor hemolysin of *Vibrio cholerae* O1 forms channels in planar lipid bilayer membranes. FEMS Microbiol Lett 150:249–254
7. Sierig G, Cywes C, Wessels MR, Ashbaugh CD (2003) Cytotoxic effects of streptolysin O and streptolysin S enhance the virulence of poorly encapsulated group A streptococci. Infect Immun 71:446–455
8. Tilley SJ, Orlova EV, Gilbert RJC, Andrew PW, Saibil HR (2005) The prepore structure of pneumolysin, obtained by fitting the alpha carbon trace of perfringolysin O into a cryo-EM map. (2005) Structural basis of pore formation. Cell 121:247–256
9. Hotze EM, Tweten RK (2012) Membrane assembly of the cholesterol-dependent cytolysin pore complex. Biochim Biophys Acta 1818:1028–1038
10. Wilke GA, Wardenburg JB (2010) Role of a disintegrin and metalloprotease 10 in *Staphylococcus aureus* alpha-hemolysin-mediated cellular injury. Proc Natl Acad Sci U S A 107:13473–13478
11. Ikigai H, Akatsuka A, Tsujiyama H, Nakae T, Shimamura T (1996) Mechanism of membrane damage by El Tor hemolysin of *Vibrio cholerae* O1. Infect Immun 64:2968–2973
12. Anderson Y, Ekdahl K (2006) Wound infections due to *Vibrio cholerae* in Sweden after swimming in the Baltic Sea, summer 2006. Euro Surveill 11:193
13. Füssle R, Bhakdi D, Sziegoleit A, Tranum-Jensen J, Kranz T, Wellesiek H-J (1980) On the mechanism of membrane damage by *Staphylococcus aureus* α-toxin. J Cell Biol 91:83–94

14. Unpublished data
15. http://www.nanoporetech.com/
16. Sanderson K (2008) Personal genomes: standard and pores. Nature 456:23–25
17. Walker B, Bayley H (1994) A pore-forming protein with a protease-activated trigger. Protein Eng 7:91–97
18. Walker B, Kasianowicz J, Krishnasastry M, Bayley H (1994) A pore-forming protein with a metal-actuated switch. Protein Eng 7:655–662
19. Bayley H (1997) Building doors into cells. Sci Am 277(3):62–67
20. Choi LS, Mach T, Bayley H (2013) Rates and stoichiometries of metal ion probes of cysteine residues within ion channels. Biophys J 105:356–364
21. Ikigai H, Ono T, Nakae T, Otsuru H, Shimamura T (1999) Two forms of *Vibrio cholerae* O1 El Tor hemolysin derived from identical precursor protein. Biochim Biophys Acta 1415:297–305

Application of Phytoplankton

Toshiyuki Takahashi

Abstract This chapter describes the phytoplankton *Chlorella* species as model organisms and their application to metal toxicity evaluation. Metallic solutions used to assess metal toxicity were eluted from steel slag with leaching conditions based on JIS K 0058-1. This study also examined the development of methods to easily estimate not only the number of algae but also their algal activities. The concentrations of metal effluents from tested slags were almost all lower than environmental quality standards for effluent and drinking water. After incubation of *Chlorella* with culture media including eluates from the respective slags tested in this study, the effects of each eluate were investigated using microscopy and flow cytometry. Analyses of algal cells treated with each eluate revealed that eluate induced neither lethality nor growth inhibition. In addition to microscopic cell counting, bioassay using flow cytometry was used to estimate vigorous and aberrant algal growth simultaneously. Instead of exhibiting cytotoxicity, the addition of each eluate from steel slag enhanced algal growth.

1 Introduction

Anthropogenic sewage, agricultural and farming drainage, and industrial drainage from mines and industrial plants can cause water pollution. Water discharge effluents from chemical industries, iron and steel industries, metallic processing and plating plants, and electric and electronics industries particularly cause metallic water pollution. To prevent several human factors that cause water pollution, laws such as the environmental quality standards (EQS) for soil pollution, Japanese Law Relating to the Prevention of Marine Pollution and Maritime Disaster and EQS for water pollutants have been instituted.

Various types of iron and steel slags have been produced as steel industrial byproducts. Some can be recycled as soil aggregate [1, 2], although other slag ultimately ends up in landfill sites [3]. New applications of slag as depuratives and sand capping materials in aquatic environments [4] have been regarded recently to

T. Takahashi (✉)
National Institute of Technology, Miyakonojo College, Miyakonojo, Japan
e-mail: mttaka@cc.miyakonojo-nct.ac.jp

H. Kanematsu, D.M. Barry (eds.), *Corrosion Control and Surface Finishing*,
DOI 10.1007/978-4-431-55957-3_19

decrease amounts of discarded slag and to reduce high costs of discarding slag. When these slags are used in aquatic environments, it is obligatory to elucidate the toxicity and impacts of the metallic eluate from slag on aquatic organisms. Methods for evaluating eluate from slags using a leaching test of JIS K 0058-1 have been based on comparisons of chemical concentrations in eluate with EQS for soil and water pollution. All values for each element of EQS have been established from the sensitivity of model organisms to each element tested. Exposure of organisms to multiple elements such as eluate from slag has not always been considered using EQS. If the concentration of each element in any eluate from slag is below some EQS, then simultaneous exposures of multiple elements to organisms might cause unexpectedly strong effects on aquatic ecosystems. In addition to comparison of chemical concentrations in eluate from slag with EQS, bioassays using aquatic organisms are important to evaluate the impacts of eluate on aquatic ecosystems. To keep slag field trials in aquatic environments from directly polluting any river, pond, or sea, *in vitro* assessment methods are of enormous importance to evaluate eluate effects on aquatic organisms.

For this study, the phytoplankton species *Chlorella* is used as a model of aquatic organisms. Its application to the evaluation for metallic toxicity of eluate from slag was investigated. This study also particularly examined the development of a method to easily estimate not only the number of algae but also their algal activities such as viability. After the incubation of algae with culture media including eluates from slags, the effects of each eluate on algae were evaluated using microscopy and flow cytometry.

2 Steel Slags and the Eluate from Each Slag Used for This Study [5, 6]

Iron and steel slags from blast furnace slag and steel making slag including converter slag and electric arc furnace (EAF) slag have been produced as steel industrial byproducts. All blast furnace slag (24,000 ton/year) can be recycled completely for use of steel making slag base and cement [1, 2] as soil aggregate, although 331 ton/year of steel making slag (11,000 tons of converter slag and 2800 tons of EAF slag) ultimately end up in landfill sites [3]. New applications such as sand capping materials of slag in coasts and lakes [4] have been recently anticipated, as described in the *Introduction* section.

Slag utilization in aquatic environments has been restricted by several laws for environmental pollution because slag contains hazardous substances. However, slags also include several essential elements for organisms such as Ca, Mg, and trace elements such as Zn and Mn [7]. For example, Ca and Zn are respectively related with signal transduction in cells and with gene expression as DNA binding proteins [8]. The toxicities of the metallic eluate from EAF slag on aquatic organisms remain poorly understood [5, 6]. The physiochemical properties and effects of converter slag on organisms have been documented often [9–14].

Table 1 Chemical compositions of EAF steel slags used for this study [mass%] [6]

	FeO	SiO_2	CaO	Al_2O_3	MgO	MnO	Cr_2O_3	ZnO	NiO	CuO
Slag A	0.74	44.1	33	5.39	7.68	4.09	3.29	0.01	0.06	0.024
Slag B	35.1	19.2	20.8	15.2	4.1	5.1	0.43	0.071	0.028	0.025

This study specifically examined EAF slags, stainless steel (designated as slag A) and normal steel slag (slag B), as metallic materials. Table 1 presents compositions of EAF slags used for this study. Here, all Fe or Cr compounds are, respectively, described as FeO or Cr_2O_3 because it is difficult to distinguish FeO and Cr_2O_3 from elemental Fe and Cr in particles of stainless steel slag. Slag A includes greater amounts of SiO_2, CaO and Cr_2O_3 than slag B, but slag A contains less FeO than slag B does. To assess several properties of each EAF slag in aquatic environments, a leaching test based on JIS K 0058-1 (Test method for chemicals in slags Part 1: Leaching test method) was used to elute metal components of slag with HCl [5, 6, 15]. After elution, the solution was filtrated with 0.45 μm pore filter to eliminate slag particles. The filtrated eluate from the slag (designated as eluate A or eluate B) was used for the following experiments. Table 2 presents concentrations of elements of eluates from EAF slags. Consequently, the chemical concentrations of metal effluents from tested slags were almost all lower than the EQS for pollutions and the EQS for effluent and drinking water.

3 *Chlorella* as Model Phytoplankton and Its Application to Evaluate Metallic Toxicity

3.1 Features of Chlorella *sp. Used for This Study*

To evaluate metallic toxicity of eluates from slags on phytoplankton, *Chlorella kessleri* was used as a model organism. Several methods used to examine algal cellular behaviors including cell numbers, their cell cycle and metabolisms, have already been established because *Chlorella* sp. has been used as a model organism among phytoplankton. All phytoplankton including *Chlorella* sp. have chloroplasts for photosynthesis in their cytoplasm. When chlorophyll is exposed to appropriate excitation lights, it irradiates red fluorescence (Fig. 1). Chlorophyll is sensitive to physiological factors including heat and acid. These factors eventually cause inactive chlorophyll because of degradation. Chlorophyll degradation has also occurred by treatment with hazardously physiological factors or agents on algae themselves. After heating algae, red fluorescence for chlorophyll decreased considerably because of biodegradation. At the same time, yellow fluorescence was greater in dead algae (Fig. 2). Consequently, the yellow fluorescence might be regarded as an index of variant algae including dying algae rather than the number of vigorous algae correlated with red fluorescence intensities [16].

Table 2 Environmental quality standards related to pollutions and others for effluent and drinking water, and concentrations of elements of each eluate [mg/L] [6]

Origin of slag		Eluate of EAF stainless steel oxidation slag (Slag A)	Eluate of EAF normal steel oxidation slag (Slag B)	Environmental quality standards				
				Soil pollution	Marine pollution	Water pollutant	Effluent standard	Drinking-water standard
Regulated substances	Total As	ND[a] (RDL[b]: 0.001)	ND[c]	0.01	0.1	0.01	0.1	0.01
	Total B	0.16	0.28[c]	1		1[d]	10[m], 230[n]	1
	Total Be	ND (RDL: 0.0005)	ND[c]		2.5			
	Total Cd	ND (RDL: 0.0001)	ND[c]	0.01	0.1	0.01	0.1	0.003
	Cr (VI)	ND (RDL: 0.005)	ND[c]	0.05	0.5	0.05	0.5	0.05
	Total Cu	0.003	ND	0.001	3		3	1
	Total Pb	ND (RDL: 0.0005)	ND[c]	0.01	0.1	0.01	0.1	0.01
	Hg	ND (RDL: 0.0001)	ND[c]	0.0005	0.005	0.0005	0.005	0.0005
	Total Ni	0.001	ND	0.001	1.2			0.02
	Total Se	0.012	0.003[c]	0.01	0.1	0.01	0.1	0.01
	Total V	ND (RDL: 0.001)	0.01[c]		1.5			
	Total Zn	0.099	0.014[c]		2	0.03[e], 0.02[f] 0.01[g]	2	1
	F^-	ND (RDL: 0.1)	0.5[c]	0.8	15	0.8[h]	8[m], 15[n]	0.8
Substances out of regulation	Total Al	ND	1.8					
	Total Ca	9.3	10.1					300[o]
	Total Fe	ND	0.23				10	0.3
	Total Mg	0.9	1.1					300[o]
	Total Mn	0.028	ND				10	0.05
	Total Si	1.8	1.9					

	Total N	0.4	0.362[c]			0.1–1[i]	100	0.04[p], 10[q]
						0.2–1[j]		
	Total P	ND (RDL: 0.1)	ND[c]			0.005–1[k]	16	
						0.02–0.09[l]		

[a]Not detected
[b]Reportable detection limit
[c]These data from a previous study reported by Takahashi et al. [5]
[d, h, m]Standard value is not applied to coastal waters
[e]Habitable river or lake for an aquatic life
[f, j, l]Habitable coastal water for an aquatic life
[g]Habitable coastal water that requires conservation in particular for nidus and nursery ground
[i, k]Habitable lake for aquatic life
[n]Standard value is applied to coastal waters
[o]Total concentrations of both calcium and magnesium are limited for water hardness
[p]Total N contents derived from nitrite nitrogen
[q]Total N contents derived from both nitrite nitrogen and nitrate nitrogen

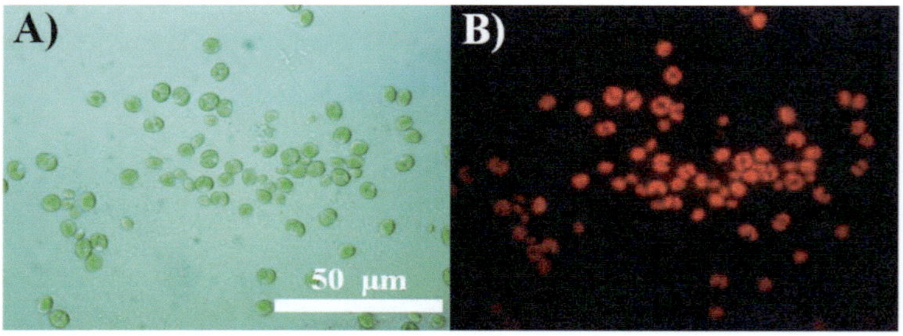

Fig. 1 Microphotographs of *Chlorella*-like algae isolated from protozoa *Paramecium bursaria*. Algae observed using a bright field (**A**) and a fluorescence microscope (**B**) are shown

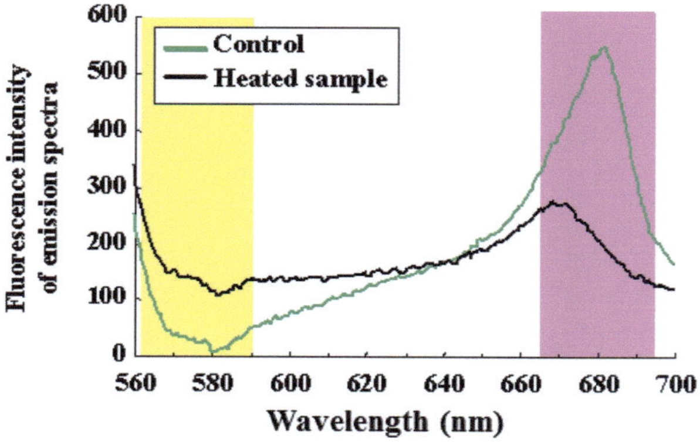

Fig. 2 Detection of living algae (*green line*, control), dying and dead algae (*black line*, heated sample) [16]. Emission spectra of *Chlorella*-like algae are shown with or without heat treatment. *Yellow* and *pink* areas represent each range of detection for the yellow or the red fluorescence channel used for the following flow cytometry

In this study, *C. kessleri* was cultured with CA medium (pH 7.2) [6] containing an eluate from each EAF slag under natural white fluorescent light at 23 ± 2 °C. After treatment of algae with each eluate for 1 week, the number and physiological features of algae were quantified respectively using hemocytometry and flow cytometry.

3.2 Evaluating Metallic Toxicity on Chlorella *sp. Using Microscopy [5, 6]*

After treatment of algae with each eluate for 1 week, the algae were quantified using a hemocytometer (Fig. 3). Consequently, algae increased dependent on concentrations, up to 30 vol.%, of each eluate. The quantities of algae were approximately

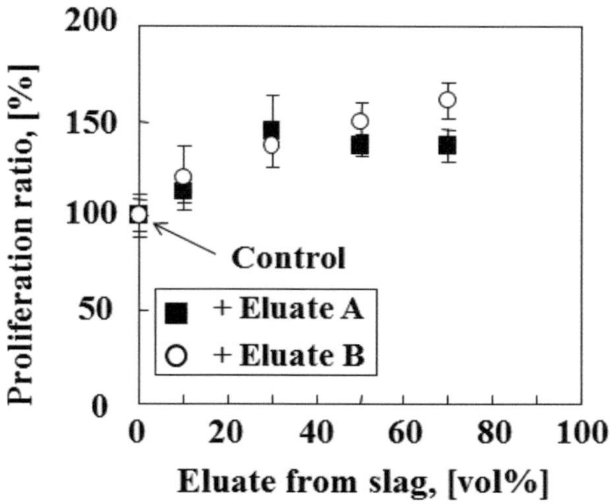

Fig. 3 Effects of respective eluates on algal growth [6]. *Dot* graphs show quantities of algae (± standard error) after treatment with each eluate. The graph shows a proliferation ratio (% of control without eluate) on the vertical and eluate contents (vol.%) on the horizontal axis. Each culture was started at an initial density of 1.0×10^4 algae/ml

constant in concentrations not less than 30 vol.% of each eluate. Addition to the quantification of algal numbers, the effects of the eluate on the morphology and population of algae were analyzed using microscopy (Fig. 4). Briefly, algae treated with or without the eluate were collected and observed using microscopy. From algal microphotographs, each algal size as a relative cell size was calculated from the pixel area of each alga using image analysis software [5]. *Chlorella* sp. can be classified morphologically into a unicellular cell to two to four autospores. The inset photograph (panel b) in Fig. 4 shows algae in each stage of the cell cycle under the control condition. Consequently, the morphological characteristics including algal size did not change between control and the treatment condition with the eluate. From the population analysis, the eluate from slag induced neither the morphological change nor population change of algae (panel a in Fig. 4). These data imply that eluates had no significant cytotoxicity.

3.3 Evaluating Metallic Toxicity on Chlorella *sp. Using Flow Cytometry [6]*

Flow cytometry (FCM), a powerful tool for assessment of cell biology, can provide single-cell optical information including cell size. Several fluorescence signals can be detected and evaluated from each cell suspension. To detect algae treated with each eluate from steel slag in this study, a flow cytometer equipped with a green

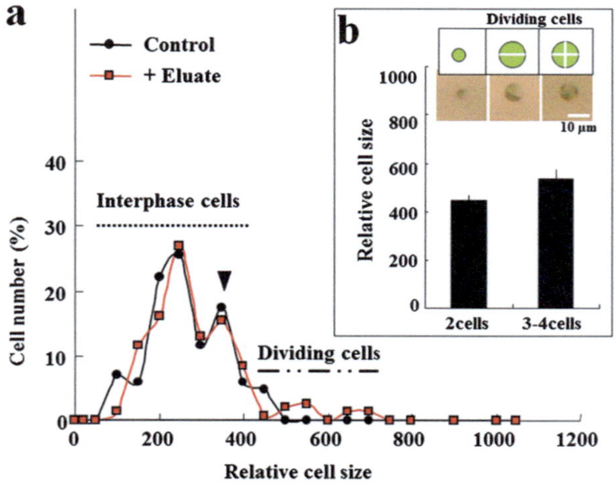

Fig. 4 Morphological observation of algae treated with eluate from slag B [5]. (**a**) Histograms show each algal size population after treatment with or without the eluate B of 70 vol.% for 1 week. A relative cell size was measured using Image J software. Cells above 400 relative cell size contained dividing cells in the panel (**a**). The *arrowhead* in panel (**a**) shows algae in the growth stage immediately before cell division. (**b**) Data show the sizes of dividing cells in panel (**a**). The *line graph* shows the average size of algae ± standard error. Inset photographs show an interphase cell and dividing cells

laser operating at 532 nm [17] was used for FCM (Fig. 5). Using this instrument, algal cells were analyzed through a rectangular capillary with a 100-μm round bore for several optical properties. Forward scatter (FSC) signals were collected to ascertain the cell size. The autofluorescence of chlorophyll was detected in the red fluorescence channel through a 680/30 nm band pass filter. A yellow fluorescence channel through a 576/28 nm band pass filter was used simultaneously (Fig. 2). To distinguish algal signals from signals of measurement noise, the following gating strategy was chosen. Signals above only the culture medium on FSC signals were removed as measurement noise. Algal signals gated by FSC were reanalyzed for chlorophyll fluorescence. Moreover, yellow fluorescence was regarded as dying or dead algae, as described previously (Fig. 2).

After treatment of algae with each eluate for 1 week, the optical features of algae including FSC and chlorophyll fluorescence were quantified using FCM. The red fluorescence intensity as chlorophyll fluorescence is shown with the yellow fluorescence intensity (Fig. 6). For comparison of vigorous algae with variant algae, algae treated with heat for 10 min at 100 °C were prepared as the index of dying or dead algae. In Fig. 6, each single dot shows the optical information of single alga. The dots of control algae correlated significantly with the red fluorescence intensity, but not with the yellow fluorescence intensity. They were found from 10^2–10^3 on the red to 10^1–10^2 on the yellow channel in the two-dimensional (2D) map of the red fluorescence intensity versus the yellow fluorescence intensity. The dots of

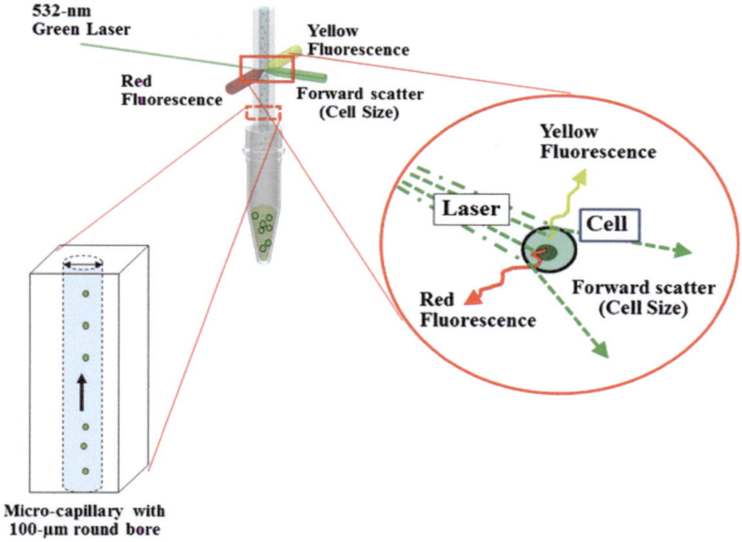

Fig. 5 Overview of FCM used for this study. Algal cells that had passed through a capillary with 100-µm round bore were analyzed. The FSC signals, the fluorescence of endogenous chlorophyll in the red fluorescence channel and that in the yellow fluorescence channel were collected simultaneously

heated algae correlated significantly with both the red and the yellow fluorescence intensity. They were found from 10^1–10^2 on the red to 10^1–10^3 on the yellow channel in the corresponding 2D map. The dot distribution of algae treated with eluate from slags was appropriately similar to that of control, although the dot distribution of algae treated with eluate from slags shifted slightly upward relative to that of control algae.

For descriptive purposes of algal distribution patterns, each graph in Fig. 6 was divided into four subareas (designated as regions I–IV). These segmentations based on algal viability were mainly set and performed. Region I, for instance, represents an area for vigorous algae. Region II shows an area for dead and variant algae like heated algae. The remaining subareas in Fig. 6 constitute region III for algae with low red fluorescence intensity. From region IV, algae were virtually absent. A few signals in control algae were distributed in region III. Therefore, these signals might be from measurement noise. To sum up, signals in region I and II shown respectively constitute an index of living algae and one of dead algae. Table 3 shows percentages in each region under each condition. From ratio analysis in the region I, the percentages of algae treated with eluate from slags were slightly greater than those of control algae. Consequently, the components of eluate from slags did not give rise to algae stresses directly. Components of eluate from slags include several metals such as copper, zinc and aluminum, which are not included in CA medium for algae. Although aluminum has been reported, particularly as a growth inhibition factor for plants [18, 19], these elements including aluminum and other metal

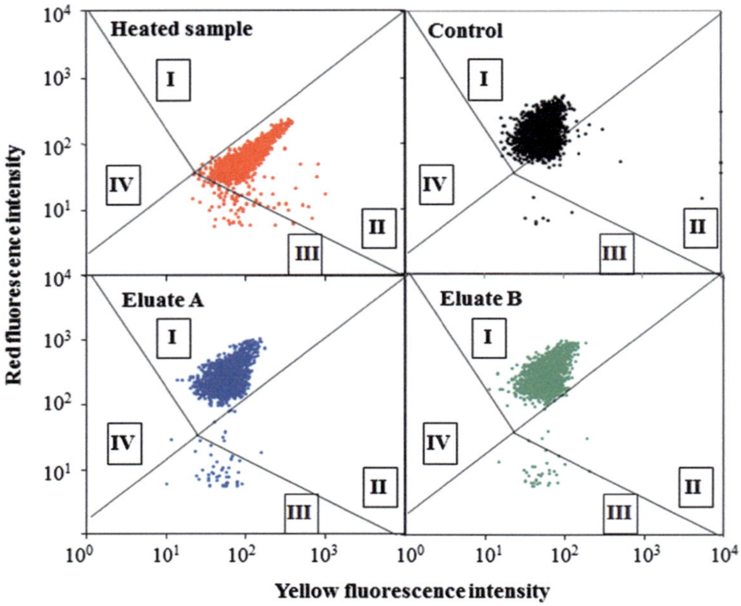

Fig. 6 Optical distribution of *Chlorella* using FCM [6]. Red fluorescence intensity of algae is shown versus the yellow fluorescence intensity. The heated sample is the heat treatment sample of *Chlorella*. Eluate A and eluate B denote solutions with respective concentrations of eluate A and eluate B of 50 vol.%

Table 3 Comparison of distribution between untreated *Chlorella* and those treated with heat or eluate from slag (% of total algal number) [6]

	I	II	III	IV
Control	96.81 ± 2.60	2.40 ± 2.91	0.73 ± 0.31	0.07 ± 0.08
Heated sample	0.28 ± 0.36	97.27 ± 0.81	2.59 ± 0.42	0.02 ± 0.03
Eluate A	98.15 ± 0.31	0.17 ± 0.08	1.67 ± 0.21	0.02 ± 0.03
Eluate B	98.13 ± 0.24	0.52 ± 0.51	1.29 ± 0.33	0.05 ± 0.05

elements did not affect algal growth directly and exert significant toxicity to algae [6]. Although great amounts of time for obtaining data are required for studies using microscopic observation, methods using FCM are more time-efficient. In the case of this metallic toxicity to algae, both the harmfulness of eluate from slags and the effect of that on algae were determined by diagnosis based on the optical distribution of algae using FCM.

This assessment system, which estimates chlorophyll fluorescence of photosynthetic pigments using FCM, is applicable to other algae, other phytoplankton and aquatic plants with chlorophyll, although this report presents data only of *Chlorella* sp. using FCM.

4 Conclusion

Effects of environmental pollution on phytoplankton, which are the bedrock of aquatic ecosystems, are related with more higher-order biota in many ways. Evaluating the impact of environmental pollution, including metallic toxicity, on phytoplankton is extremely important. For evaluating metallic toxicity of eluate from steel slag in this study, bioassay techniques using phytoplankton *Chlorella* sp. were discussed. This study also specifically examined the development of a method to estimate not only the number of algae but also their algal activities. To achieve these, the effects of each eluate from the slags on *Chlorella* were investigated using microscopy and FCM as representative cell biological techniques.

Results revealed that eluate from the steel slags tested in this study induced neither lethality nor growth inhibition. From data based on microscopic algal counting and observation, the addition of each eluate did not show cytotoxicity: rather it enhanced algal growth. In relation to methods evaluating metallic toxicity for algae, bioassay using FCM, aside from microscopic techniques, can distinguish vigorous algae from aberrant algae simultaneously. Taken together, this assessment method using FCM in addition to microscopy can be a powerful tool for the rapid and simple evaluation of the effects of metallic toxicity as environmental pollution on algae.

References

1. Shinozaki H, Takahashi K (2010) Iron and steel making slag & its applications for road constructions. Bull Iron Steel Inst Jpn (Ferrum) 15(4):177–182
2. Takahashi K, Yabuta K (2009) Road temperature mitigation effect of "Road Gool" a water-retentive material using blast furnace slag. JFE Tech Rep 13:58–62
3. Nippon Slag Association (2012) Iron and steel slag. Tekko Sulagu Toukeinenpou, pp 2–10
4. Miyata Y, Sato Y, Shimizu S, Oyamada K (2009) Environmental improvement in the sea bottom by steelmaking slag. JFE Tech Rep 13:41–45
5. Takahashi T, Ogura Y, Ogawa A, Kanematsu H, Yokoyama S (2012) An effective and economic strategy to restore acidified freshwater ecosystems with steel industrial byproducts. J Water Environ Technol 10(4):347–362
6. Takahashi T, Yokoyama S (2015) Bioassay of components eluted from electric arc furnace steel slag using microalgae chlorella. Tetsu-to-Hagané 101(9):506–514
7. Koolman J, Röhm KH (2007) In: Color atlas of biochemistry, pp 362–363
8. Sadava D, Heller HC, Orians GH, Purves WK, Hillis DM (2008) In: LIFE, 8th edn, Chapter 3
9. Chlou CS, Chang CF, Chang CT, Shle JL, Chen YH (2006) Mineralization of reactive black 5 in aqueous solution by basic oxygen furnace slag in the presence of hydrogen peroxide. Chemosphere 62:788–795
10. Kang WH, Hwang I, Park JY (2006) Dechlorination of trichloroethylene by a steel converter slag amended with Fe (II). Chemosphere 62:285–293
11. Xue Y, Wu S, Hou H, Zha J (2006) Experimental investigation of basic oxygen furnace slag used as aggregate in asphalt mixture. J Hazard Mater 16:261–268

12. Chaurand P, Rose J, Briois V, Olivi L, Hazemann JL, Proux O, Domas J, Bottero JY (2007) Environmental impacts of steel slag reused in road construction: a crystallographic and molecular (XANES) approach. J Hazard Mater 139:537–542
13. Stimson J, Chae GT, Ptacek CJ, Emelko MB, Mesquita MM, Hirata RA, Blowes DW (2010) Basic oxygen furnace slag as a treatment material for pathogens: contribution of inactivation and attachment in virus attenuation. Water Res 44:1150–1157
14. Miki O, Kato T, Kusui T (2010) Bioassay test on steelmaking slag leachate using Vibrio fischeri and Tigriopus japonicus. J Jpn Soc Water Environ 33:141–146
15. Yokoyama S, Suzuki A, Izaki M, Umemoto M (2009) Elution behavior of electronic arc furnace oxidizing slag into fresh water. Tetsu-to-Hagané 95(5):434–443
16. Takahashi T (2014) Direct evaluation of endosymbiotic status in Paramecium bursaria using a capillary flow cytometer. Cytom Part A 85:911–914
17. Millipore Corporation (2013) Muse™ cell analyzer user's guide, pp 1–124
18. Bose J, Babourina O, Rengel Z (2011) Role of magnesium in alleviation of aluminum toxicity in plants. J Exp Bot 62:2251–2264
19. Zheng L, Lan P, Shen RF, Li WF (2014) Proteomics of aluminum tolerance in plants. Proteomics 14:566–578

The Application of Fish

Masahiro Yamaguchi

Abstract Most heavy metals have toxic effects on aquatic organisms. Therefore, toxicity tests of these metals using living organisms are required to estimate their toxicity. Fish are representative of aquatic organisms and have been used for the toxicity tests. Model species such as fathead minnows (*Pimephales promelas*), zebrafish (*Danio rerio*, formerly referred as *Bracydanio rerio*), medaka (*Oryzias latipes*), and rainbow trout (*Oncorhynchus mykiss*) have been heavily used. However, the intensity of toxicity varies at developmental stages even in the same species. In addition, temperature, pH, hardness, and salinity of test water also affect the toxicity of heavy metals. Thus the influences of these factors should be considered. The toxicity of heavy metals includes both acute and chronic cases. However, not much data exists for the chronic toxicity tests because these toxicity tests require a long time and great effort compared with acute toxicity tests. In the future, an accumulation of chronic toxicity test data and elucidation of the mechanisms for the toxicity expressions will be needed.

1 Introduction

Various kinds of metals, originating from either natural or artificial substances, can contaminate water. In order to evaluate the toxicity of these metals, bioassays are required using aquatic organisms. Some species of algae, crustacean, or fish have been used for the toxicity tests.

Among these organisms, fathead minnows, zebrafish, and medaka, have been widely used as test fish. They have some advantages; (1) they are model animals in various research fields such as developmental biology and neuroscience, and the methods for keeping and breeding have been already established, (2) they are vertebrate and have a similar body plan as humans. Thus the results obtained from the toxicity tests would be applicable to humans to some extent, and

M. Yamaguchi (✉)
National Institute of Technology, Suzuka College, Suzuka, Japan
e-mail: yamaguchi@chem.suzuka-ct.ac.jp

H. Kanematsu, D.M. Barry (eds.), *Corrosion Control and Surface Finishing*,
DOI 10.1007/978-4-431-55957-3_20

(3) especially in zebrafish and medaka, genome information has been available. Therefore, we can analyze the gene(s) and/or protein(s) expression pattern(s) in the toxicity tests. In this chapter, the application of fish to the evaluation of metallic toxicity is discussed.

2 Methods for Toxicity Tests Using Fish

Ecotoxicity includes acute and chronic toxicity. Acute toxicity is defined as the toxicity in which the influence appears within a short period after exposure to pollutants. On the other hand, chronic toxicity emerges after long-time exposure to pollutants. Various methods could exist to estimate ecotoxicity. However, the standardized methods should be adopted to compare the independently obtained results. The Organization for Economic Cooperation and Development (OECD) provides guidelines for toxicity tests, and they are the most reliable and widespread protocols. Among the guidelines, some illustrating the tests using fish are TG203, TG210, TG212, TG215, TG229, TG230, TG234, and TG236 [25] (Table 1). Based on these toxicity tests, some parameters such as NOEC (no observed effect concentration; the concentration of a pollutant that will not harm the tested organisms), LOEC (lowest observed effect concentration; lowest concentration of a pollutant that harms the tested organisms), LC_{50} (lethal concentration, 50 %; the concentration of a pollutant that kills half of the tested organism population), and EC_{50} (effective concentration, 50 %; the concentration of a pollutant that affects half of the tested organism population) are determined.

When the influences of heavy metals on aquatic organisms are considered, it is natural to assume they are exposed to a trace amount of substances over a long period of time. Therefore, the parameters obtained from chronic toxicity tests could

Table 1 OECD guidelines concerning fish toxicity tests

	Name of tests	Duration of tests	Main endpoints
TG203	Acute toxicity test	96 h	LC_{50}
TG210	Early-life stage toxicity test	Until free feeding	LOEC, NOEC
TG212	Short-term toxicity test on embryo and sac-fry stages	Until yolk-absorption completed	LOEC, NOEC
TG215	Juvenile growth test	28 days	ECx
TG229	Fish short term reproduction assay	21 days	Egg production
TG230	21-day fish assay	21 days	Vitellogenin and secondary sexual characteristics
TG234	Sexual development test	60 days	Vitellogenin and proportion of males, females, intersex and undifferentiated fish
TG236	Embryo acute toxicity (fet) test	96 h	LC_{50}

be informative. However, these values are less reliable than those of acute toxicity tests, because the chronic toxicity tests require a long time and lots of effort. Thus the number of tested organisms is small. However, some methods to speculate the chronic toxic values have been developed. For example, the method using the index of ACR (acute to chronic ratios; the ratio between chemical concentrations exerting acute toxicity, such as LC_{50}, versus a chronic toxic value such as NOEC) is useful. AF (application factor) is the inverse of ACR. Kenega [20] reported that the values of ACRs are less than 100 in 86 % of analyzed chemicals. However, Tabata [31] showed that the AF of lead, chromium (VI), and cadmium is 0.003. Therefore, the assumption that ACR is less than 100 could underestimate the risks of these heavy metals.

3 Species and Developmental Stages of the Sample Fish

The OECD guidelines recommend certain species of sample fish. For example, TG203 recommends zebrafish, fathead minnows, carp (*Cyprinus carpio*), medaka, guppy (*Poecilia reticulate*), bluegill (*Lepomis macrochirus*), and rainbow trout as sample fish. These species show different tolerances to heavy metals. For instance, LC_{50} of zinc during 96 h for an acute toxicity test with fathead minnows, bluegill, and zebrafish were 0.551 mg/L [8], 2.86–3.87 mg/L [5], and 25 mg/L [6], respectively, suggesting that the toxicity of zinc differs depending on the species of test fish. From the viewpoint of the protection of aquatic organisms, most susceptible species should be used for toxicity tests. However, a most susceptible species for all pollutants does not exist. Instead, most susceptible species differ depending on the tested substances. Thus, the use of various fish species is preferable to estimate the toxicity for most of the aquatic organisms. In addition to fish, the use of creatures in other kingdoms such as algae and crustacean is also preferable.

The toxicity of heavy metals or chemicals differs depending on the stages of development even in the same species. The developmental process of fish could be divided into embryonic stage (from the fertilization to hatching), pre-larval stage (from the hatching to the absorption of yolk), post-larval stage (from the absorption of yolk to the phase in which the number of fin rays reaches the same with those of an adult fish), juvenile stage (from the phase in which the number of fin rays reaches the same with those of an adult fish to the phase when scale is synthesized), pre-adult stage (from the stage when scale is synthesized to the first maturity), and adult stage (after the first maturity). During embryogenesis, the sensitivity to heavy metals or chemicals is relatively low, because the chorion entraps these substances. Generally, the sensitivity increases after hatching, and again decreases gradually thereafter. Thus, it is possible that the toxicity tests with post-hatching stages, such as TG203 detects metallic toxicity more sensitively than the one with embryonic stages such as TG236. Eaton et al. [11] analyzed cadmium toxicity using seven different fish species and found that larvae were more sensitive than embryos in all species. In addition, McKim et al. [23] examined copper toxicity using eight different fish species and also found that larval stages were more sensitive than the embryonic stages. However, it was reported that copper was toxic to the same degree in the embryonic stage and larval stage in rainbow trout [28]. This suggests that the embryonic stage is not necessarily an insensitive stage. Furthermore, the pattern of the change in sensitivity to metals is different depending on the type of metals, even in same fish species [28]. Therefore, it would be important to examine the toxicity in various developmental stages.

When the adult fish were used with toxicity tests, the amount of required reagents and wastewater was large. In addition, it was claimed that the application of adult fish to toxicity tests is not preferable ethically, and that toxicity tests with adult fish should be replaced by those with embryos [3].

4 The Quality of the Test Water

4.1 Temperature of Water

Each fish species has a specific water temperature appropriate for development, propagation, and living. TG203 designated a water temperature for each test fish species. A suitable water temperature for most fish is 20–25 °C, but for rainbow trout it is 13–17 °C. Thus more than 25 °C could damage rainbow trout even if harmful substances do not exist.

Water temperature also could influence metallic toxicity. For instance, the metallic toxicity tends to increase with a rise of water temperature for goldfish (*Carassius auratus*); acute toxicity test of chromium (VI) and zinc for goldfish in 24 h revealed that the rise of water temperature from 5 to 30 °C accompanied a decrease of LC_{50} [30]. It is known that 20–28 °C is suitable for keeping goldfish. Thus metallic toxicity is not necessarily lowered when the temperature of the test water is most suitable for sample fish. Seasonal fluctuation of water temperature is relatively large in temperate zones, and daily fluctuation is also large in a small river. Therefore, the influence of water temperature on metallic toxicity should be considered especially in such places. However, in the case of some fish species such as bluegill and rainbow trout, there is no relationship between water temperature and the toxicity of heavy metals. This suggests that the existence or absence of the relationship depends on the fish species.

4.2 pH

Potential of hydrogen (pH) of natural environmental fresh water such as a river or lake is usually neutral. However, it is known that acid rain decreases pH and eutrophication increases pH. Extremely low and high pH values are not suitable for aquatic organisms. In addition, it was reported that pH influences metallic toxicity.

When the concentration of H^+ rises, the binding of H^+ to ligands of dissolved organic substances would be accelerated, thus antagonizing the binding of heavy metal ions to dissolved organic substances, resulting in the increase of metallic toxicity because of the rise of the concentration of free heavy metal ions. However, it is not only free ions that are toxic to aquatic organisms. Heavy metals in a natural aquatic environment are found in many different forms such as free ions, complexes, colloids, and absorbed species on the surface of suspended particles, and each form of heavy metal has different toxic effects on aquatic organisms. Because pH influences metal speciation, it has a complex effect on metallic toxicity.

Acute toxicity tests of zinc for rainbow trout and brown trout (*Salmo trutta*) revealed that a rise of pH elicits an increase of toxicity [2, 12]. The difference of speciation for zinc would not account for the difference of toxicity between pH5 and pH7, because most zinc exists as free ions both in pH5 and pH7. It is possible that a rise of pH (i. e., the decrease of H^+ concentration) would reduce antagonistic effects of H^+ on free zinc ions against positive ion-binding sites of sample fish, resulting in the increase of zinc toxicity. On the other hand, the rise of pH from 7 to 9 elicits a moderate increase of zinc toxicity. Most of the zinc exists as $ZnCO_3$ or Zn (OH)x in pH9. Thus the adhesion of such a zinc precipitate to body surfaces or mucus of test fish (rather than the binding of free zinc ions to putative positive-ion binding sites of test fish) would account for the toxicity of zinc in pH9.

In the case of copper, it was also reported that a rise of pH is accompanied by an increase of the toxicity in rainbow trout [15]. On the other hand, acute toxicity tests of aluminum for smallmouth bass (*Micropterus dolomieui*) showed that the rise of pH weakens its toxicity [19]. When pH is elevated, toxic Al^{3+}, $Al(OH)^{2+}$, and Al $(OH)^{4-}$ would decrease, and less toxic hydroxide precipitate would increase. These changes of speciation could explain the weakened toxicity.

As mentioned above, metallic toxicity is largely influenced by pH. However, the effects of pH vary depending on the type of species and the type of metals present. The pH of test water should be controlled carefully with consideration of such effects.

4.3 Hardness

It is well-known that an increase of water hardness decreases metallic toxicity. For instance, LC_{50} of zinc during 96 h of an acute toxicity test with brown trout is 0.14 mg/L when the hardness is 10 mg/L $CaCO_3$, and it rises to 1.0 mg/L when the hardness is 204 mg/L $CaCO_3$ [12]. The same effect of the hardness on the metallic toxicity is also reported in rainbow trout [29]. However, the effect of the hardness differs depending on the types of metals. The increased hardness weakened the toxicity of cadmium very much, but its effect on the toxicity of lead is relatively small. The effect on the toxicity of copper, zinc, or nickel is intermediate between that of cadmium and lead [4]. In the case of cadmium, the hardness almost does not affect its speciation. Thus a rise in the hardness would cause decreased incorporation of cadmium, resulting in decreased toxicity. It is also possible that some positive ions such as Ca^{2+} or Mg^{2+} antagonize Cd^{2+} by blocking absorption of Cd^{2+} from the gills. The metallic toxicity-reducing effect of Ca^{2+} is greater than that of Mg^{2+} [7]. In addition, it is thought that carbonate (in water) captures metallic ions and composes complexes, resulting in decreased toxicity because the concentration of free metallic ions is reduced [32].

The hardness in a natural fresh water environment varies depending on the regions in the world. In general, the hardness is high in Europe and North America, and low in Japan. However, it varies in each river, lake, or pond, even in the same country. Therefore, the metallic toxicity should also vary in each aquatic environment.

4.4 Salinity

Seawater contains 3.5 % salt. Some fish species living in brackish water can survive in wide ranges of salinity (euryhaline species). However, most saltwater and freshwater fish cannot survive in fresh water and seawater, respectively. Thus an excess or a lack of salinity itself causes lethal damage to test fish, and salinity of test water should be controlled carefully according to the species of test fish. Currently there is no model species of saltwater fish designated as a recommended species by OECD guidelines. In order to estimate metallic toxicity in seawater, it is necessary to establish a saltwater species suitable for toxicity tests. Recently some studies used Java medaka (*Olyzias javanicus*) as test fish in blackish and seawater [9, 21].

Some reports showed that the rise of salinity decreases metallic toxicity. However, there are not so many reports investigating the relationship between the salinity and the metallic toxicity in fish. We conducted acute toxicity tests of zinc with medaka larvae and found that sodium chloride and sodium dihydrogenphosphate decrease the toxicity of zinc [26]. As in the case of Ca^{2+} and Mg^{2+}, it is possible that Na^{+} antagonized Zn^{2+} by inhibiting the intake of zinc from the gills. However, the effect of sodium dihydrogenphosphate is more intense as compared to that of sodium chloride. Therefore, it is possible that $H_2PO_4^{-}$, HPO_4^{2-}, or PO_4^{3-} forms complexes or suspended particles with zinc, which are less toxic than free zinc ions. We also reported that sodium dihydrogenphosphate decreases the toxicity of lithium to medaka larvae in a dose-dependent manner [17]. However, we should keep in mind that medaka is relatively tolerant to salt water.

5 Mechanisms of Action for the Toxicity of Heavy Metals

Most heavy metals are essential trace elements that are required for organisms, but are also toxic substances. Heavy metals such as iron, zinc, copper, manganese, molybdenum, selenium, chromium, and cobalt are known as essential trace elements for animals, and they function by binding to specific proteins in an animal's body. For example, iron is incorporated in hemoglobin, and zinc is an essential component of transcription factors having a zinc-finger domain. Lack of these essential trace elements causes physiological damage such as anemia caused by a deficiency of iron. On the other hand, in general, some types of heavy metals such as cadmium and lead are thought not to be required for organisms. However, these "unessential elements" have similar physical and chemical characteristics to the essential elements. Therefore, they could be taken into the body by the same channels through which essential elements are absorbed.

The mechanisms for acute toxicity of heavy metals against aquatic organisms are still unclear. One possible explanation is that there are positive ion-binding sites (called "biotic ligands") on the surface of the gills, and heavy metal ions compete with other positive ions (such as H^+ and Ca^{2+}) for binding to the biotic ligands. Thus heavy metals would inhibit absorption of such essential ions and cause severe damage. This model is named the Biotic Ligand Model (BLM) [22, 27]. According to the BLM, a decrease of metallic toxicity by an increase of hardness (of the water) could be explained by the decline of binding of heavy metals to the biotic ligands which is caused by an increase of Ca^{2+} concentration. Furthermore, it could be hypothesized that the amount and/or ratio of biotic ligands which is occupied by heavy metals determines the degree of toxicity. This hypothesis has not been verified yet, but there is a report supporting it mathematically [18].

On the other hand, acute toxicity tests with rainbow trout revealed that the sensitivity for cadmium decreases gradually from fertilization to hatching, but for zinc it is unvaried and for copper it increases in the same developmental stages [28]. BLM cannot explain this phenomenon unless it is assumed that there are specific developmental stages at which specific species of heavy metal ions preferentially bind to the biotic ligands. In addition, it should be mentioned that not only free ions are toxic to aquatic organisms. As mentioned above, the toxicity of zinc is strengthened in alkaline test water, even if the ratio of precipitates such as $ZnCO_3$ and/or Zn(OH)x increases and that of free ions decrease drastically. According to BLM, heavy metals cause a deficiency of essential ions like Ca^{2+} by inhibiting their absorption, and this accounts for lethal damage. However, calcium is stored in bone vertebrae and released by bone resorption when the plasma concentration of Ca^{2+} is reduced. Thus the lack of calcium would not necessarily induce acute toxicity. On the other hand, it is known that heavy metal ions can interact with proteins of various species by binding to the thiol group of cysteine residues, and inhibiting their functions. Thus the behavior of heavy metals in the body, as well as their permeability at the gills, would affect the toxicity.

In order to reveal the mechanisms for the toxicity of heavy metals, pathological and biochemical analyses of the tissues are required. In addition, analyses of swimming behaviors of test fish are also helpful to inspect how the toxicity of heavy metals is expressed. Drummond [10] analyzed swimming ability, response to stimuli, and respiration patterns of fathead minnows, and classified the action of toxic substances into four groups; narcosis, respiratory uncoupler, respiration irritant, and acetylcholinesterase inhibitor [10]. He did not analyze heavy metals, but the methods of his analyses are possibly applicable for evaluating the mode of action of heavy metals.

The mechanisms for chronic toxicity would be somewhat different from those of acute toxicity. The lack of calcium mentioned above could have a chronic, rather than an acute effect. When we consider the action of chronic toxicity of heavy metals, their narcotic or respiration uncoupler effects would be less important. Rather, the amount of accumulated heavy metals over a long period of time, the sites where they accumulated, and their speciation are important. However, there are relatively few reports investigating these points.

6 Future Prospects

6.1 Acute Toxicity Test

A large number of acute toxicity tests have been conducted (about various kinds of metals) with a lot of species of aquatic organisms, and there is an enormous stock of data. These data are very useful to conserve aquatic organisms. Most of these tests adopt survival as the endpoint. However, if the swimming ability of fish is damaged in a natural environment, then they cannot feed and/or escape from their predators, and would die even if they could survive during a relatively short test period for an acute toxicity test. Therefore, they could be regarded as dead "ecologically". The use of this "ecological death" as an endpoint could allow for more sensitive toxicity tests. Recently some studies reported acute toxicity tests using deficiency of behaviors as endpoints.

On the other hand, finer toxicity tests would be allowed if the symptoms of severe damage could be caught before the test fish fall into a critical situation. Nowadays some toxicity tests adopting cellular or molecular events as endpoints are performed. For example, it was reported that the amount and activity of the enzyme cytochrome P450 is upregulated in copper sulfate-treated silver carp (*Hypophthalmichthys molitrix*) [13]. It was also reported that secretion of adreno-corticotropic hormone (ACTH) from the pituitary gland is activated in heavy metal-exposed fish, resulting in the acceleration of cortisol secretion from the internal gland [1]. Thus quantification and monitoring of plasma ACTH and/or cortisol concentration would be helpful to grasp the symptoms of damage to the test fish.

Heat shock protein (HSP) is another useful marker and the amount of HSP70 in the gills increases in heavy metal-exposed larvae of rainbow trout [33]. Thus it can be said that various molecules could be used as a worthwhile marker to analyze metallic toxicity, although some molecules or genes show opposite responses to different toxic substances. Thus their molecular characteristics and functions should be investigated precisely. Recently the data of genome and proteome have become available easily (especially for zebrafish and medaka) and their use could contribute to establishing more sensitive toxicity tests. In order to analyze gene and protein expression, Northern blotting, RT-PCR and in situ hybridization, and Western blotting and immunohistochemistry have been carried out, respectively as conventional methods. Nowadays however, the transgenic fish can be produced easily in which the gene coding green fluorescent protein under the control of a specific gene promoter is introduced. Using such transgenic fish, the gene expression pattern can be visualized in living fish. Thus it provides a good system to analyze the gene expression and the degree of disorder in the same individuals.

6.2 Chronic Toxicity Test

Environmental standards should be set based on the values of chronic toxicity tests. This is because most toxicity from heavy metals in natural water environments is chronic, rather than acute. However, there is relatively little data for chronic toxicity tests, and the results of the chronic tests are also somewhat unreliable. This is because these tests require a considerable amount of time and effort, and their results tend to be different depending on the conditions of each test. Therefore there is a need to accumulate data for chronic toxicity tests of heavy metals with various fish species. However, there are several problems to be solved such as establishing methods to evaluate chronic toxicity.

At present, OECD provides some guidelines for chronic toxicity tests dealing with fish such as TG229, TG230, and TG234. However, it still remains controversial as to whether a dozen days, in which test fish are exposed to toxic substances as defined in these guidelines, are enough for being the "long-term" test. Also there is the question of what should be set as the endpoint. Norris et al. [24] compared plasma concentration of ACTH and cortisol in brown trout collected from a site contaminated with cadmium and zinc with those from an uncontaminated site in the same river. They found that there are no significant differences between the two groups. But they also found that when fish are subjected to a continuous confinement stressor in 5-gal buckets, the rise of plasma cortisol level delays in fish residing at a contaminated site significantly, even if ACTH secretion is elevated initially compared with the control fish. Also the plasma cortisol level declines during the test period compared with the control fish. These data suggest that the response to confinement stress by the hypothalamus-pituitary-internal gland axis in fish chronically exposed to cadmium and zinc is depressed and these fish could not sustain stress responses. The concentration of cadmium in the contaminated site of this study was 1.3 μg/L, less than 1/1000 of LC_{50} in the 48 h acute toxicity test with rainbow trout embryos. This suggests that if we set an appropriate endpoint (for example, the sensitivity against confinement stress) the toxicity caused by a trace amount of heavy metal could be detected. In order to conserve aquatic organisms, the methods concerning their exposure with lower concentrations for a longer amount of time should be established as well as the current OECD guidelines.

6.3 *Mechanisms of Action for the Toxicity of Heavy Metals*

It is not fully understood how the toxicity of heavy metals is expressed. Heavy metals disrupt transports of ions across the gills or other body surfaces, resulting in an abnormal composition of ions. On the other hand, heavy metals incorporated into the body would bind to various proteins and hamper their functions. It is considered that metallic toxicity is caused by these combined effects. Thus in order to understand how the toxicity is expressed, it is required to reveal the system by which various species of ions are transported into and out of bodies and cells. However, the isolation and functional analysis of ion channels and ion pumps on the surface of fish gills have not been entirely carried out. In 2003, type 3 Na^+/H^+ exchanger (NHE3) was firstly identified in chloride cells of Osorezan dace (*Tribolodon hakonensis*) living in extremely acidic Lake Osorezan in Japan [14]. It has since been isolated in some other fish species, and in 2014, it was reported that NHE3 in zebrafish gill mediates Na^+ absorption from ion-poor fresh water by its Na^+/H^+ and Na^+/NH_4^+ exchange activities [16]. Integration of the theory of environmental toxicology like BLM with the knowledge about physiological and biochemical functions of biotic ligands will be the challenge for the future.

The composition and behaviors of ions in saltwater fish are totally different from those in freshwater fish because saltwater fish are always exposed to an inflow of ions into their bodies. The toxicity tests using saltwater fish species are required to consider the conservation of aquatic organisms in the marine area. However, a model saltwater fish species has not been established as yet. It will be crucial to collect the data of toxicity tests using saltwater fish species.

Acknowledgement I would like to say thank you very much to Dr. H. Kai (National Institute of Technology, Suzuka College) and Mr. A. Nakata for their critical reading of the manuscript.

References

1. Barton BA (2007) Stress in finfish: past, present, and future-a historical perspective. In: Iwama GK, Pickering AD, Sumpter JP, Schreck CB (eds) Fish stress and health in aquaculture. Cambridge University Press, Cambridge, pp p1–p33
2. Bradley RW, Sprague JB (1985) The influence of pH, water hardness, and alkalinity on the acute lethality of zinc to rainbow trout (*Salmo gairdneri*). Can J Fish Aquat Sci 42:731–736

3. Braunbeck T, Boettcher M, Hollert H, Kosmehl T, Lammer E, Leist E, Rudolf M, Seitz N (2005) Towards an alternative for the acute fish LC50 test in chemical assessment: the fish embryo toxicity test goes multi-species – an update. ALTEX 22:87–102
4. Brown VM (1968) The calculation of the acute toxicity of mixtures of poisons to rainbow trout. Water Res 2:723–733
5. Cairns J Jr, Scheier A (1957) The effects of temperature and hardness of water upon the toxicity of zinc to the common bluegill (*Lepomis macrochirus* Raf.). Not Nat Acad Nat Sci 299:1–12
6. Cairns J, Scheier A, Loos JJ (1965) A comparison of the sensitivity to certain chemicals of adult zebra danios, *Brachydanio rerio* (Hamilton-Buchanan) and zebra danio eggs with that of adult bluegill sunfish, *Lepomis macrochinus*. Acad Nat Sci Phila Not Nat 381:1–9
7. Calamari D, Marchetti R, Vailati G (1980) Influence of water hardness on cadmium toxicity to Salmo gairdneri rich. Water Res 14:1421–1426
8. Carlson AR, Nelson H, Hammermeister DE (1986) Evaluation of site-specific criteria for copper and zinc: an integration of metal addition toxicity, effluent and receiving water toxicity, and ecological survey data. Natl Technical Information Service, Springfield, VA. EPA/600/S3-86/026: 1-3 Order No. PB85-121101
9. Daryoush K, Ismail A (2012) Acute toxicity test of Zn, on Java medaka (*Oryzias javanicus*) fish as an indicator of estuary pollution. Sci Res Essays 7:3302–3306
10. Drummond RA, Russom CL (1990) Behavioral toxicity syndromes: a promising tool for assessing toxicity mechanisms in juvenile fathead minnows. Environ Toxicol Chem 9:37–46
11. Eaton JG, McKim JM, Holcombe GW (1987) Metal toxicity to embryos and larvae of seven freshwater fish species—i. cadmium. Bull Environ Contam Toxicol 19:95–103
12. Everall NC, Macfarlane NAA, Sedgwick RW (1989) The interactions of water hardness and pH with the acute toxicity of zinc to the brown trout, *Salmo trutta* L. J Fish Biol 35:27–36
13. Henczová M, Deér AK, Filla A, Komlósi V, Mink J (2008) Effects of Cu2+ and Pb2+ on different fish species: liver cytochrome P450-dependent monooxygenase activities and FTIR spectra. Comp Biochem Physiol C Toxicol Pharmacol 148:53–60
14. Hirata T, Kaneko T, Ono T, Nakazato T, Furukawa N, Hasegawa S, Wakabayashi S, Shigekawa M, Chang MH, Romero MF, Hirose S (2003) Mechanism of acid adaptation of a fish living in a pH 3.5 lake. Am J Physiol 284:R1199–R1212
15. Howarth RS, Sprague JB (1978) Copper lethality to rainbow trout in waters of various hardness and pH. Water Res 12:455–462
16. Ito Y, Kato A, Hirata T, Hirose S, Romero MF (2014) Na+/H+ and Na+/NH4+ activities of zebrafish NHE3b expressed in Xenopus oocytes. Am J Physiol Regul Integr Comp Physiol 306:R315–R327
17. Kai H, Yamaguchi M, Ohta M, Arizono K, Ishibashi Y (2014) Study of metallic toxicology of medaka on coexisting component and pH (in Japanese). In: Abstracts of 23rd Symposium on Environmental Chemistry, Kyoto University, Kyoto, 14–16 March 2014)
18. Kamo M, Hayashi T (2011) Biotic ligand model: historical overview and perspectives (in Japanese with English abstract). Jpn J Environ Toxicol 14:25–38
19. Kane DA, Rabeni CF (1987) Effects of aluminum and pH on the early life stages of smallmouth bass (Micropterus dolomieui). Water Res 21:633–639
20. Kenaga EE (1982) Predictability of chronic toxicity from acute toxicity of chemicals in fish and aquatic invertebrates. Toxicol Chem 1:374–358
21. Koyama J, Kawamata M, Imai S, Fukunaga M, Uno S, Kakuno A (2007) Java medaka: a proposed new marine test fish for ecotoxicology. Environ Toxicol 23:487–489
22. Meyer JS, Santore RC, Bobbitt JP, DeBrey LD, Boese CJ, Paquin PR, Allen HE, Bergman HE, Di Toro DM (1999) Binding of nickel and copper to fish gills predicts toxicity when water hardness varies, but free-ion activity does not. Environ Sci Technol 33:913–916
23. McKim JM, Eaton JG, Holcombe GW (1987) Metal toxicity to embryos and larvae of eight species of freshwater fish—II: copper. Bull Environ Contam Toxicol 19:608–616

24. Norris DO, Donahue S, Dores RM, Lee JK, Maldonado TA, Ruth T, Woodling JD (1999) Impaired adrenocortical response to stress by brown trout, Salmo trutta, living in metal-contaminated waters of the Eagle River, Colorado. Gen Comp Endocrinol 113:1–8
25. OECD (1998) Guideline for testing of chemicals. Section 2: effects on biotic systems. OECD, Paris
26. Ohira M, Nakagawa G, Ohta M, Ishibashi Y, Arizono K, Yokoyama S, Kai H, Yamaguchi M (2015) Acute toxicity test of $ZnCl_2$ with or without NaCl or NaH_2PO_4 using medaka larvae: its effect on mortality and cell proliferation (in Japanese with English abstract). J Environ Saf 6:31–42
27. Santore RC, Di Toro DM, Paquin PR, Allen HE, Meyer JS (2001) Biotic ligand model of the acute toxicity of metals. 2. Application to acute copper toxicity in freshwater fish and Daphnia. Environ Toxicol Chem 20:2397–2402
28. Shazili NAM, Pascoe D (1986) Variable sensitivity of rainbow trout (Salmo gairdneri) eggs and alevins to heavy metals. Bull Environ Contam Toxicol 36:468–474
29. Sinley JR, Goettle JP, Davies PH (1974) The effects of zinc on rainbow trout (*Salmo gairdneri*) in hard and soft water. Bull Environ Contam Toxicol 12:193–201
30. Smith MJ, Heath AG (1979) Acute toxicity of copper, chromate, zinc, and cyanide to freshwater fish: effect of different temperatures. Bull Environ Contam Toxicol 22:113–119
31. Tabata K (1979) On the relationship between acute and long-term toxicity of water pollutants to aquatic organisms (in Japanese with English abstract). Bull Tokai Reg Fish Res Lab 98:1–19
32. Wang W (1987) Factors affecting metal toxicity to (and accumulation by) aquatic organisms — overview. Environ Int 13:437–457
33. Williams JH, Petersen NS, Young PA, Stansbury MA, Farag AM, Bergman HL (1996) Accumulation of hsp70 in juvenile and adult rainbow trout gill exposed to metal-contaminated water and/or diet. Environ Toxicol Chem 15:1324–1328

Life Cycle Assessment (LCA) of Surface Treatment Products

Hotaka Kai

Abstract In this chapter, life cycle assessment (LCA) for surface treatment products is discussed from the view points of the outline of LCA, the history of LCA in the steel industry, and overview and issues in applying LCA to surface treatment. The LCA calculates the resource consumption and amount of generated waste during the life cycle of a product or service, and assesses the way in which this burdens or affects the environment. The results of LCA give us the amount of environmental burdens or affects in regards to services and products. Also it provides us with some decisions for lowering the environmental burden or affect caused by services and products. Applying LCA to surface treatment is expected to be a very complex and highly labor-intensive task, because LCA requires collecting and summarizing extremely reliable, but diverse, inventory data. Furthermore, inventory data would be open to the public in the LCA reports. However this means that some important data and private, technical know-how for services and products would be open too.

1 Introduction to LCA

1.1 Overview

A life cycle assessment (LCA) considers the entire life cycle of a product or service—from the acquisition of raw materials to the manufacturing of materials, production of the product, actual use of the product, and final disposal—in other words, from the cradle to the grave. The LCA calculates the resource consumption and amount of generated waste during the life cycle of a product or service, and assesses the way in which this burdens or affects the environment. A prominent characteristic of LCA is that hidden environmental burdens and effects of

H. Kai (✉)
Department of Chemistry and Biochemistry, National Institute of Technology, Suzuka College, Suzuka, Japan
e-mail: kai@chem.suzuka-ct.ac.jp

H. Kanematsu, D.M. Barry (eds.), *Corrosion Control and Surface Finishing*,
DOI 10.1007/978-4-431-55957-3_21

manufacturing and disposal on the environment are considered in addition to the actual use of the product or service [1]. The result of a LCA can be used by a company to improve product development and production processes, by enabling an understanding of the effect of the product on the environment. In addition, the company can emphasize its efforts in reducing its environmental burdens by disclosing the results of the LCA to the public [2]. The LCA is therefore used as a decision making tool to reduce environmental burdens and effects on various aspects of human activity, and to assist with the choice of products and services with lower environmental burdens and effects.

1.2 Overview and Rules for Using the LCA Method [3–7]

The Introduction of International Standards ISO14040:2006 and 14044:2006 defines the LCA as a method to better understand and address possible impacts associated with manufactured and consumed products. There are four phases involved in conducting a LCA; the first phase determines and defines the goal and scope of the assessment; in the second phase an inventory analysis is conducted; the third phase involves making an impact assessment; and in the fourth phase the results are interpreted (Fig. 1: LCA phases). These four phases are presented below.

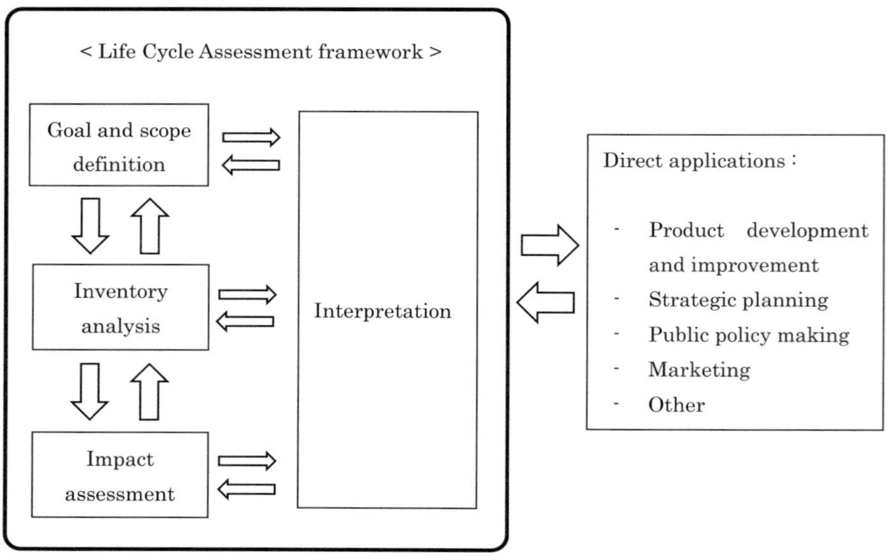

Fig. 1 LCA phases

1.2.1 Determining the Goal and Scope of the Assessment

During this phase the following is ascertained: (1) the reasons for conducting a LCA are clarified, in addition to the potential use of the results and to whom they will be disclosed, (2) the characteristics of the product and its functional units (the contents of the product or service to be assessed and quantified numbers describing the capabilities and effects of the product or service); (3) the system boundaries are designated (the scope of the product or service to be assessed, and anything related to the assessed product or service that is excluded from the LCA must be clarified); and (4) details of data collected for the LCA (the source and reliability of data collected for the LCA) are finally summarized. The use of factors (1) to (4) can significantly alter the criteria and data of the product or service to be collected in the LCA, and therefore this phase is critical.

1.2.2 The Inventory Analysis

In the inventory analysis, the input and output data required to accomplish the LCA that have been previously defined when determining the goal and scope of the LCA, are determined and calculated. To be precise, the amount of resources and energy consumed by the target product or service (input data), the waste matter from manufacturing, and the use and disposal of the product or service (output data) are collected and added up. The collection of reliable data gives the LCA result increased reliability, and therefore the collection of input and output data during inventory analysis is very important. However, it is difficult to collect all the necessary data unless the product is manufactured in-house, and therefore data typically needs to be obtained from associated companies, and by undertaking literature searches of LCA examples of other products and data in commercial software (Fig. 2: method of collecting inventory data). The use of software is desirable when making calculations, because diverse data are used in inventory analyses. In this respect, a considerable amount of LCA software has been developed, and commercial LCA software is available (Table 1: LCA software abroad. Table 2: LCA software in Japan).

1.2.3 The Impact Assessment

The environmental load (environmental load materials) can be determined by functional units such as manufactured products and services. This is firstly compiled as input and output data in the inventory analysis. In the impact assessment, these data are then categorized into impact categories, which are items generally recognized as causing environmental issues, such as global warming, ozone layer depletion, eutrophication, and soil acidification. Classification of the environmental burden by impact category illustrates the way in which the environment will be

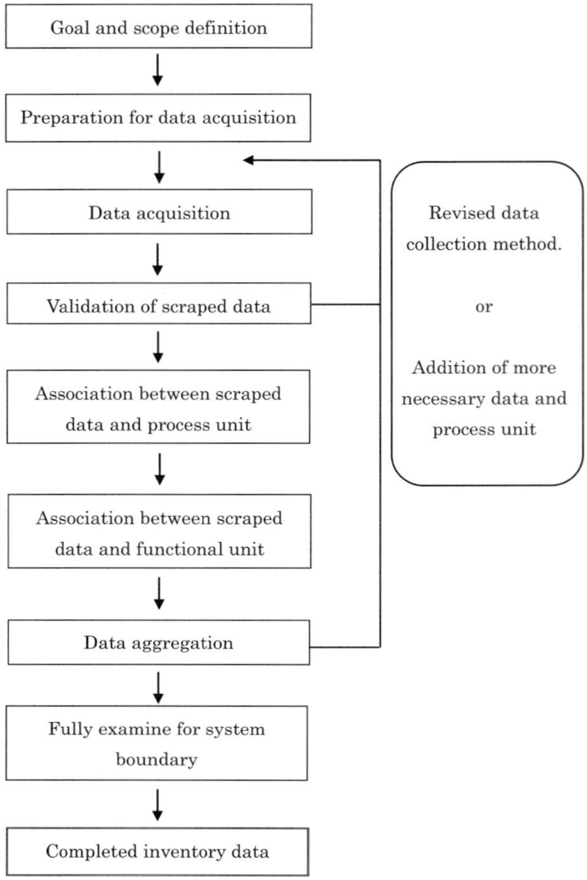

Fig. 2 Method of collecting inventory data

affected by the product. Such a classification of the generated environmental burden converts the results into objective data that are easy for a decision-maker to understand. There is a difference between the effects of an environmental burden material on a number of impact categories, and the effect of different environmental burden materials on one impact category. For example, it is known that carbon dioxide and methane are both environmental burden materials that affect global warming, but methane has a greater effect. Therefore, a factor is multiplied to the generated environmental burden materials, and the individual environmental burden scores are added up in each impact category. This process is called characterization, and the factor used is known as the characterization factor. Characterization can convert inventory analysis results by environmental burden material into results by an impact category, In addition it shows which environmental burden material has the greatest effect within the assessed impact categories. The international standards of LCAs require classification and characterization. It is also possible to use weights within the assessment. Such a process weighs the environmental

Table 1 LCA software abroad

Software	Sima Pro	TEAM	GaBi	CMLCA
Country Who's developed	Netherlands	France	Germany	Netherlands
Inaugural year	1990	1993	1990	1996
Method for inventory analysis	Build-up	Build-up	Build-up	Matrix
Data source	Ecoinvent	Ecoinvent	Ecoinvent	Not include inventory data, impact category in the software
	ETH-ESU96	APME	ETH-ESU96	
	BUWAL250	ETH&BUWAL250	BUWAL250	
	EPS		Original data	Import from EcoIndicator99, LIME, etc to the software
	TMR's Data			
Integrated method on impact assessment	LIME	Eco-indicator	Eco-indicator user setting data	Import Eco-indicator LIME etc
	ExternE	EPS2000		
	Eco-indicator	CVCH		
	EPS2000			
	EDIP	EcoPoint		
	OMPACT2002 +			
	EcoPoint			
Characteristic	LCA methods adhere to the ISO	Making LCA's tree flexible	LCA methods adhere to the ISO	Data input and output by matrix calculation
	A quick calculation and simple interface	A quick calculation and simple interface	Possible for flow caption reflected calculation result	Easily calculation of complex system including recurrence loop and recycle roop
	SPOLD format	SPOLD format a lot of analytical functions (sensitivity analysis, Monte Carlo approach)	A quick calculation and simple interface	Cannot making LCA's process tree in the software → import from self making excel's table
			DfE, flow analysis, possible for input cost data	Easily for selectable impact assessment and uncertainty analysis

effects after characterization and provides the results of the environmental effects as a single index. There are two weighing procedures: firstly one that calculates the damage, where weighing is based on objective and scientific quantification of

Table 2 LCA software in Japan

Software	MiLCA	JEMAI LCA-pro	Simple-LCA
Country Who's developed	Japan (JEMAI, AIST)	Japan (JEMAI, AIST)	Japan (JEMAI, AIST)
Inaugural year	2010	2000	2007
Method for inventory analysis	Build-up	Build-up	Build-up
Data source	A lot of scientific reports, analytical results (inventory data base which is concerned about more than 3000 service and products in the software)	A lot of scientific reports, analytical results (BUWAL, NEDO, JEMAI, CMC, plastic association, etc.) Japanese electric data transportation data	Manufacture, process, transfer unit and many LCA data calculated by AIST Ver4 as background data in the software Possible to input data as foreground data in inventory analysis
Impact assessment	Characterization, damage estimate, integration by LIME2	Characterization, damage estimate, integration by LIME	Characterization (warming, acidification, eutrophication, disposal field, resource consumption, energy consumption)
Integrated method on impact assessment	LIME2, T-E2A	LIME, eco-indicator, EPS (users can select)	–
Characteristic	LCA methods adhere to the ISO Easily and directly for a lot of case study Many relational database in the software	LCA methods adhere to the ISO setting many type of LCA stage (nation, area, transportation)	Easy for beginners

ultimate damage (end point) in each impact category; and secondly one in which the assessor weighs the environmental categories, where the categories are weighed directly using scientific insight before damage is incurred (midpoint). Both methods are susceptible to the interference of subjective judgment to some extent, so transparency of the weighing process is crucial.

The most common impact assessment procedure in Japan is the life-cycle impact assessment method based on endpoint modeling (LIME). LIME makes life-cycle impact assessment possible based on Japanese environmental conditions and environmental ideology, using the three steps of "characterization", "damage assessment", and "integration". LIME1 (published in 2005) was developed through the first-term LCA National project (1998–2003). This is the first method used in Japan that reflects the environmental conditions within Japan, and is at the cutting edge of

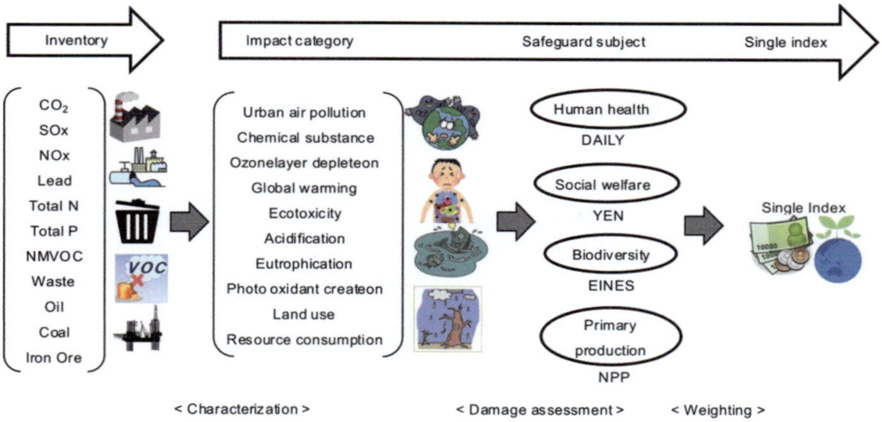

Fig. 3 Outline of LIME 1

environmental science. LIME2 was developed within the second-term LCA National project (2003–2006), and represents an improvement of LIME1. Categories of "Indoor air pollution" and "noise" were added to cover a total of 15 impact categories. Furthermore, since 2011 researchers have been working to develop LIME3, which is an improvement of LIME2, under the Funding Program for the Next Generation World-Leading Researchers since 2011 [8]. LIME3 aims to enable impact assessments outside of Japan. Similar to the inventory analysis, the impact assessment is provided via software, and is generally integrated with the inventory analysis so that the impact assessment is automatically conducted once the inventory analysis data have been provided (Fig. 3: Outline of LIME 1).

1.2.4 Interpretation of Results

Interpretation is carried out at the end to check whether the conclusions derived from the inventory analysis and impact assessment are rational, and to convert the results obtained into those that are suitable for making an objective judgment. In other words, the results of the inventory analysis and impact assessment are summarized according to the first procedure—the definition of the goal and scope of the LCA—and the reliability of the results are determined. Furthermore, information is provided in a form that makes the data easy to comprehend and use by decision makers and third parties.

2 Use and Application of the LCA in Surface Treatment

2.1 *Introduction*

Surface treatments are applied to materials, including plastics, to add new functionalities, characteristics, or properties such as strength, hardness, softness, an improved look, or rust resistance. Materials that have been surface-treated can alter the initial product into one that has a new functionality, longer lifetime, endures a higher load, or has increased stability. Therefore, surface treatment contributes significantly to manufacturing.

Diverse surface treatment methods are available for use in relation to the various objectives of the particular surface treatment, and in this respect there are many existing ways used to classify surface treatment methods. Thus it is difficult to make distinct classifications. Table 3 shows a method of classifying surface treatment methods by objectives. It can be seen that the process involved in each method differs widely. For example, the process involved in applying surface treatment methods changes significantly when the material undergoing the treatment, such as metals or plastics, is also altered. Examples of the input and output data necessary for LCA inventory analysis, when considering the LCA of a surface treatment based on the fundamental LCA "cradle-to-grave" principle, are as follows: data related to the materials to be surface-treated are received in addition to the resources, reagents, and energy necessary to implement the actual surface treatment (this includes those necessary for establishment of the entire surface treatment process); data related to the burden involved in the post-treatment and testing processes

Table 3 Processing method and dominant purpose [9]

Processing method	Dominant purpose
Cleaning	Removing grease
Derusting	Removing scale, rust stripping
Polishing	Smoothing and glazing
Etching	Construction of surface shape
Shot peening	Fatigue endurance, removing scale
Printing	Surface decorating
Painting	Corrosion resistance, decorating
Lining	Corrosion resistance, resistance to abrasion
Wet plating	Decorating, corrosion resistance, resistance to abrasion
Chemical conversion coating	Surface for painting, corrosion resistance, sliding property
Anodizing	Corrosion resistance, resistance to abrasion, coloring
Dry plating	Corrosion resistance, sliding property, optical property
Ion implantation	Electric property, resistance to abrasion, heat resistance
Hot-dip plating	Corrosion resistance
Surface heat treatment	Resistance to abrasion, resistance to fatigue, sliding property
Melting treatment	Resistance to abrasion, heat resistance, corrosion resistance
Thermal spraying	Resistance to abrasion, heat resistance, corrosion resistance

necessary to ensure the quality of the surface-treated material; and data pertaining to the post-processing of byproducts and waste water that may be generated during the surface treatment process. It is evident that compiling a LCA of surface treatment is a very complex and highly labor-intensive task. However, it is considered that as surface treatment contributes significantly to manufacturing processes that make our lives convenient and comfortable, a LCA of such treatment would be extremely valuable in relation to enabling future sustainable development.

2.2 History of LCA in the Steel Industry and Surface Treatment Technology

The LCA is a method for use in quantitatively assessing the environmental burden, or the impact on the environment that occurs throughout the entire life of a process or product beginning with the manufacture of raw materials, to the processing of the product, its transport, use, abandonment, treatment, and disposal. The concept of the LCA was established in the 1980s in relation to beverage containers. Reports of the application of LCA in relation to surface processing technology, includes that of the surface treatment of metal materials, particularly steel. For example, the World Steel Association, which was established in 1967 and consists of around 170 steel-making and iron and steel institutes, began working on applying the LCA to the steel industry in 1994. It has now established a LCA procedure conforming to standards ISO14040 and ISO14044, and is additionally compiling a life cycle inventory dataset. The World Steel Association currently compiles accurate and comprehensive datasets on environmental aspects of 15 steel products and updates these data approximately every 5 years (the next update is scheduled for 2015). This promotes the use of the LCA within the steel industry [10].

During the late 1990s in Japan, investigation began to determine whether the LCA was an effective method for use in assessing the environmental burden of a material. A system used to conduct LCAs in the steel industry has since been adopted by all related industries in Japan. In addition, life cycle inventory analysis and the assessment of environmental impacts are carried out using actual statistical data on steel products and on other materials, including tin and aluminum. However, discussions are still ongoing with respect to issues and measurements used for the LCA within the steel industry. The Japan Iron and Steel Federation, which is the core system investigating the use of LCAs in the steel industry, aims to ensure that the LCA methodology used by the World Steel Association is regarded as an ISO (International Organization for Standardization) standard. This would mean that it would also achieve global standardization by the second half of fiscal year 2016, which would enable the possibility of, for example, including the environmental value of automobile scrap recycled after abandonment in an LCA [11, 12]. Steel, which has a high recycling rate and low environmental burden, will be fairly assessed, and the environmental assessment of the Japanese steel industry and consumers using Japanese steel products is expected to benefit from its high scrap recycling rate.

Many cleaning processes are used throughout the various manufacturing steps of: manufacturing of raw materials, processing, transport, use, abandonment, further processing, and disposal. In this respect, LCAs related to cleaning metal parts have been reported in Japan. For instance, in processing it is necessary to remove oil coatings used to reduce friction heat, coatings used to protect machine surface quality and prevent equipment damage, and to remove impurities and dust attached to devices and equipment. A wide range of chemical products are employed as detergents in the cleaning processes involved in metalwork, and these produce extremely large amounts of total waste materials. In this respect, although waste water in manufacturing is generated separately from the main product, its effect on the environment cannot be neglected. Thus it is necessary to consider assessing how the by-product waste water affects the environment when developing and using LCA methodology.

In summary, LCAs appear to be used increasingly within the steel industry. However, only a limited amount of research exists in relation to the use of LCAs with surface treatment technologies. One of the reasons, as already mentioned, is that diverse surface treatment methods that have different objectives are used, and the processes involved in each method differ significantly. Therefore, it is difficult to define the functional unit, and an enormous amount of input data is required. As a result, there is a lack of studies related to the development of LCAs that sufficiently consider the surface treatment of steel products.

2.3 Overview and Issues in Applying LCAs to Surface Treatment

As previously mentioned, there are diverse surface treatment methods with various objectives, and these can be classified into wet processes (surface treatments using a solution) and dry processes (surface treatments that do not use a solution), as introduced in Table 3. There are several benefits to wet processes, such as the ability to treat large areas and conduct mass productivity, and it is relatively simple to set up the simple processing equipment that is open to the atmosphere. In addition, the processes involve low equipment costs. However, the disadvantages are that there is a limit to the amount of metals that can be extracted because of the method of electrolysis used, and it is also necessary to treat the waste liquid and wastewater. This section discusses issues necessary for a high quality LCA applied to wet processes, in relation to surface treatment using only a chemical reaction, which is the simplest form of surface treatment. Key factors are emphasized in relation to how the system boundary and functional units should be set, and how the data needed for input and output data should be collected for inventory analysis.

Chemical conversion processing is a type of surface treatment that consists only of a chemical reaction. In this respect, metal is immersed in a solution where it dissolves slightly, and an insoluble corrosion product, or an oxide film forms simultaneously on the surface. Examples of such films are corrosion resistant films, undercoat films, and lubrication films. Examples of chemical conversion processing are phosphating and chromic acid treatment. The following is an example of the use of a LCA in the phosphating of zinc plates.

The materials required are the zinc plates to be treated and zinc phosphate ($Zn_3(PO_4)_2$), which is the reagent necessary for chemical conversion treatment, and these items are reflected in the input data section of the inventory analysis. The system boundary is then set, representing the assessment scope of the LCA, and is determined based on its objective. For example, it may be necessary to consider whether to include the manufacturing processes involved in making the zinc plates from the original raw material, which is zinc ore, in the scope.

It is then important to define the functional units that follow, as these need to be set concretely and realistically using target values and numbers. Examples of the quantities that require clear definition are the dimensions and weight of the zinc plates to be processed, the thickness (in micrometers) of the film to form on the surface of the zinc plates, and the production output in a fixed timeframe, such as the day, month, or year.

The fundamental concept of the LCA is to assess the entire "cradle-to-grave" life cycle of a product or service. Therefore, using only the information pertaining to the material to be surface treated, and the treatment conditions such as the film thickness, is insufficient for a LCA. For instance, the dimensions of the zinc plates determine the size of the treatment tank required, and the amount of solution used in the treatment tank. As the surface treatment of zinc plates is impossible without the use of a treatment tank, the step required to manufacture the treatment tank also needs to be included in the assessment scope. The amount of necessary zinc phosphate reagent required is also determined by the treatment tank size, and the amount of solution in the treatment tank is determined by the thickness of the film on the surface of the zinc plates. The zinc phosphate in the solution will be consumed faster with an increase in the number of zinc plates requiring phosphating; hence zinc phosphate needs to be added regularly. Therefore, the amount of solution in the treatment tank gradually decreases, and water needs to be added to maintain the amount of solution fixed. It is thus necessary to make calculations under a fixed functional unit condition of the reagents, including zinc phosphate and the water resources necessary to carry out the surface treatment in a given treatment tank. Moreover, most of the surface treatment process is carried out using a number of machines, and is not performed by hand. Consequently, it is necessary to sufficiently summarize the electricity consumption used in the surface treatment process within its designated functional unit, and this amount must then be used as input data in the inventory analysis. In addition, the cleaning and waste water treatment processes are also required in the LCA scope, as mentioned in Sect. 2.2.

In summary, a good quality LCA requires collecting and summarizing extremely reliable, but diverse, inventory data. In Sect. 2.1, it was stated that the LCA of surface treatment is expected to be a very complex and highly labor-intensive task. There is currently a lack of available research related to the LCA of surface treatment technology, and this is inevitably related to the difficulties involved in defining surface treatment technology processes and setting the functional unit. However, we believe that the primary reason for the lack of past research is related to the immense burden involved in collecting the diverse and reliable input data required to make high quality LCAs, as previously discussed. The industry itself cannot recommend LCA because LCA also requires disclosure of input data to some extent, and private technical know-how regarding surface treatment technology has to be protected.

References

1. Inaba A (2005) Affairs of life cycle assessment. JEMAI, Tokyo
2. Itsubo N, Tahara K, Narita A (2007) Outline of life cycle assessment. JEMAI, Tokyo
3. Itsubo N, Inaba A (2005) Methods of life cycle impact assessment. JEMAI, Tokyo
4. Matsuno Y (2013) J Life Cycle Assess Jpn 9(4):349–351
5. Inaba A (2014) J Life Cycle Assess Jpn 10(1):67–68
6. Tahara K (2014) J Life Cycle Assess Jpn 10(2):183–184
7. Nakano K (2014) J Life Cycle Assess Jpn 10(4):492–493
8. Life-cycle impact assessment method based on endpoint modeling. http://lime3.jimdo.com/
9. Nihira N, Mitsuo A (2001) The first – surface treating technique. pp 13–16
10. Nippon steel and sumitomo metal (2013) LCA activities on world steel Association (The seminar of Life Cycle Assessment Society of Japan and the Institute of Life Cycle Assessment, Japan, 2013). http://www.j-lca.sakura.ne.jp/seminar/pdf/20130315/4-WSA-ono.pdf#search=‘%E9%89%84%E9%8B%BC+LCA’
11. Nikkan Kogyo Shimbun (Business & Technology Daily News) (13 Jan 2015). http://nsri.nssmc.com/assets/files/japanmetal-20150113_01.pdf
12. Nikkan Kogyo Shimbun (Business & Technology Daily News) (13 Jan 2015). http://nsri.nssmc.com/assets/files/japanmetal-20150113_10.pdf

Part V
Environmentally Friendly Surface Finishing

What Is Environmentally Friendly Surface Finishing?

Hideyuki Kanematsu and Dana M. Barry

Abstract As already described, environmentally friendly surface finishing is presently very important and the requirement will be become much more important in the future. This is because we are facing a serious environmental crisis in nature as well as the severe environmental politics in industries. We have to balance them and find a new and better way for the future. In this chapter, we describe what environmentally friendly surface finishing is. Also we classify this process type into some cases. Particularly in this chapter, we focus on Hazard Analysis and Critical Control Point (HACCP) and explain its concept.

1 Introduction: What Is Environmentally Friendly Surface Finishing?

Environmental friendliness is becoming more important, since we seem to face a critical environmental crisis at this point, putting aside the authenticity of the situation. As a result, we have to take environmental friendliness into consideration when designing new materials or conventional materials. Since the surface is always the interface between bulk materials and the environment, materials' surfaces should always be carefully designed.

From this viewpoint, four types of environmental friendly surface finishing processes are described. The first and second ones are closely related to strict environmental regulations and laws, as discussed in Part II. As the premise of research and development, there are solid and strict environmental regulations and

H. Kanematsu (✉)
National Institute of Technology, Suzuka College, Suzuka, Japan
e-mail: kanemats@mse.suzuka-ct.ac.jp

D.M. Barry
Clarkson's Departments of Chemical & Biomolecular Engineering and Electrical & Computer Engineering, Clarkson University, Potsdam, NY, USA

Center for Advanced Materials Processing (CAMP), Clarkson University, Potsdam, NY, USA
e-mail: dmbarry@clarkson.edu

H. Kanematsu, D.M. Barry (eds.), *Corrosion Control and Surface Finishing*,
DOI 10.1007/978-4-431-55957-3_22

laws. In order for surface finishing industries to survive, it is very important and critical for them to be able to meet these legal requirements and standards. Therefore, many new environmentally, friendly processes have been developed. In this chapter, these surface finishing processes are classified into some categories conceptually. Also this chapter focuses on Hazard Analysis and Critical Control Point (HACCP), an inspection process for products. The concept and examples developed by the authors are provided too.

2 Non-harmful Environmentally Friendly Processes

Non-harmful environmentally friendly processes are classified into two types. One of them is the closed system, as shown in Fig. 1. Generally, the substances and materials that have been used for conventional processes have had many merits and priorities from the viewpoint of performance. They have their own unique reasons why they have been used. The closed system makes the process completely closed. It often means zero emission to the outside environment. In some cases, the harmful gasses or liquids produced as byproducts or those that remain in the process might be reused for other purposes within the process. At any rate, no harmful materials would be discharged outside of the system (when using a closed system). The concrete examples are shown and explained in chapter "Process Changes".

Another type (of environmentally friendly, surface finishing process) is the substitution of more environmentally friendly materials. As shown in Part II, there are already many materials, particularly metals, appearing on the hazardous substance lists. To name a few, we can mention lead, cadmium, hexavalent chromium, mercury, etc. So far, they have been replaced with more friendly substances.

Figure 2 shows metal elements available for electroplating. This is the general tendency at this point. Some metals might be useful at the research stage or in very special cases.

The details are introduced and explained in chapter "Process Changes". The concept for the process is explained in Fig. 3 schematically. Another possibility is to completely replace materials with different ones. However in such a case, the

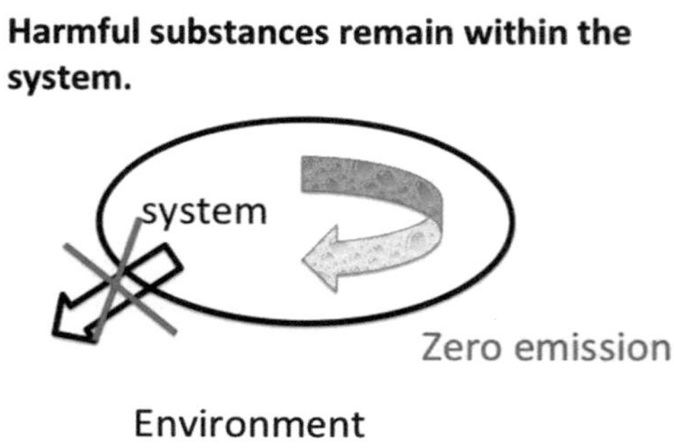

Fig. 1 The concept of a closed system

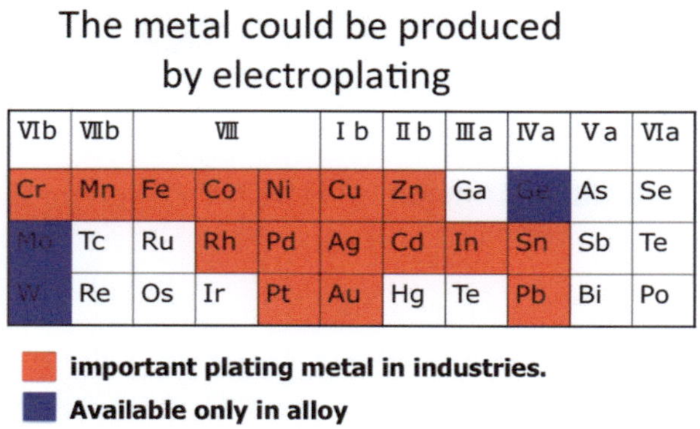

Fig. 2 Metals used for electroplating in practical industries

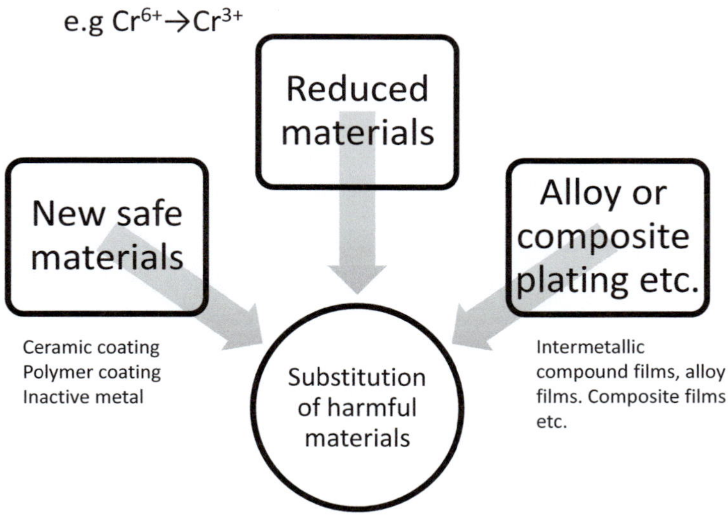

Fig. 3 Basic concept for substitution type environmentally friendly surface finishing

merits that the conventional surface finishing process had, might be completely lost. For example, cadmium plating might be replaced with zinc alloy plating very easily from the viewpoint of corrosion resistance. However, cadmium plating has also been well-known for its good lubricating ability. Therefore, it has been often used for sliding surfaces. There are still many more concrete cases other than the cadmium plating. Since the hexavalent chromium plating was issued, the reduced trivalent chromium has been investigated and developed. However, some other films or composites may work well as substitutes. All of these trials have been carried out directly to solve the environmental problems. The detailed explanations will be given in chapters "Substitution with More Friendly Elements", "Change from Metals to Nonmetals", and "Process Changes". However in this chapter, we would like to introduce some unique, environmentally friendly processes, even though people might not recognize them as environmentally friendly surface finishing processes at a first glance.

3 Surface Finishing Assured by HACCP (Hygiene Related Surface)

HACCP stands for Hazard Analysis and Critical Control Point and is an inspection process for products, particularly for food [1, 2]. It started with a NASA project. In the 1960s, an appropriate inspection process was being investigated in order for astronauts to be free from health damages while on board and carrying out space activities. Conventionally, only final products were sampled and checked just before packaging and shipping. On the other hand, there are some CCPs (Critical Control Points) throughout the whole process in the case of the HACCP system. Therefore, we can say that the HACCP system provides us with better and safer inspections than the conventional methods. Figure 4 shows the difference between the two inspection systems schematically.

To pass inspection and comply with the HACCP requirements in the twenty-first century, we have to change the basic concepts for surface finishing. Figure 5 shows the schematic illustration about the kinds of surface finishing performance that are assured by HACCP (HACCP surfaces).

Conventionally, surface finishing provides three main functions: wear resistance, corrosion resistance, and color tone. In addition to the conventional

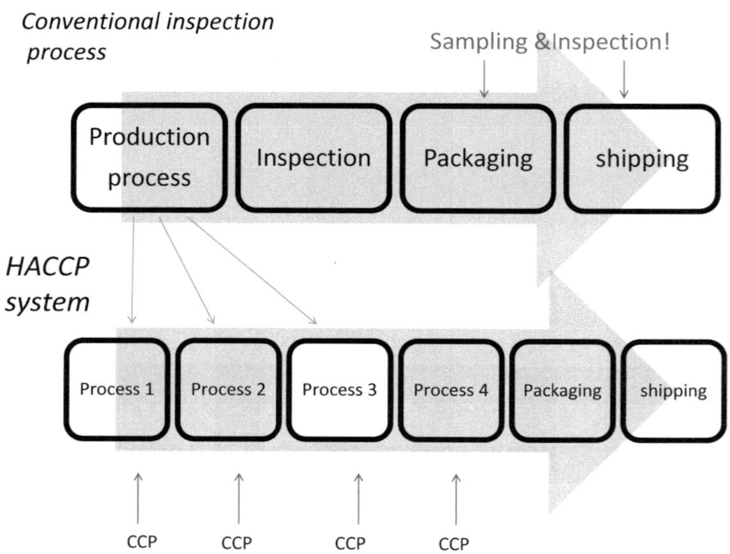

Fig. 4 HACCP system and a conventional one

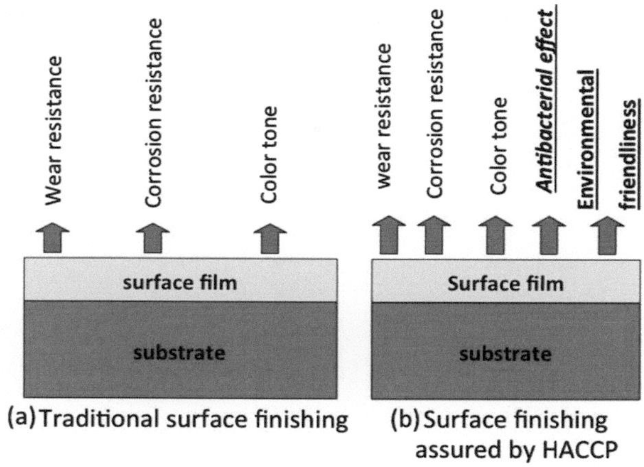

Fig. 5 Performances provided by the surfaces assured by HACCP

properties, antibacterial effect and environmental friendliness have to be provided for HACCP surfaces. For the latter, the dissolution of harmful components from a surface finishing film may trigger detrimental contamination and lead to negative effects for humans. It might also violate environmental laws in some cases.

For the former (antibacterial effect), SIAA (Society of Industrial Technology for Antimicrobial Articles) in Japan already defined the antibacterial effect and standardized the evaluation method as ISO22196 [3]. According to their definition, an antibacterial effect is a state where the growth of bacteria on materials' surfaces would be suppressed [4]. It should be differentiated from sterilization and the killing of bacteria completely. Sterilization might lead to contamination of the environment with very harmful chemicals. On the other hand, the antibacterial effect (Kokin) would not be harmful for organisms and would never contaminate the environment beyond the criteria of the environmental laws. From this standpoint, the surface finishing processes that realize Kokin (antibacterial effect) are important to industries pursuing environmentally, friendly surface finishing.

To understand the concept described above, we can mention the effect of metallic ions on bacteria for example. Substances bringing about the antibacterial effect can be classified into three categories, as shown in Fig. 6 schematically. One of them is the organic chemical group. It could destroy the outer part of bacteria cells, leading to the appearance of an antibacterial effect. However, these chemicals might also be harmful to organisms including human beings. This would not be an environmentally friendly surface finishing process. The second group contains photocatalysts. A very well–known example is titanium oxide. When light irradiates the substance, electrons are excited (with higher energy) to the valence band and form positive holes in the former. The positive holes tend to be stabilized by getting electrons from substances in the environment. Hydroxide ions often provide electrons to substances to become radicals. Radicals generally have strong oxidation power and could kill bacteria. The third group has antibacterial metals. The group includes silver, copper, etc., and conventional antibacterial metals. For the

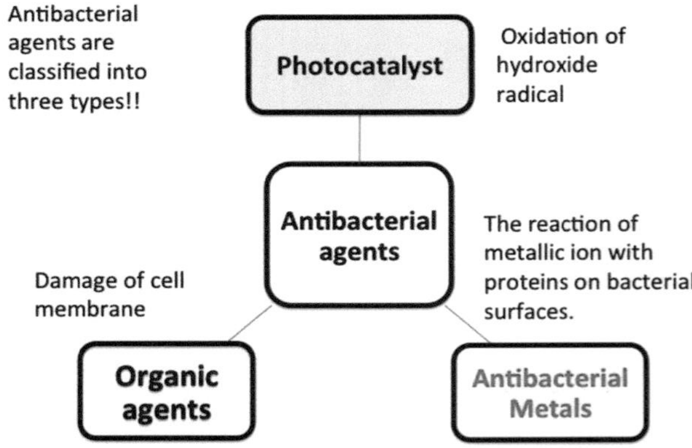

Fig. 6 Three types of antibacterial agents

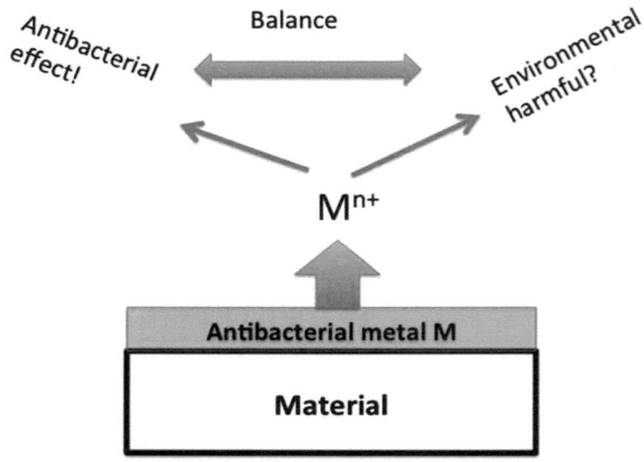

Fig. 7 Ambivalence of antibacterial metals from a viewpoint of the HACCP concept

substances in the group, metallic ions (dissolved from material surfaces) tend to bind to amino acids on the surface of bacterial cells. Therefore, the binding of antibacterial metallic ions with the amino acids inhibits the metabolism action of bacteria. This action leads to the control of bacterial growth. Thus the antibacterial effect appears. Since the dissolution might lead to an increase of metallic ions in the environment, one might need to control the dissolution characteristics of materials' surfaces very carefully. Even though the increase of dissolution would be favorable from the antibacterial standpoint, it might violate environmental laws. Surface finishing processes assured by HACCP should take all of these factors into consideration (Fig. 7).

The concept of an antibacterial surface is connected to the broader and more important concept, an antifouling surface. Here we are talking about biofouling, where there are concerns about attachments of organisms to materials' surfaces. Biofouling is classified into two main groups: microfouling, based on bacterial activity, and macrofouling which is based on larger organisms such as oysters,

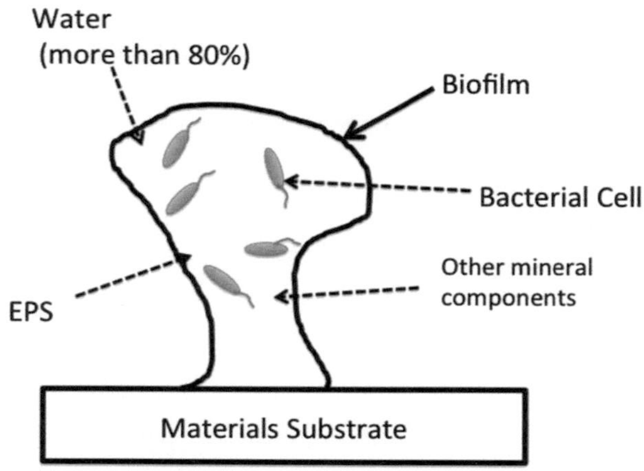

Fig. 8 Biofilm and its composition

barnacles, etc. As for the former (microfouling), bacteria tend to form sticky thin films called biofilm. Biofilm is generally composed of 80 % water, exopolymeric substances (EPS), bacteria themselves and other extra minor components (Fig. 8) [5]. From a sanitary viewpoint, biofilm is very detrimental, since the bacteria could survive through the attacks of bactericidal agents, antibiotics, larger organisms, creatures such as natural enemies, strong flows, oligotrophic environments, and so on. This means that the contaminated and probably environmentally harmful surfaces would remain on the materials.

The antibacterial surface and antifouling surface need specially designed surface finishing. To solve many industrial problems from hygiene, corrosion, scale buildup on pipes, etc. to new novel sensors, fuel cells, and environmental cleanup, new surface finishing processes are required. Most of them should be able to produce environmentally friendly surfaces for materials. In the next section, we mention our own research about such an environmentally friendly surface finishing process.

4 Concrete Examples of Environmentally Friendly Surfaces from the Standpoint of HACCP

As described in the previous section, some metals, particularly silver and copper, show very strong antibacterial effects. However, the important place is the interface between materials and environments. Therefore, it is very important to give the antibacterial effect to the material's surface. Silver plating and copper plating could show clear antibacterial effects. However, they just dissolve into the environment as ions, when silver or copper plating takes place. In such a case, the dissolved ions would react with the environmental factors to form some insoluble compounds. Also the dissolved ions might accumulate in the environment or exceed the

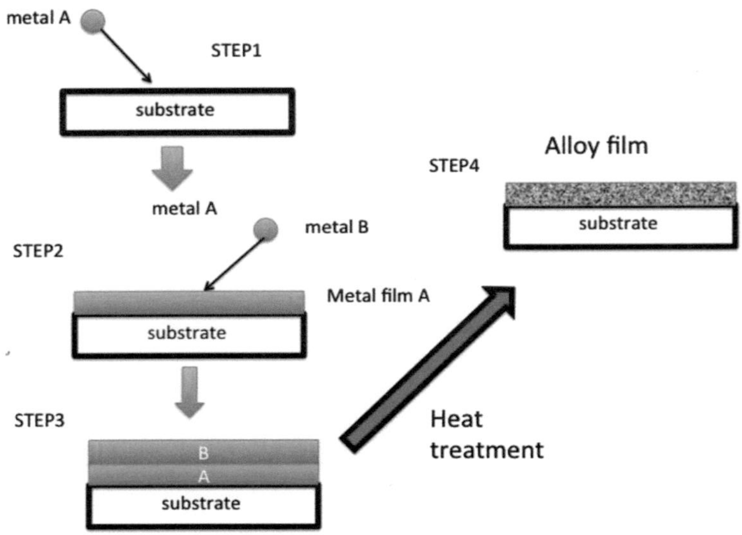

Fig. 9 The principle of the heat treatment and surface finishing

permissible levels defined by environmental laws. In the former case, it would lead to the cessation of the antibacterial effect, while the latter case might lead to serious environmental problems in the future. From the standpoint, one of the authors focused on the intermediate metallic compounds and devised a surface treatment process based on the combination of heat and surface treatment. Figure 9 shows the basic principle of the production process [6].

As shown in Fig. 9, the metal A is first plated or coated (by any process) on a substrate. Then the metal B is plated/coated on top of film A. Next a heat treatment is applied. If the melting temperature of metal A would be lower than that of metal B, the temperature should be slightly over the melting point of metal A. Then some of the metal A would melt, and the atoms of metal B would diffuse into the metal A at a very high speed, since one of the phases (phase of metal A) is a melted form. To enhance the antibacterial effect for tin plating used in food industries, tin was chosen as metal A. And as metal B, silver and copper were chosen independently. As a result, the process was used to produce tin-silver or tin-copper intermetallic compounds. The evaluation tests for the antibacterial effect (ISO 22196) confirmed that tin film did not show any antibacterial effect. On the other hand, the intermetallic compounds for the alloy films clearly showed antibacterial effects. This suggests that silver or copper could dissolve into the environment at a low concentration level, and that the silver or copper ions could make the antibacterial effect appear on materials' surfaces. Fortunately, these well-known antibacterial metals could show antibacterial effects at very low levels [7].

As for antifouling surfaces, we confirmed that silane compound coatings dispersed by nano-antibacterial powders such as silver, copper, etc. or organic compounds (for those antibacterial metals) showed high antifouling properties on materials [8].

5 Future Scope

In this chapter, we explained the meaning and basic concept for environmentally friendly surface finishing. And as a concrete example, a unique case – HACCP (the assured environmentally friendly surface finishing process) was introduced. There are still many typical environmentally friendly surface finishing processes. However, those typical processes will be described in other chapters. In this chapter, two cases were introduced. One of them was the production process composed of heat treatment and surface treatment to produce films that were intermetallic compounds between Sn-Ag and Cu-Ag. The intermediate compounds could show antibacterial effects and control dissolution simultaneously. The other example was a composite film based on the silane compound. Nano silver or copper powders were dispersed into a silane compound. The organic compounds of those antibacterial metals were also dispersed into silane compounds. The antibacterial, nano-powder dispersed, silane compounds were coated on various materials. These coated materials showed antifouling properties on their surfaces. They all assured HACCP and environmental friendliness.

These cases explained in this chapter were the authors' inventions and trials. However, analogous and similar processes should be continued much more in the future.

References

1. Mortimore S, Wallace C (2013) HACCP: a practical approach, 3rd edn. Springer, New York
2. Okuda K (2010) In: Kanematsu H (ed) "HACCP and quality control", antibacterial materials for safety, security and reliability. Yoneda Shuppan Co, Chiba, pp 7–15
3. International Organization for Standardization (ISO) (2007) Measurement of antibacterial activity on plastics and other non-porous surfaces
4. SIAA (Society of Industrial Technology for Antimicrobial Articles) (2008) Kohkin, hidden ingenuity to global industry. http://www.kohkin.net/dcms_media/other/SIAAbrochure-en.pdf
5. Kanematsu H, Barry DM (eds) (2015) Biofilm and materials science. Springer
6. Kanematsu H, Ikigai H, Yoshitake M (2009) Evaluation of various metallic coatings on steel to mitigate biofilm formation. Int J Mol Sci 10(2):559–571
7. Kanematsu H (2013) Alloy plating by heating stacked single layers and the possibility of its application in the future. Prod Finish 78(3):1–14
8. Kanematsu H, Sasaki S, Miura Y, Kogo T, Sano K, Wada N, Yoshitake M, Tanaka T (2015) Composite coating to control biofilm formation and effect of alternate electro-magnetic field. Mater Technol 30(B1):21–26

Substitution with More Friendly Elements

Hideyuki Kanematsu and Dana M. Barry

Abstract As described in detail already, environmentally friendly surface finishing processes are available by substituting the original process or plating itself with something less harmful. As for the process changes, we described the details and gave examples in chapter "Process Changes". In that chapter, we mainly mentioned physical vapor deposition (PVD). On the other hand, we focused on the substitution of plating materials from aqueous solutions. As concrete examples, we talked about some aqueous processes for the substitution of hexavalent chromium plating. This chapter mainly describes alloy plating processes, trivalent chromium processes and composite plating.

1 Introduction

As we have already seen, there are quite a few substances that regulations and laws prohibit or control their use. Many of them are unfortunately (for us metallurgists) metals. As shown in Fig. 1, we have to decide how to develop surface finishing technology so that there is a balance between the functionality and environmental friendliness based on political regulations. As we have seen so far, many metals such as lead, cadmium, mercury, hexavalent chromium ions, etc., are only useful from the functional standpoint. If the metals would be judged as toxic substances, then they should not be used. However, there are some problematic metals whose toxicity and detrimental effects are under dispute and investigation. They may be

H. Kanematsu (✉)
National Institute of Technology, Suzuka College, Suzuka, Japan
e-mail: kanemats@mse.suzuka-ct.ac.jp

D.M. Barry
Clarkson's Departments of Chemical & Biomolecular Engineering and Electrical & Computer Engineering, Clarkson University, Potsdam, NY, USA

Center for Advanced Materials Processing (CAMP), Clarkson University, Potsdam, NY, USA
e-mail: dmbarry@clarkson.edu

H. Kanematsu, D.M. Barry (eds.), *Corrosion Control and Surface Finishing*,
DOI 10.1007/978-4-431-55957-3_23

Fig. 1 The design principle for the future surface finishing processes

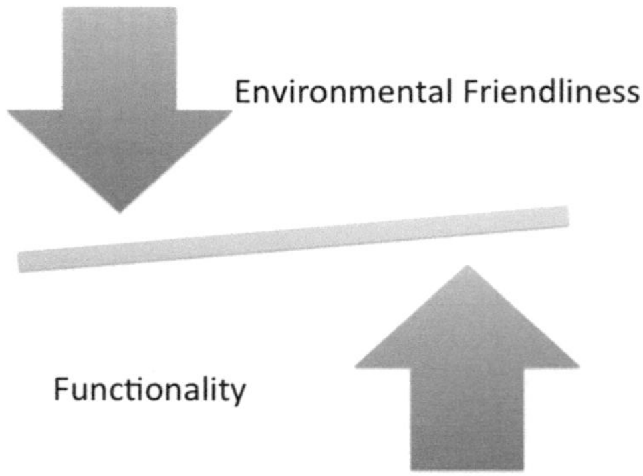

judged as toxic substances or controlled for their uses to some extent. In such a case, industries should prepare ways to substitute them with less harmful ones.

In this chapter, we deal with these cases. Particularly, we focus on hexavalent chromium ion, even though cadmium, lead, and other problematic metals might cause similar cases now or in the near future. There are many possibilities for all of them. Unfortunately, we have to restrict them to some concrete examples due to limitations of space. However, the basic concept for all of them could be applied to different cases in the future. Here in this chapter, we would like to mainly introduce some substitutions for hexavalent chromium.

2 Substitution for Hexavalent Chromium Surface Finishing

Since hexavalent chromium ion is a hazardous substance [1], it should be substituted with other more environmentally friendly ones. You could also use a completely closed system so no harmful substances would be discharged. As for the closed system shown in Fig. 1 (in chapter "What is Environmentally Friendly Surface Finishing?"), we could mention the PVD and chemical vapor deposition (CVD) processes. Actually, chromium plating from aqueous solutions could be replaced with processes in gas phases, as shown in Fig. 2. It includes Diamond Like Coating (DLC), a recently very useful and popular process. However, we have already described the meaning, concept, and the general information in chapter "Process Changes". Therefore for this chapter, we mention some substituted plating examples from aqueous solutions and discuss them.

Figure 2 shows some plating processes to substitute for hexavalent chromium plating. Ni-based and Co-based alloy plating is the first class. Generally speaking, alloy plating is composed of the base metal (nickel or cobalt) and the second alloy element such as phosphorous, boron, tungsten, or tin. Usually chromium plating has many merits such as high wear resistance, high corrosion resistance and high

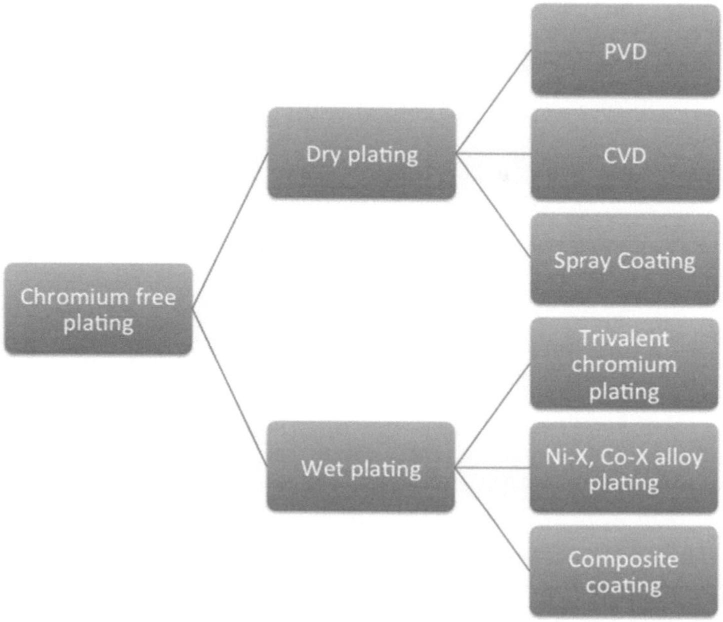

Fig. 2 Various chromium free plating processes [2]

lubricant ability, etc. It has been very useful for industrial plating so far. People have been confused about if chromium itself might be harmful or not. However, we do not have any concrete reports for its negative effects at this point. The problem is actually caused by hexavalent chromium ion in most cases. The confusion might be caused by the phrase, "Chromium free coating." It might suggest that chromium would also be a hazardous substance. Therefore, we will use the word, chromium ion free coating, instead.

Originally, hexavalent chromium has been a poisonous substance. Since it has strong oxidation power, it can destroy the cells of organisms. When humans touch the substance directly, they can develop dermatitis, a growth, etc. It has been very well- known that workers touching hexavalent chromium ions for a relatively long time often developed a hole in the nasal septum. This case could be considered a sort of chronic ulceration. The hexavalent chromium plating causes a "chromium mist" problem and gives the workers detrimental effects. In such a case, they might develop lung cancer through disorders of their respiratory system. If it would violate ailment systems for workers, they might get colorectal cancer or stomach cancer. These cases show that hexavalent chromium would be considered a carcinogen. Such a poisonous carcinogen, even though it might have some useful functions, should be replaced with some other environmentally friendly substances.

2.1 Electroless Nickel Plating for Hexavalent Chromium Ion Free Plating [3]

One of the promising plating processes to substitute for hexavalent chromium plating is electroless nickel plating. Electroless plating is literally the plating without any external power supply. The electrons needed for the plating process are provided by a series of chemical reactions. Figures 3, 4, and 5 show three representative plating processes from aqueous solutions schematically. And we can understand the difference between electroless and electroplating schematically.

Figure 3 shows the general concept for electroplating and the configuration of components. The work is immersed into the plating bath and coupled with the other counter electrode, so that the work should be the cathode and the counter electrode should be the anode. The metal ions in the bath are precipitated onto the work, having the electrons provided by an external electron source.

Figure 4 shows the schematic principle for displacement plating. In this case, you don't need any external electron source to make the metal ions in the solution precipitate onto the work. Instead, the metal component of the work would dissolve to give the metal ions in the solution electrons. The chemical reactions can be written in the following way schematically. Eq. 1 is the cathode reaction for the precipitation and plating film formation, while Eq. 2 is the anode reaction to give the plating metal electrons.

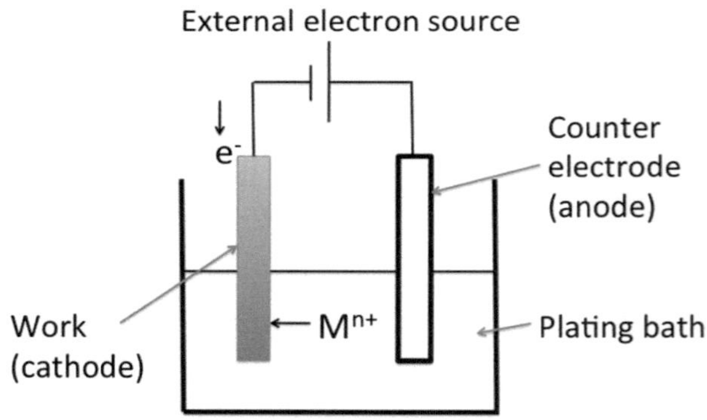

Fig. 3 The general concept for electroplating

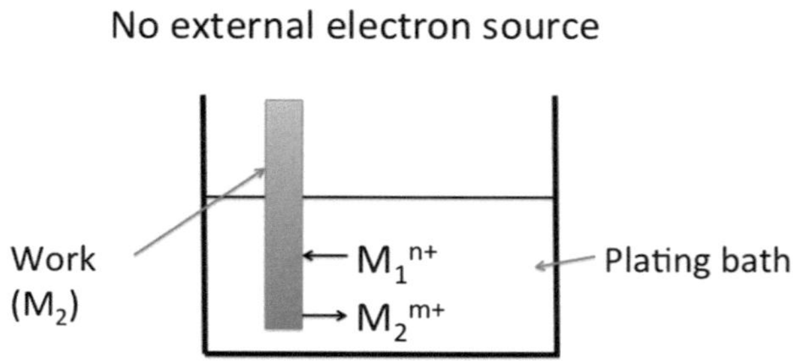

Fig. 4 The schematic illustration for displacement plating

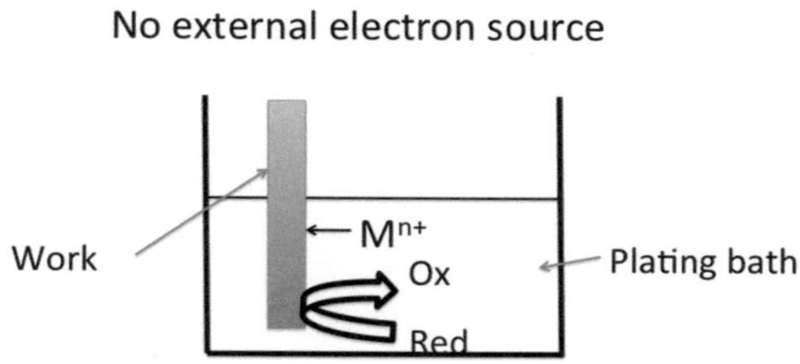

Fig. 5 The schematic principle for chemical plating

$$M_1^{n+} + ne^- \rightarrow M_1 \tag{1}$$

$$M_2 \rightarrow M_2^{m+} + me^- \tag{2}$$

This type of plating is often called displacement plating and is a type of electroless plating.

Finally, the third possibility can be mentioned. It may be called "chemical plating," since the electrons for the precipitation of the objective plating metal are not provided by any outer electron sources, but by the inner chemical reactions (in the bath). See Fig. 5. This is also electroless plating and the main one for this category. A reducing agent (Red) exists in the bath. When it is oxidized chemically (Ox), then the electrons are released and available for the metal ions and then precipitation occurs. The chemical reactions can be described schematically as follows.

$$\begin{aligned} &M^+ + e^- \rightarrow M \\ &Red + H_2O \rightarrow Ox + 2H^+ + 2e^- \end{aligned} \tag{3}$$

As a typical example, we can mention nickel electroless plating. The process generally utilizes hypophosphorous acid solution. When iron is immersed into the solution, the metal surface works as a catalyst and promotes the dehydrogenation reaction of hypophosphite ion as follows.

$$[H_2PO_2]^- \rightarrow [PO_2]^- + 2H \tag{4}$$

The produced substance is meta phosphite ion and it reacts with water to make phosphite ion. The produced atomic hydrogens combine with each other to form hydrogen gas. Some parts (of them) react with nickel ion to precipitate a nickel coating.

$$2H \rightarrow H_2 \uparrow \tag{5}$$

$$Ni^{2+} + 2H \rightarrow Ni + 2H^+ \tag{6}$$

Fig. 6 The concept for electro-alloy deposition

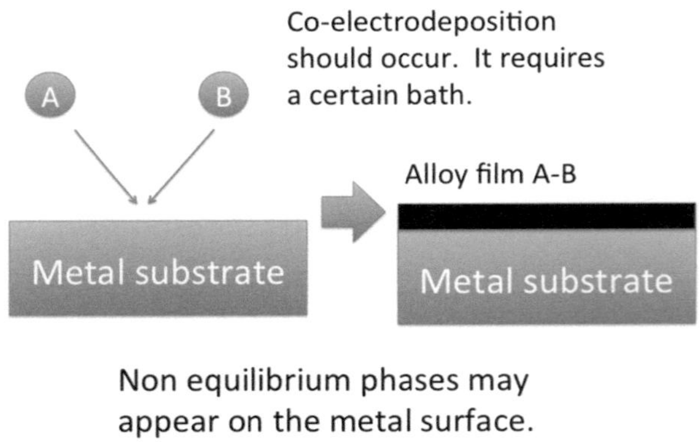

By using electroless plating, we can produce a substitute coating for hexavalent chromium ion free plating. Concretely, Ni-P, Ni-B, Ni-P-B and such alloy films have been investigated and produced so far. These electroless nickel alloy films would be available by choosing appropriate bath components.

On the other hand, the electroplated nickel alloy coating can be used as the substitute for hexavalent chromium ion free coating. Concretely, Ni-P, Ni-W, Ni-B, Ni-Mo, Ni-Sn, Co-W, and Co-Sn can be mentioned. Basically, the alloy film can be made by co-electrodeposition. Figure 6 shows the basic concept for co-electrodeposition schematically. Since the co-electrodeposition is basically the fundamental mechanism, the bath solution should be available to achieve it. It might sometimes force us to use certain hazardous chemicals to make the bath solution. In such a case, some appropriate effluent processing or the adsorption of some, or a closed system might be needed. From the standpoint, it might cause other environmental problems. Therefore, one needs to be cautious in regards to this work.

To avoid such complicated additional problems, one of the authors proposed a completely different process to produce alloy films [4]. It is actually a combination of heat treatment and coating (including plating) and named Heating Stacked Single Layers Process (HSSL process) by the author and his colleagues. It was already explained as an example for the surface finishing assured by HACCP in the previous chapter (chapter "What Is Environmentally Friendly Surface Finishing?"). However, it was originally developed as a substitute for hexavalent chromium ion free coating. As an example of an application to hexavalent chromium ion free coating, we mention the production process of Nickel-tin alloy film.

The process is composed of multi-steps. At the first stage, a film would be made on a steel surface for example (The substrate should not be restricted only to steel. It is just an example at this point.). The coating process would not be restricted only to electroplating nor electroless plating. Any coating method would be available. From the viewpoint, the surface finisher could generally have broad choices at this point. Then the second film would be made on the first coating film. The film would be available also by any coating processes. At this point, the multi-layer

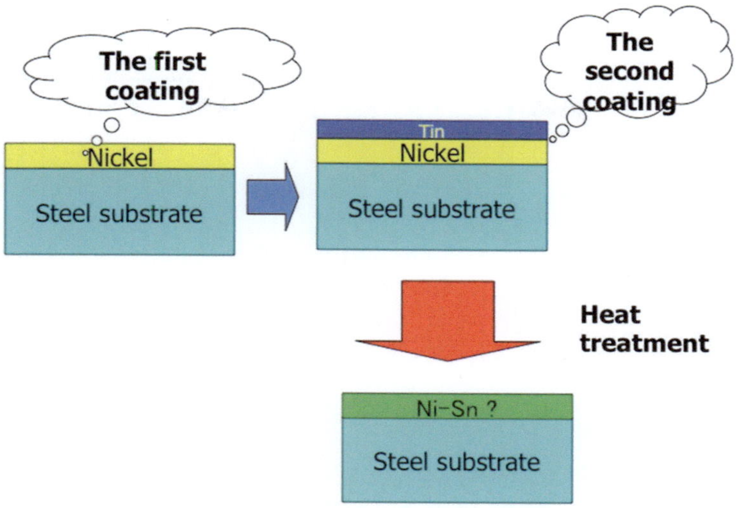

Fig. 7 The basic concept for the HSSL process

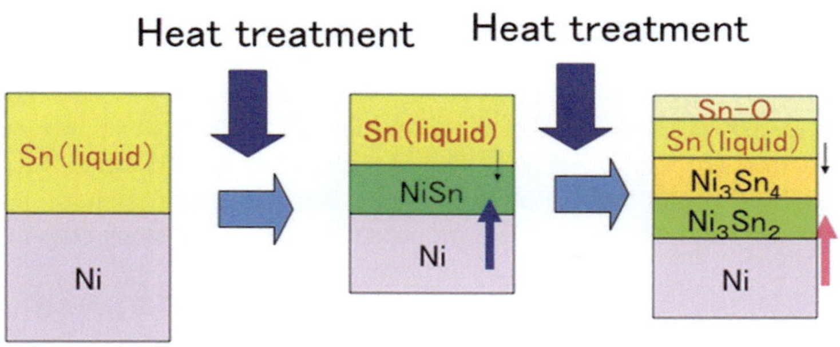

Fig. 8 The schematic illustration for the alloy film formation by the HSSL process

(stacked single layers) would be formed. Then the heat would be applied to the stacked single layers to form the alloy film for nickel and tin (Fig. 7).

Figure 8 shows the schematic illustration for what will be going on during the heating. Since the melting point of nickel differs from that of tin (nickel is a high melting point material, while tin has a low melting point.), the heat could make tin melt first. The situation would make nickel diffuse into the tin phase more easily and diffusion would occur. Various phases between nickel and tin could be formed at relatively low temperatures. However, all of them are stable phases found in the phase diagram of nickel and tin. On the other hand, the phases formed by co-electrodeposition might be unstable thermodynamically, which might lead to instability during their use. The author and colleagues already confirmed that the coating film has high wear resistance, corrosion resistance, and a certain unique

Fig. 9 The general concept for composite films in this book

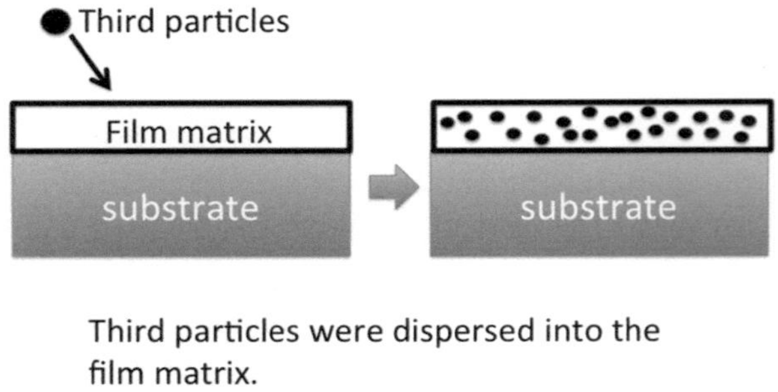

color tone. This process could be applied to many other two alloy systems for the same purpose.

3 Composite Coatings as a Substitute for Hexavalent Chromium Plating

If the composite coating would be defined as the combination film of two or more substances, it would be incorporated into the category of an alloy film. In this chapter, however, we would like to restrict the composite film as a dispersion type film in a matrix as shown in Fig. 9 schematically [5]. As an example, we have already mentioned (in chapter "What Is Environmentally Friendly Surface Finishing?"), a silane based film with some dispersed nano-metallic powders. However, other systems have been proposed for the hexavalent chromium ion free process.

Concretely, there are some cases with the utilization of SiC powder or teflon powder. They are dispersed into a plating film to improve lubricant ability, etc. More possibilities will be designed and developed in the future.

4 Trivalent Processes [6]

Trivalent chromium plating already has a relatively long history. Some processes have been available commercially. Usually, chromium sulfate or chromium chloride is used as the main component in the bath. A graphite electrode or composite one is used. Some additives are added to the bath to prevent the oxidation of trivalent chromium at the anode. In some cases, the lead anode is used. It is surrounded with walls as a compartment filled with sulfuric acid. It is well-known as a shielded anode. The arrangement is also made to prevent the oxidation of trivalent chromium at the anode. Such a trivalent chromium plating process has already achieved similar characteristics and performances to those of the conventional hexavalent chromium plating process from the viewpoint of hardness, etc. under a similar thickness of the film.

The trivalent chromium plating process has a very strong and remarkable advantage over the hexavalent one from the standpoint of environmental friendliness. Workers could deal with the process under healthy conditions. As for the plating performance, the trivalent chromium process has higher cathode efficiency and throwing power. Therefore, we can say in general that the trivalent chromium plating process is more energy efficient.

5 Future Scope

In this chapter, we focused on substitute films for hexavalent chromium ion free plating and explained the general sketch for the topic at this point. The substituted coating films were classified into three categories – electroless/electroplated alloy films, composite coating, and trivalent chromium plating. The processes should not be restricted to the cases that this chapter mentioned. New ideas will be proposed and developed in the near future. And the same ideas and principles will be applied to other plating and coating systems.

References

1. U.S Department of Labor, Occupational Safety and Health Administration, Hexavalent Chromium, Paperback, CreateSpace Independent Publishing Platform, March 15, 2014, the USA
2. T. Morikawa, T. Nakade, Y. Sato, M. Yokoi (1999) Substitution for chromium plating. Rust Prevent Control 43(11)
3. Nawafune H, Matsuoka M (2014) Modern electroless plating. Nikkan Kogyo, Co.
4. Kanematsu H (2013) Alloy plating by heating stacked single layers and the possibilities of its application in the future. Prod Finish 78(3):1–14
5. Makhlouf ASH, Scharnweber D (2015) Handbook of nanoceramic and nanocomposite coatings and materials. Elsevier, Amsterdam
6. Wikipedia (2015) Chrome plating. https://en.wikipedia.org/wiki/Chrome_plating

Change from Metals to Nonmetals

Dana M. Barry and Hideyuki Kanematsu

Abstract Treatments to the surfaces of metals as well as other items are very important in most industries. They are used to enhance substrates' appearances and to improve their functions such as reflectivity, hardness, and corrosion resistance. This chapter describes some surface finishing processes. It discusses the change from metals to nonmetal substrates. Also the topics of resins, composites, and plastics are introduced.

1 Introduction

Industrial treatments are made to material surfaces for various reasons. One is to enhance the appearance of a finished item to attract potential buyers. Another is to increase the safety of products. Other purposes for surface treatments are to improve corrosion resistance, wear resistance, mechanical functions, and electrical properties. This chapter describes some surface finishing processes. It discusses the change from metals to nonmetal substrates. Also the topics of resins, composites, and plastics are introduced.

The trend of switching from metals to nonmetals is due in part to three main concerns: costs, environmental regulations, and products. If the prices of health care, operating systems, metals, chemicals, etc. increase and labor costs remain low in competitive countries, then US finishers' profits will decrease. As a response to these potential problems, finishers have reduced the workforce to cut costs, partnered with foreign-based companies to sell products in those markets, and

D.M. Barry (✉)
Clarkson's Departments of Chemical & Biomolecular Engineering and Electrical & Computer Engineering, Clarkson University, Potsdam, NY, USA

Center for Advanced Materials Processing (CAMP), Clarkson University, Potsdam, NY, USA
e-mail: dmbarry@clarkson.edu

H. Kanematsu
National Institute of Technology, Suzuka College, Suzuka, Japan
e-mail: Kanemats@mse.suzuka-ct.ac.jp

H. Kanematsu, D.M. Barry (eds.), *Corrosion Control and Surface Finishing*,
DOI 10.1007/978-4-431-55957-3_24

focused on the finishing of components that are difficult to prepare globally. Also some finishing companies have left the industry.

Another industrial concern is environmental regulations. Over the years domestic and international environmental regulations have become stricter. As previously discussed (chapter "Regulations by the Environmental Protection Agency in the US"), the US Environmental Protection Agency (EPA) was established to write and enforce regulations for laws such as the Clean Air Act, the Clean Water Act, the Toxic Substances Control Act, and the Resource Conservation and Recovery Act (which relates to the disposal of solid and hazardous waste and impacts most industry sectors in the US). The EPA continues to update and write additional regulations that take into account advancements in technology and shifts in priorities. Therefore, industries and other organizations must abide by these rules or face punishments such as stiff fines or even the total ban of their products. The US regulations mostly focus on managing manufacturing processes and limiting the amounts of specific contaminants (examples: heavy metals) released into the air, water, soil, etc.

Regulations in the countries of Europe differ from those of the US. They are based on life-cycle thinking (taking into account the whole system) and producer responsibility [1]. Life cycle thinking implies that every step involved to produce the final product can impact humans and the environment and therefore must be taken into account. Some of these steps include material selection, product design, manufacturing process, etc. Also the original producer of the material is ultimately responsible for the chain of impacts that result. Generalizing from these statements, one can conclude that products exported to European markets may require new materials and/or new technologies in order to comply with Europe's regulations. A possible example is a product from the plating industry which has existed for many years. This surface finishing method uses metals, chemicals, chemical baths, etc. and generates chemical waste that must be disposed of.

The final products from industries are of concern and importance because they must be acceptable to domestic and international customers. Continuing to be able to please customers is a real industrial challenge. This is because products and their finishing processes are changing from metals to nonmetals as a result of stricter environmental regulations, increased concerns about global warming, a desire for a greener environment, etc.

The next section introduces resins which are used in the nonmetals (composites and plastics) that have replaced finished metals in some products. Following this section on resins, the chapter continues with a description of composites and plastics and their future trends.

2 Resins

Resins are most often considered to be hydrocarbons secreted from many plants, especially coniferous trees. See Fig. 1 [2]. They are viscous liquids containing volatile fluid terpenes and some dissolved nonvolatile solids that make them thick

Fig. 1 Drops of resin

and sticky. An example of such a liquid is amber, a hard fossilized resin from ancient trees.

Natural plant resins are valuable for their chemical properties and uses like the production of varnishes (transparent, hard, protective films for the finishing of wood and other materials). Varnishes are generally a combination of a resin, a drying oil, and a solvent. After resin varnishes are applied to a surface, they usually become hard as soon as the solvent evaporates. Shellac is a varnish that uses a natural resin from insects. It is mostly applied to surfaces that remain indoors. The shellac resin is secreted by a female lac insect called *Kerria lacca*, found in the forests of Thailand [3]. Lacquer, a coating made from the sap of the lacquer tree *Rhus verniciflura*, is typically more durable than shellac [4].

A modern meaning of the word "resin" applies to most any component of a liquid that will set into a hard protective finish. Many of these components are synthetic resins. Synthetic resins have an important property that is similar to the natural resins. They both are able to harden when heat and/or pressure are applied. However, synthetic resins are made using chemical polymerization. This process results in polymers that are more stable and homogeneous than natural occurring resin. Polymers are long chain-like molecules that contain many simple repeating units. Acrylic is a synthetic resin used in latex or water-based paints [5]. This resin forms the paint's binder which allows the coating to hold its color and last longer. Various types of paint are used to provide protective coatings for metals, wood, etc.

Generally synthetic resins are grouped into two categories: thermoplastic synthetic resins and thermosetting synthetic resins [6]. Thermoplastic resins can be softened and reformed by applying heat and pressure even after being set in a certain shape. Examples of this type of synthetic resin include polyethylene and polypropylene. These resins (polymers) are used in packaging films, molded containers, and for automotive components.

Thermosetting synthetic resins are generally softened during the forming stage and then set into their final shape. Thermosetting materials are formed during a chemical reaction where a resin and a hardener (catalyst) are mixed to form a hard product by undergoing a non-reversible chemical reaction. Epoxy is this type of resin. It is used to make hard, chemical resistant coatings and adhesives. It has good chip resistance too [7].

A specific example for epoxy is fusion bonded epoxy (FBE) coating [8]. It is widely used to protect steel pipes. In this case, the resin and hardener components of a dry powder stock of FBE do not react until the coating application temperature is reached. Then the powder contents melt and form a liquid which flows onto a steel surface where it becomes a solid coating by chemical crosslinking and the use of heat. This is known as fusion bonding and the chemical crosslinking reaction is not reversible.

Polyurethanes are mostly thermosetting polymers [9]. They are used to make foam products (example: seats), wheels, tires, hoses, synthetic fibers, etc. and for surface coatings and sealants. Also their foams have some resistance to the damaging effects of visible light.

3 Composites: Description and Future Trends

A composite is a material made from two or more materials that have different physical or chemical properties. However when these items are combined, they form a composite with different characteristics from the individual components. These components remain separate and distinguishable in the final structure. See Fig. 2 [10]. Concrete is a simple composite material that is made with loose stones and cement. It is more robust than cement.

To be more specific, composites are made of constituent materials that are categorized as matrix and reinforcement. The matrix material surrounds and supports the reinforcement materials by maintaining their positions. On the other hand, the reinforcements provide mechanical and physical properties to enhance those of the matrix. As for the previous example of concrete, the stones are the reinforcement while the cement serves as the matrix. To further strengthen the concrete and prevent stretching (due to a tensile load), steel bars can be added to it. This modified form (of concrete) is known as reinforced concrete.

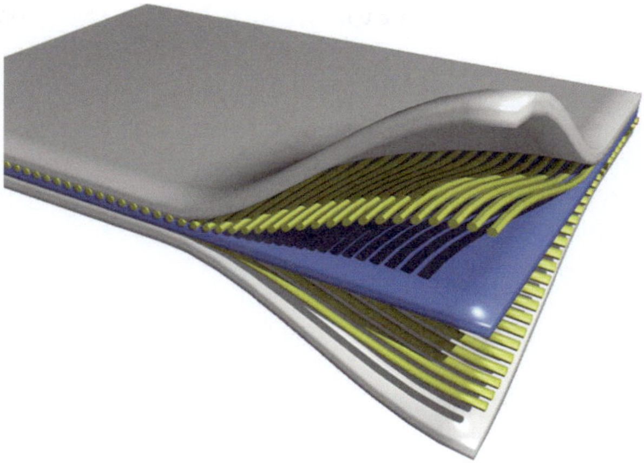

Fig. 2 Simple illustration of a composite displaying several components

Other sources of reinforcement are fibers like those of carbon and metal. Fibers can add rigidity and help prevent crack propagation. A common composite is fiberglass, which consists of small glass fibers embedded within a polymeric material that is typically an epoxy or polyester [11]. Epoxy resin is used as a structural matrix in the aerospace industry. It is also used as glue.

The fabrication of composites involves wetting, mixing, and saturating the reinforcement with matrix. Then heat or a chemical reaction is used to bind the matrix together into a rigid structure.

Composites have replaced finished metals in many products. They typically consist of fibrous reinforcement such as glass and carbon or other materials that are encapsulated in a hardened matrix of a resin system. Carbon composite is used in launch vehicles and heat shields for the re-entry phase of spacecraft. Composites have advantages over finished metals. They can be stronger and stiffer than metals per unit weight, highly corrosion resistant, electrically insulating, and much more.

Composite materials have been used extensively in the military aircraft B-2 “stealth” bomber [12]. Most of this aircraft’s body is made from composite materials designed to absorb radio waves. Currently composites are being used in aircraft and boat propellers, helicopter rotors, and in wind turbines. They are also replacing the use of finished metals in materials applications for missiles, spacecraft, automobile parts, golf clubs, and bicycles.

The trend shows that new applications will be developed for the use of composites. They are lightweight compared to metals and have excellent properties that can be tailored for specific functions. There is ongoing research to develop composite chassis and body panels for automobiles and to use composite materials for airplane fuselages and wings, etc.

4 Plastics: Description and Future Trends

Plastics are generally organic polymers of high molecular mass. They are usually synthetic and derived from petrochemicals. Also they are malleable and can be molded into solid objects of various shapes. Most plastics contain other organic or inorganic compounds that are blended in. These include fillers which improve performance and may reduce production costs. One such additive might be a fire retardant to lower the flammability of the material. Another could be a colorant to enhance its appearance. Figure 3 displays a plastic vase [13].

Some plastics are biodegradable while others must be recycled. Plastics can be either thermoplastics or thermosetting polymers. The thermoplastics are plastics, like polyethylene, that can be heated and still be molded over again to take on different shapes. On the other hand, the thermosetting plastics undergo a chemical reaction that is not reversible. Therefore, they solidify into one final shape.

Properties (examples: hardness and resistance to heat) of plastics are determined by the organic chemistry of the polymers that they contain. The common plastics polyethylene, poly(vinyl chloride), and polypropylene are presented.

Fig. 3 A plastic vase

4.1 Polyethylene

Polyethylene (PE), also called polyethene, is the most common plastic and mainly used for plastic bags and bottles. It is a thermoplastic, with long hydrocarbon chains, made from the monomer ethylene (ethene). Many forms of it exist such as low density polyethylene (LDPE) and high density polyethylene (HDPE). Most of them have excellent chemical resistance to acids and bases, etc. and generally have the chemical formula $(C_2H_4)_n$ [14]. Figure 4 is a model of polyethylene [15].

4.2 Polyvinyl Chloride

Polyvinyl chloride (PVC), also called poly(vinyl chloride) and polychloroethylene, is a widely produced synthetic polymer. It is used in pipe construction, for doors, windows, etc. and has the chemical formula $(C_2H_3Cl)_n$. This polymer can be made softer and more flexible by the addition of plasticizers. In its softer form, the plastic can be used in plumbing and imitation leather. Figure 5 shows a model of polyvinyl chloride [16].

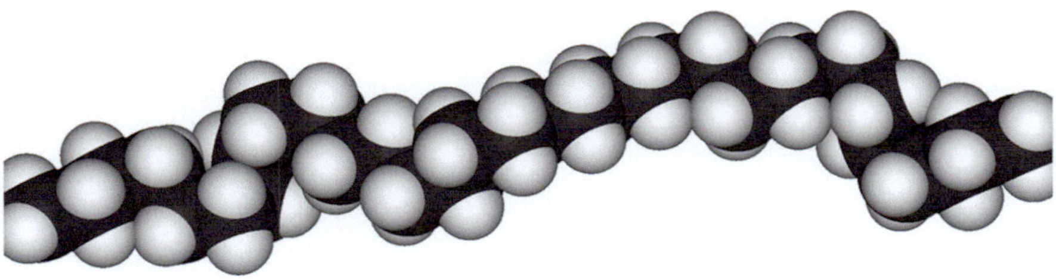

Fig. 4 Model of polyethylene

Fig. 5 Model of polyvinyl chloride

4.3 Polypropylene

Polypropylene (PP), also called polypropene, is a thermoplastic polymer that has a variety of uses. Some include applications for laboratory equipment, automotive components, ropes, thermal underwear, reusable containers, and plastic hinges. Polypropylene is normally tough and flexible and has good resistance to fatigue. It has the chemical formula $(C_3H_6)_n$. Most polypropylene used commercially is isotactic and has some level of crystallinity. Both isotactic and syndiotactic forms can be made using a Ziegler-Natta catalyst [17]. Isotactic polypropylene has all the methyl groups (CH_3 group) positioned on the same side of the polymer chain, while the syndiotactic form has the methyl groups located (at alternating positions) on both sides of the polymer chain. Figure 6 displays a ball and stick model of syndiotactic polypropylene [18].

In recent years, plastics have been used to replace some metal products. They have lighter weight, low corrosion, and are electrically non-conductive. Plastics are easy to color and offer design flexibility. Metal components have been replaced by amorphous plastics like polyvinyl chloride and by crystalline plastics such as polypropylene and polyethylene [12]. Processes for plating on plastics to give a bright metallic surface (example: automobile grill) have been developed too.

There is great potential for plastics to continue to replace metals in many products. Developments in plastic engineering should result in improved material properties which could lead to a wider use of plastics. One area of interest is the development of plastics to conduct heat, possibly by adding thermally conductive additives. Also ongoing research is being carried out to expand the use of plastics in the automotive industry.

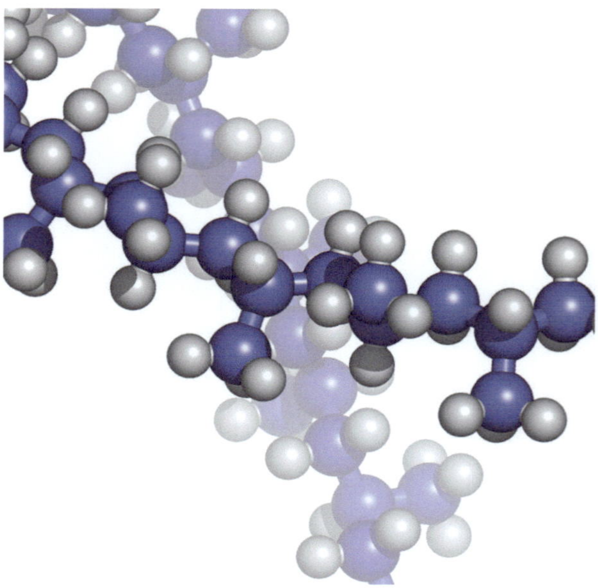

Fig. 6 Ball and stick model of syndiotactic polypropylene

5 Conclusion

Treatments to the surfaces of metals as well as other items are very important in most industries. They are used to enhance substrates' appearances and to improve their functions such as reflectivity, hardness, and corrosion resistance. This chapter discussed the change from metals to nonmetal substrates. This trend is due in part to high prices involved with the use of metals, stricter environmental regulations both domestically and internationally, increased concerns about global warming, a desire for a greener environment, and more.

The topics of resins, composites, and plastics were also introduced. These items have replaced finished metals in some products. The trend shows that new applications will be developed for them, especially in the automotive and aerospace industries.

References

1. Chalmer PD (2008) The future of finishing. Published by the National Center for Manufacturing Science (NCMS), 23. http://www.google.com/url?sa=t&rct=j&q=&esrc=s&frm=1&source=web&cd=1&ved=0CCAQFjAAahUKEwjS7P_7xP7GAhVCcj4KHX18ABs&url=http%3A%2F%2Fwww.nmfrc.org%2Ffutfinrev2.doc&ei=G9W3VZLzEsLk-QH9-IHYAQ&usg=AFQjCNHrhkAJYlmcmSP_yWd8tSYjXzIcFg&bvm=bv.98717601,d.cWw
2. Sharma A (Rgraphics) (2012) File: Grate Giver. CC – BY – SA 3.0. Search Category: Resins. https://commons.wikimedia.org/wiki/File:Grate_Giver.jpg
3. https://en.wikipedia.org/wiki/varnish
4. https://en.wikipedia.org/wiki/Toxicodendron_vernicifluum
5. https://en.wikipedia.org/wiki/Acrylic_resin
6. http://www.wisegeek.com/what-is-synthetic-resin.htm
7. (2011) Basics on coatings chemistry: what the generic types mean practically. Coatings manual HRSD, Appendix C. http://www.hrsd.com/pdf/Coatings%20Manual/2011/APPENDIX%20C.pdf
8. http://en.wikipedia.org/wiki/Fusion_bonded_epoxy_coating
9. https://en.wikipedia.org/wiki/Polyurethane
10. PerOX (2009) File: Composite 3d.png. This work is in the public domain. https://commons.wikimedia.org/wiki/File:Composite_3d.png. https://en.wikipedia.org/wiki/Composite_material
11. https://en.wikipedia.org/wiki/Composite_material
12. Chalmer PD (2008) The future of finishing. Published by the National Center for Manufacturing Science (NCMS), 61. http://www.google.com/url?sa=t&rct=j&q=&esrc=s&frm=1&source=web&cd=1&ved=0CCAQFjAAahUKEwjS7P_7xP7GAhVCcj4KHX18ABs&url=http%3A%2F%2Fwww.nmfrc.org%2Ffutfinrev2.doc&ei=G9W3VZLzEsLk-QH9-IHYAQ&usg=AFQjCNHrhkAJYlmcmSP_yWd8tSYjXzIcFg&bvm=bv.98717601,d.cWw
13. Photo by Dana M. Barry
14. https://en.wikipedia.org/wiki/polyethylene
15. User: Benjah-bmm27 (talk/contributes) (2006) File: polyethylene -3D-vdW.Png. This work is in the public domain. https://commons.wikimedia.org/wiki/File:Polyethylene-3D-vdW.png. https://en.wikipedia.org/wiki/polyethylene

16. User: Benjah-bmm27 (talk/contributes) (2006) File: PVC-3D-vdW.png. This work is in the public domain. https://commons.wikimedia.org/wiki/File:PVC-3D-vdW.png. https://en.wikipedia.org/wiki/Polyvinyl_chloride
17. https://en.wikipedia.org/wiki/polypropylene
18. User: Dubaj-commonswiki (talk/contributes) (2006) File: Syndiotactic polypropene.png. This work is in the public domain. https://commons.wikimedia.org/wiki/File:Syndiotactic_polypropene.png. https://en.wikipedia.org/wiki/polypropylene

Process Changes

Hideyuki Kanematsu and Dana M. Barry

Abstract As described in detail already, environmental friendliness is needed for sustainable development. There are two basic ways to make the surface finishing process more environmentally friendly. One of them is the substitution of hazardous materials with harmless substances. The other one is the utilization of a closed system for surface finishing/coating. From the viewpoint, physical vapor deposition (PVD) and chemical vapor deposition (CVD) seem to be the most favorable processes, since both have already been established. In this chapter, we describe both processes and discuss the effectiveness and possibility of changes from the conventional plating process to a dry process.

1 Introduction

Environmental friendliness is needed not only for products, but also for processes. Hazardous materials used or produced in the production process should be transformed into harmless substances or retained in a closed system (Fig. 1). The latter is a topic and problem-solving issue for wastewater treatment. The latter is composed of many different processes. Several of them are physical vapor deposition (PVD) and chemical vapor deposition (CVD). Both are called dry coating processes. In this chapter, the dry processes such as PVD and CVD will be introduced and explained as environmentally friendly coating processes.

H. Kanematsu (✉)
National Institute of Technology, Suzuka College, Suzuka, Japan
e-mail: kanemats@mse.suzuka-ct.ac.jp

D.M. Barry
Clarkson's Departments of Chemical & Biomolecular Engineering and Electrical & Computer Engineering, Clarkson University, Potsdam, NY, USA

Center for Advanced Materials Processing (CAMP), Clarkson University, Potsdam, NY, USA
e-mail: dmbarry@clarkson.edu

H. Kanematsu, D.M. Barry (eds.), *Corrosion Control and Surface Finishing*,
DOI 10.1007/978-4-431-55957-3_25

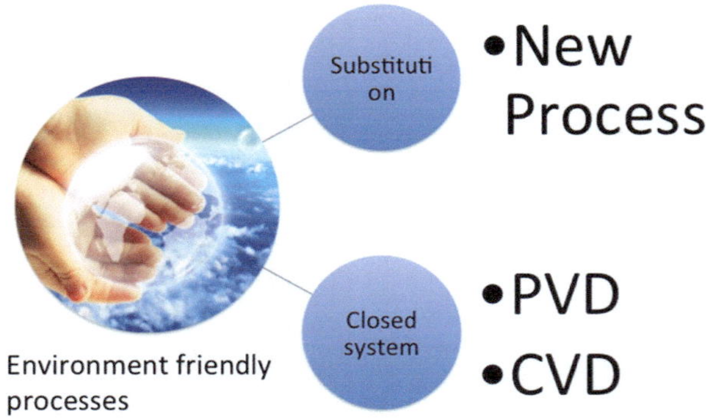

Fig. 1 Two possibilities of process changes for surface finishing

2 Vacuum Technology

Both PVD and CVD belong to vacuum technology. So what is a vacuum?

Usually, it means the space without any substances. On the other hand, when we come to deal with the system from the viewpoint of surface finishing technology, it can be defined as the system whose pressure is lower than atmospheric pressure. Therefore, it includes a very broad range of states from diminished pressure to an ultrahigh vacuum. Physically speaking, the mean free path of gas molecules increases with the degree of a vacuum. The number of collisions among gas molecules is very low in a high vacuum state.

How can we make a vacuum state in a system? Usually, we need vacuum pumps for this purpose. There are many kinds of vacuum pumps and the performance/ characteristics differ from pump to pump. Therefore, the appropriate pump would be fixed, according to the required degree of vacuum for a specific purpose.

Table 1 shows the kinds of pumps that could be used for various degrees of vacuum. The corresponding vacuum gauges are mentioned in the table.

For a relatively low vacuum (down to 0.1 Pa), oil-sealed rotary vacuum pumps are used. To obtain the vacuum of $0.1–10^{-3}$ Pa, the combination of a diffusion pump and the rotary pump is used. To accelerate the pumping rate, booster pumps may be effective. For the super-high vacuum (lower than 10^{-6} Pa), the turbo-molecular pump is often used. As for the vacuum gauge, a Geissler tube, Pirani gauge (PG), etc., are often used for a medium vacuum, while mercury manometers are used for a low vacuum. For a high and super-high vacuum, ionization gauges are used.

Table 1 Vacuum pumps and gauges often used in dry processes

	The degree of vacuum	
Super-high vacuum (<10^{-8} Pa)	High vacuum (10^{-1}–10^{-3} Pa)	Low vacuum (<10^{-1} Pa)
	Pumps to realize vacuum	
Turbo-molecular pump	+ Diffusion pump	Oil-sealed rotary
	Booster pump	Vacuum pump
	Vacuum gauge	
Ionization gauge	Geissler tube	Mercury manometer
	Pirani gauge	

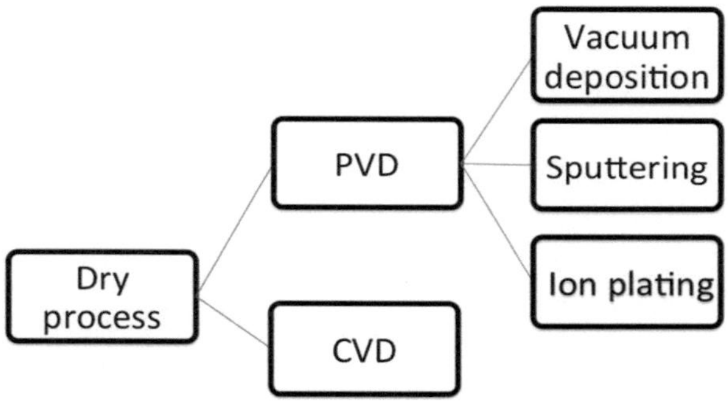

Fig. 2 Main dry process types for practical applications

3 Dry Processes and Their Characteristics

The deposition process of various materials from gas phases has been utilized for surface finishing (coating) and it is called a dry process. As shown in Fig. 2, it is classified into two main categories: physical vapor deposition (PVD) and chemical vapor deposition (CVD), even though there are many kinds of dry processes such as spray coating, ion implantation, ion carburization, ion nitriding, laser treatment and various other plasma treatments, etc. While the wet process such as electroplating has had a major problem about effluent processing from the viewpoint of environmental friendliness, the dry process is generally considered to be environmentally friendly. The process typically has the following characteristics [1].

#1. It would be available for any substrates: metals, inorganic materials and organic materials.
#2. Compounds can be coated easily.
#3. New chemical synthesis of matter is possible in a non-equilibrium state and the excitation state.
#4. High adhesion.
#5. Controllability of coating thickness at the atomic or molecular levels.
#6. Easy formation of multi-layers.
#7. Availability of gradient coating.

#8. Changeability of substrate temperatures in a broad range.
#9. Easy exhaust gas treatments.
#10. Beneficial use of basic ingredients.

As described above, even though there are many types of dry processes, universalistic dry processes are restricted to certain ones at this point: vacuum deposition, sputtering and ion plating (examples of the PVD process) and many kinds of CVD processes. General descriptions of them are provided.

4 Vapor Deposition, Sputtering, and Ion Plating

As shown in Fig. 3, vapor (vacuum) deposition can be defined as the process where a material's source is heated and then its evaporated vapor condenses as a film coating onto a substrate. Table 2 shows the metals, their material source for vapor deposition, melting points, etc. that are often used in practice.

Deposition sources are put into boat type or basket type containers or attached to a filament made of W, Mo, or Ta. The source is generally heated by resistance heating, an electron beam, high frequency, laser, etc. Evaporated metal gases are deposited on substrates. However, if the degree of vacuum is low, then the evaporated gases would be interrupted very often and the film formation would be insufficient. Plus the residual gases might be mixed into the film. Generally, the degree of vacuum for the vapor deposition would be lower than 10^{-2} Pa. Typically, the film formation rate is proportional to the evaporation rate and decreases in proportion to the square of the distance between the evaporation source and the substrate.

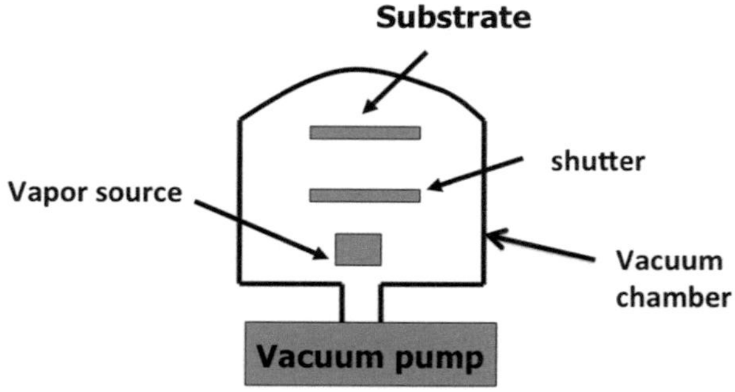

Fig. 3 Schematic illustration of the apparatus for vapor deposition

Table 2 Metals used for vapor deposition in practice and their process temperatures [1]

Metals	Melting points (K)	Evaporating temperature (K)
Ag	1234	1301
Al	933	1421
As	1087	546
Au	1335	1676
B	2573	2409
Be	1551	1485
Bi	544	922
C	3923	2874
Co	1768	1793
Cr	2163	1637
Cu	1356	1537
Fe	1808	1698
Ga	303	1315
In	430	1185
Mg	924	710
Mn	1517	1247
Mo	2883	2797
Nb	2741	2980
Ni	1726	1803
Pb	601	995
Pt	2042	2322
S	388	377
Sb	903	806
Si	1683	1803
Sn	505	1521
Sr	1042	806
Ta	3269	3329
Ti	1948	2021
U	1405	2206
V	2163	2133
W	3683	3502
Zn	692	618
Zr	2125	2662

5 Sputtering Process

The sputtering phenomenon can be described as follows. When high energy particles such as positive argon ions collide with solids, then surface atoms, molecules, or clusters of atoms are generally ejected from their surfaces. This phenomenon is called “sputtering”. Utilizing this phenomenon, films can be formed on materials’ surfaces. Figures 4 and 5 show some sputtering apparatus schematically. Figure 4 shows two types of sputtering: (a) is called DC bipolar sputtering.

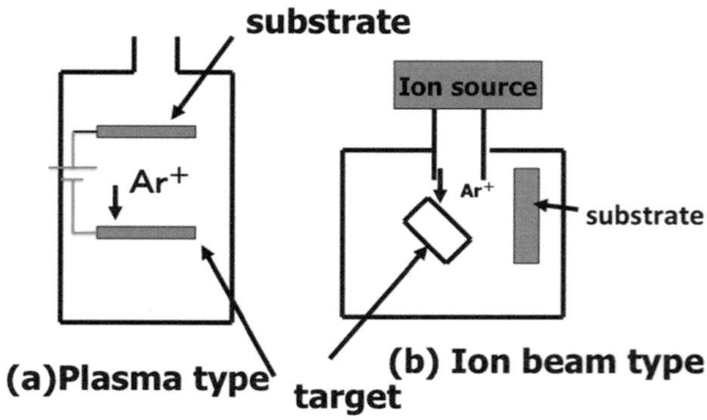

Fig. 4 Plasma sputtering and ion beam types

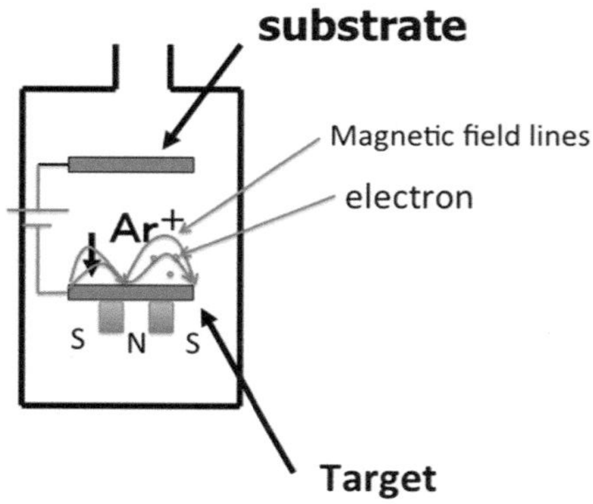

Fig. 5 Schematic illustration for magnetron sputtering

First of all, the vacuum chamber is evacuated to form a high degree of vacuum. Then argon gas is introduced up to 10^{-1}–10^{-2} Pa. DC high voltage current (1–1.5 kV) is applied to the electrodes and a glow discharge occurs as a result. The discharge produces positive argon ions that are accelerated toward the target (cathode). When the accelerated ions collide with the surface, the surface atoms are ejected and deposit on the substrate. And finally the thin film forms. The metal source is evaporated by resistor heating for (a). On the other hand, (b) is called an ion beam sputtering process. An ion produced in the separated cell is irradiated on the target to produce a sputtering process. Figure 5 shows magnetron sputtering. In this case magnets are placed behind the target so that a magnetic field forms on the target's surface. The magnetic field makes electrons, captured around the surface, follow helical paths around magnetic field lines. This arrangement increases the number of collisions with argon atoms which creates extra argon ions. Therefore, magnetron sputtering has a higher film formation rate than that of the usual DC sputtering.

Being compared with the vapor deposition, the sputtering process has the following characteristics. #1–#5 are general merits, while #6 and #7 are demerits [1].

#1. Most of high melting point materials including metals, alloys, compounds, organic materials, etc. can be formed on materials as thin films.
#2. The composition of alloy films can be controlled easily.
#3. Compound films are easily formed by introduced gases.
#4. High adhesion force.
#5. Good reproducibility.
#6. Relatively low film formation rate.
#7. Argon gas and reactive gases in the vacuum chamber can easily enter the produced film.

The process is mainly used in the semiconductor industries.

6 Ion Plating

Ion plating is a representative PVD process. Figure 6 shows the schematic principle for ion plating. The material source is evaporated in the same or similar way as vapor deposition. However, gas is introduced to the chamber and a plasma state is realized by various ways. When nitrogen or acethylene is introduced into the system, their plasmas would lead to the formation of metal nitride or carbide films. The advantages of ion plating are as follows [1].

#1. An adhesion force of ion plating is stronger than that of vapor deposition.
#2. A high adhesion force can be obtained at relatively low temperatures.
#3. Film formation is available in a broader range of production conditions.
#4. Compound or alloy films can be formed by adjusting gases.
#5. Work pieces of irregular shapes can be coated pretty well, as compared with vapor deposition.

There are many types of ion plating processes and they can be classified in various ways. From the viewpoint of plasma formation, we can mention several

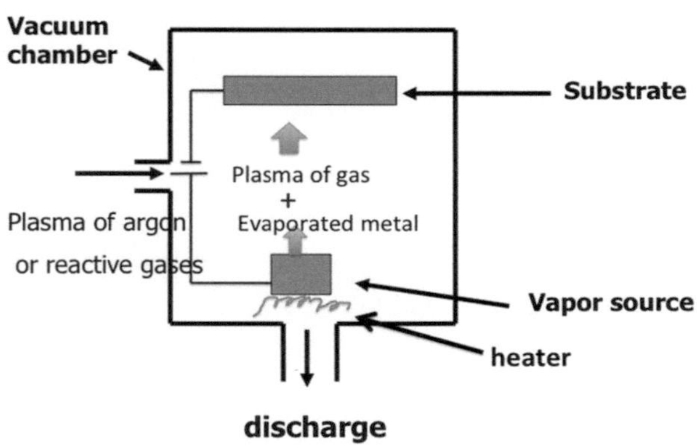

Fig. 6 Schematic illustration for a type of ion plating

types: DC discharge excitation, high frequency discharge excitation, a hollow cathode discharge, an arc discharge type, etc. Various alloy and compound films are formed by ion plating processes and used in industries for cutting instruments, mold tools, IC, automobile parts, electric equipment, glass frames, decoration items, etc.

7 CVD Process

CVD stands for chemical vapor deposition. Raw materials as gases are introduced into the system and the desired substance is formed as a coating on a substrate. Figure 7 shows the schematic figure for a CVD process. Usually the source gas or gases contain elements that would constitute the deposited films. A carrier gas is used in the system and the chemical reaction(s) occurs on the substrate to form a surface coating.

Being compared with PVD, CVD has the following advantages [1].

#1. More versatile films: metals, alloys, ceramics, etc.
#2. Not only the film-like matter, but also needle-like, fiber-like, powders, whiskers, etc. could be formed.
#3. Higher adhesion force.
#4. Simplicity of processes.
#5. Not so expensive.

On the other hand, they have the following disadvantages [1].

#1. Due to high temperatures and the use of corrosive gases, the component materials would be damaged and the life cycle would be relatively short.
#2. Deformation of work pieces and residual stresses would occur after the process.

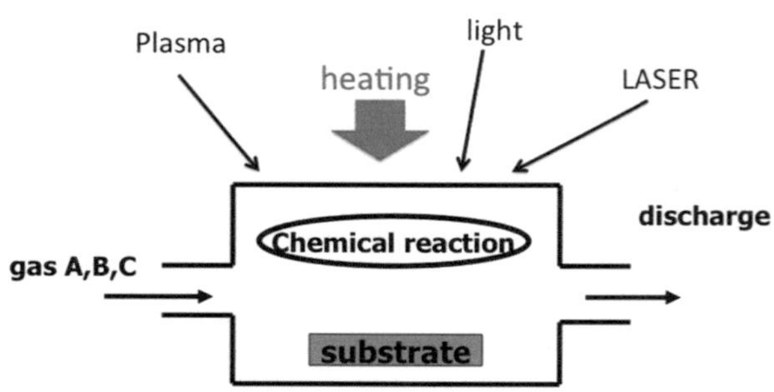

Fig. 7 Schematic illustration for the Chemical Vapor Deposition principle

8 Conclusions

In this chapter, we explained why some representative dry processes such as the closed system would be more environmentally friendly than other methods. The closed system could help us avoid effluent problems and enhance the safety of workers. The closed system might also be realized for many wet processes such as plating. However, the dry processes would provide more benefits. The costs might be higher than those of the wet processes, but the expenses are compensated by their environmental friendliness.

Reference

1. Surface treatment technology – fundamentals and Application edited by Surface Finishing Society, Nikkan Kogyo Newspaper Co. Feb. 2000 ISBN 4-526-04522-5

The Application of Corrosion Protection

Hideyuki Kanematsu and Dana M. Barry

Abstract As described in detail already, environmentally friendly technology is very important for sustainable developments in the future. Surface Finishing industries are facing a critical point, since many of them have to change their processes completely or find substituted coating substances in order to be environmentally friendly and sustainable. In this chapter we provide and describe some metallic coatings from the viewpoint of corrosion protection, since metallic ions are able to interact with human cells with possible detrimental effects.

1 Introduction

Many environmental laws have been established to regulate and control the use of metallic ions. Also some heavy metals have been mentioned as harmful elements. However a discussion, about the harmful effects of heavy metal elements, could continue forever. Apart from the long lasting discussion, surface finishing industries in the country or community should obey the regulations and laws, and seek new ways to survive. Some conventional metals for surface finishing have been judged as toxic substances. As for metallic coatings, the toxic substances can be classified into two groups for surface finishing, as shown in Fig. 1. One of them is the toxic substance in effluent. Cyanide and arsenic can be mentioned as examples. Cyanides have been used often as chemical substances to hold metallic components as stable and easy-to-use electrolytes for plating. The cyanide has to be treated properly after

H. Kanematsu (✉)
National Institute of Technology, Suzuka College, Suzuka, Mie, Japan
e-mail: kanemats@mse.suzuka-ct.ac.jp

D.M. Barry
Clarkson's Departments of Chemical & Biomolecular Engineering and Electrical & Computer Engineering, Clarkson University, Potsdam, NY, USA

Center for Advanced Materials Processing (CAMP), Clarkson University, Potsdam, NY, USA
e-mail: dmbarry@clarkson.edu

H. Kanematsu, D.M. Barry (eds.), *Corrosion Control and Surface Finishing*,
DOI 10.1007/978-4-431-55957-3_26

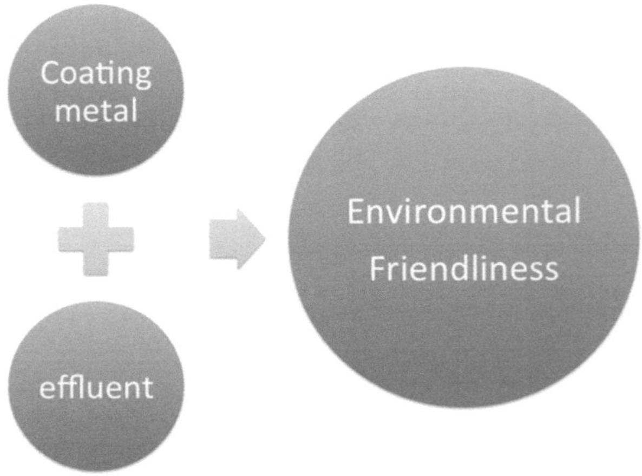

Fig. 1 Two types of toxic substances in plating processes

the plating processes. This means that the effluent should be free from all toxic cyanide.

Some organic additives for plating belong to this category too. On the other hand, the plating materials might cause environmental problems, since the metals may produce toxic effects. In this chapter toxicity (relating particularly to metal surface finishing) will be presented and countermeasures will be explained from the viewpoint of corrosion protection.

2 What Kind of Case Leads to Toxicity in Metal Surface Finishing?

In this section, we introduce some typical cases in surface finishing processes where toxic substances are involved and issued.

2.1 Chromium

First of all, we have to mention chromium. As already described in chapter "Substitution with More Friendly Elements", the toxicity of chromium is controlled by the oxidation state and solubility. Chromium itself does not have any toxicity. Generally, the toxicity of metals appears when they dissolve in aqueous solutions. It means that the toxicity is a function of solubility in aqueous environments. Chromium does not corrode so easily. It has some ionic forms. Trivalent chromium ion does not show any serious toxicity. Actually, it is sometimes needed for organisms. However, hexavalent chromium has very high toxicity. As for the solubility, both types of ions are easy to dissolve at almost the same extent. However, the absorbance characteristics for hexavalent chromium are much higher than that for trivalent chromium. Table 1 shows the potential detrimental effects of hexavalent chromium ion.

The surface finishing process where hexavalent chromium ion would be issued (used) is mostly for chromate conversion treatment. Chromate is a chemical conversion coating process. This means that the surface finishing is by chemical reactions between chemical agents and materials. The chromate conversion treatment uses a chromate bath composed of hexavalent chromium ions. Currently trivalent chromate and some topcoats are used.

Table 1 Detrimental effects of the hexavalent chromium ion [1]

Effects	Symptoms
Respiratory effects	Asthma, chronic bronchitis, chronic irritation, chronic pharyngitis, chronic rhinitis, congestion and hyperemia, polyps of the upper respiratory tract, tracheobronchitis, and ulceration of the nasal mucosa with possible septal perforation
Skin effects	Dryness, erythema, fissuring, papules, scaling, small vesicles, and swelling
Carcinogenic effects	Lung cancer, nasal cancer, sinus cancer
Renal effects	Kidney damage, glomerular injury
Hepatic effects	Elevated liver enzyme levels, hepatic failure
Gastrointestinal effects	Ulcer formation, hypertrophic gastritis, abdominal pain and vomiting before death
Cardiovascular effects	Cardiopulmonary arrest, early hypoxic changes in the myocardium, cardiogenic shock, complicated by pancreatitis and gut mucosal necrosis and hemorrhage, hypotension, ventricular arrhythmias, severe respiratory distress, and metabolic acidosis
Hematological effects	Decreased hemoglobin content and hematocrit, and increased total white blood cell counts, reticulocyte counts, and plasma hemoglobin, intravascular hemolysis, anemia and thrombocytopenia
Reproductive and developmental effects	Higher risk of spontaneous abortions, developmental toxicant in rats and mice. Adverse developmental effects in animals include greater incidence of post-implantation loss, decreased fetal body weight, reduced ossification and decreased number of live fetuses
Genotoxic and mutagenic effects	DNA damage in vitro, DNA oxidative damage from the urinary samples of the patient who ingested 2–3 g of potassium dichromate in a suicide attempt, the oxidative damage pathway in the mechanism of toxicity of chromium in occupationally exposed individuals
Other effects	Dizziness, headache, and weakness when working over the chromate tanks, erosion and discoloration of the teeth, papilloma of the oral cavity and larynx

Table 2 Lead toxicity and the concrete effects and targets [2]

Effects or targets	Concrete symptoms
Neurological effects	Damaged nervous system (The nervous system is the most sensitive)
For children	Ataxia, coma, convulsions, death, hyperirritability, stupor
For adults	Remained childhood neurological effects, including ADHD. The same neurological symptoms with children, but the threshold is higher. Lead encephalopathy, Precursors of encephalopathy, such as dullness, irritability, poor attention span, muscular tremor, and loss of memory, decreased libido, depression/mood changes, headache, diminished cognitive performance, diminished hand dexterity, diminished reaction time, diminished visual motor performance, dizziness, fatigue, forgetfulness, impaired concentration, impotence, increased nervousness, irritability, lethargy, malaise, paresthesia, reduced IQ scores, and weakness
Renal effects	Aminoaciduria, glycosuria, and hyperphosphaturia (a Fanconi-like syndrome), a toxic stress on the kidney, some chronic advanced renal disease or decrement in renal function, etc.
Hematological effects	Decreased heme biosynthesis by inhibiting *d*-aminolevulinic acid dehydratase (ALAD) and ferrochelatase activity. an increase in blood and plasma d-aminolevulinic acid (ALA) and free erythrocyte protoporphyrins, hemolytic anemia and Frank anemia
Endocrine effects	Impeded vitamin D conversion into its hormonal form, 1, -25-dihydroxyvitamin D, which is largely responsible for the maintenance of extra- and intra-cellular calcium homeostasis. impaired cell growth, maturation, and tooth and bone development, chronically high BLLs: blood lead levels. (mostly in children with BLLs >62 μg/dL) and chronic nutritional deficiency, especially with regard to calcium, phosphorous, and vitamin D
Gastrointestinal effects	Severe cramping abdominal pain
Cardiovascular (hypertension) effects	Hypertension, increased the risk for hypertensive heart disease and cerebrovascular disease as latent effects
Reproductive effects	Decreased sperm count totals and increase abnormal sperm frequencies, increased frequency of spontaneous abortions and miscarriages and stillbirths
Developmental effects	Premature births and low birth weights, congenital abnormalities, post birth effects on growth or neurological development
Other potential effects	Increased likelihood of osteoporosis (weakened bones later in life) in animals exposed to lead, impaired dental health

2.2 Lead

Lead is also clearly well-known as a toxic substance. Lead and inorganic lead compounds are generally insoluble. Therefore, they seldom cause acute poisoning. However, some organic compounds such as tetraethyl lead can move across cell membranes and accumulate in the body. In such a case, it causes many problems and symptoms as shown in Table 2.

Lead had been used for plated vessels and pipes in oil industries, since the corrosion resistance in such an environment is excellent. However, lead plating is not used nowadays due to the environmental problem. Hot Dip Galvanizing processes sometimes use lead intentionally. It might be mixed into the process inevitably because the zinc ore contains some lead. In any case, lead has been involved with zinc plating. Now, there are some test trials to replace lead with other metals. Bismuth is one of the choices.

Lead has been used as solder and for solder plating. In the past, the most frequently used solder was a tin-lead alloy. However, the movement for lead free solder has accelerated the development of substitutes. Concretely speaking, Sn-Ag-Bi-Cu, Sn-Ag-Cu, Sn-Ag, Sn-Bi, Sn-Ag-In, Sn-Ag-Bi, etc. can be mentioned.

2.3 *Cadmium*

Cadmium plating has been used for components of automobiles, aircrafts, etc. It provides high corrosion resistance (particularly in sea water), high solderability, good appearance, low risk for hydrogen embrittlement, high lubricating ability, etc. In Japan its use has already been forbidden due to its environmental harmfulness. However, there are still some countries where the production of cadmium plating is available. To produce cadmium plating a cyanide bath is often used. Also the plating often needs a chromate coating as the topcoat. Therefore, cadmium plating may cause various environmental problems from some standpoints. Usually, the alloy plating, Sn-Zn plating, has been used for the substitute plating. Around the time of World War II, the alloy plating was tentatively used very often. However, that should not be attributed to the environmental problem, but just to the lack of cadmium in those days. Nowadays, the alloy plating appeared again due to the environmental problem. For the general public, the primary source of exposure to cadmium is through our diet. Smoking may increase the amount of exposure too. Also nutritional deficiencies can increase the risk of cadmium toxicity. Generally, cadmium poisoning can lead to cancer and failure of the lungs and kidneys. Chronic cadmium exposure primarily affects the kidneys and secondarily the bones. Acute inhalation of fumes containing cadmium affects the lungs. Unfortunately, no proper treatment exists for cadmium toxicity [3]. Therefore, the development of substitutes for plating (such as Sn-Zn alloys) has been needed so far.

One of the authors proposed a new way to produce the alloy film. Figure 2 shows the schematic illustration of how to produce the alloy film.

This is actually the application of a Heat Stacked Single Layer Process to the Sn-Zn alloy system. While the conventional method utilizes the electrochemical co-deposition of tin and zinc, the Heat Stacked Single Layer Process utilizes the combination of multi-layer plating (or multi-layer coating) and heat treatment. The produced alloy films depend on the conditions of heat treatment and plating thickness.

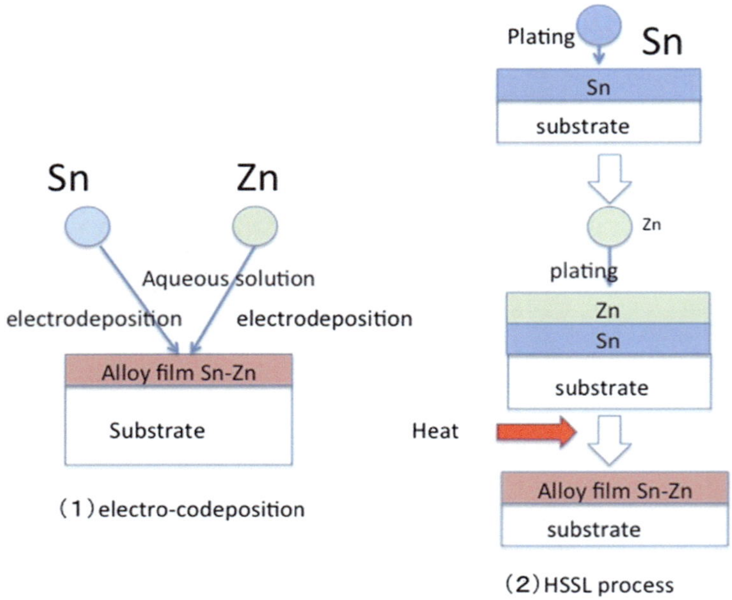

Fig. 2 HSSL process for tin and zinc alloy film [4–6]

All of the metals mentioned above were obviously classified into the toxic substance group. On the other hand, there are some problematic metals from the environmental standpoint. As plating metals, zinc and nickel can be mentioned as examples. Both plating processes have been well-known due to their high corrosion resistance and other excellent performances. Generally speaking, the toxicity is not a physical property, but is actually related to a range of exposure. Figure 3 shows the concept schematically. When the intake amount would be too large, it would lead to the appearance of toxic results. When the intake amount would be relatively small, it would lead to some detrimental result(s).

2.4 Zinc

The toxicity of zinc has been discussed so far. And it still continues to be a matter of debate. It might affect human being's Gastrointestinal (Digestive), Hematological (Blood Forming), and Respiratory (From the Nose to the Lungs) functions [7]. There are still concerns not only for human beings, but also about the detrimental effects on fish in the marine environment [8].

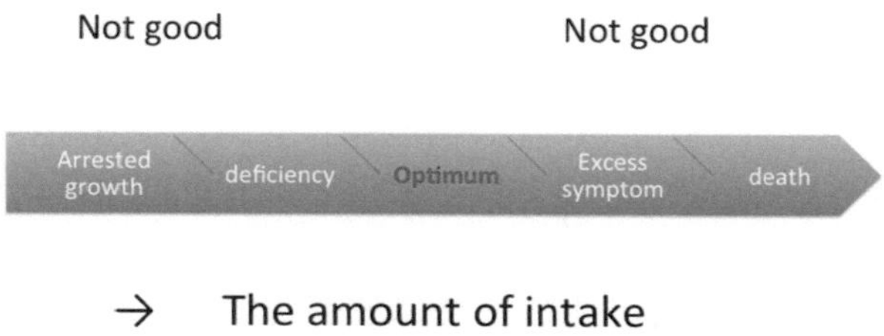

Fig. 3 The appearance of toxicity for all metals in most cases

2.5 *Nickel*

Nickel also continues to be a matter of debate in regards to its toxicity. The same concept shown in Fig. 3 could be applied to nickel plating. The most well- known characteristics of nickel, from a harmful environmental standpoint, are allergy problems. However, Cardiovascular (Heart and Blood Vessels), Dermal (Skin), Immunological (Immune System), and Respiratory (From the Nose to the Lungs) issues could also be mentioned [9].

3 The Countermeasure by Utilizing the Corrosion Characteristics

As shown in Fig. 3, the toxicity would appear in a range and should be related to the amount of intake. And the intake by organisms would be related to the solubility of the toxic substance. Therefore, the control of corrosion characteristics could adjust the appearance of toxicity. Here we mention an example of tin-zinc alloy film. When we come to think about the alloy system, it becomes toxic when the content of zinc would increase, since the toxicity of tin may be low or can be ignored. The adjustment of concentration in the alloy film can be achieved by the arrangement of heat treatment conditions and thickness for each layer, when we would carry out the HSSL process. When the concentration of zinc in the alloy film would decrease, the amount of dissolved zinc ion would decrease. In this way, the adjustment of corrosion characteristics for alloy plating films would lead to the increase of environmental friendliness.

References

1. Agency for Toxic Substances & Disease Registry (2008) Chromium toxicity, what are the physiologic effects of chromium exposure? http://www.atsdr.cdc.gov/csem/csem.asp?csem=10&po=10
2. Agency for Toxic Substances & Disease Registry (2007) Lead toxicity, what are the physiologic effects of lead exposure? http://www.atsdr.cdc.gov/csem/csem.asp?csem=7&po=10

3. Agency for Toxic Substances & Disease Registry (2008) Cadmium toxicity cover page. http://www.atsdr.cdc.gov/csem/csem.asp?csem=6&po=0
4. Kanematsu H, Kobayashi T, Oki T (2001) Tin-zinc alloy film formed from stacked single layers by heating in furnaces. The second Asian conference on heat treatment of materials, pp 241–246
5. Kanematsu H, Masuo Y, Omura H, Oki T (2001) Rapid formation of tin-zinc film from stacking layers by lasers irradiation. The second Asian conference on heat treatment of materials, pp 230–235
6. Kanematsu H, Murata K, Ogawa A, Miyano Y (2008) Sn-Zn alloy films produced by HSSL and their environmental characteristics. 2008 Sur/Fin proceeding
7. Agency for Toxic Substances & Disease Registry (2011) Zinc. http://www.atsdr.cdc.gov/substances/toxsubstance.asp?toxid=54
8. Clearwater SJ, Farag AM, Meyer JS (2002) Bioavailability and toxicity of dietborne copper and zinc to fish. Comp Biochem Physiol C Toxicol Pharmacol 132(3):269–313
9. Agency for Toxic Substances & Disease Registry (2011) Nickel. http://www.atsdr.cdc.gov/substances/toxsubstance.asp?toxid=44

Future Scope

Hideyuki Kanematsu and Dana M. Barry

Abstract For this closing book chapter, the authors have summarized their concepts (about corrosion control and surface finishing using environmentally friendly approaches) that were presented and discussed within the book. Using this information, they propose future trends and preparations (for emotional and technological/economical, etc. viewpoints) to maintain a balance for the merits of civilization and a sustainable global community by taking human and environmental protection into account. Consider this book as a useful guide and great resource for designing future innovative surface finishing/corrosion prevention technologies, etc. with a focus on environmental friendliness.

At the final stage of this book, we would like to look toward the future. In the twenty-first century, humans are facing serious environmental problems. These problems are going to control the use of surface treatments for materials and their manufacturing processes, as we have already mentioned in this book. The tendency to promote environmental protection seems to be very natural, when we come to think about the current situation. Natural resources may be exhausted in the near future. The global warming problem may be occurring seriously and drastically, based on concrete examples such as melting glaciers displayed on television and other forms of media. However, there are some who argue that these negative ideas are not real and just a plot for our future. Even though the future prediction by so-called extreme environmentalists might seem unrealistic and too negative, the tendency for sustainability and environmental protection seems to be real. Yes. We need to prepare for our future and for future generations.

H. Kanematsu (✉)
National Institute of Technology, Suzuka College, Suzuka, Mie, Japan
e-mail: kanemats@mse.suzuka-ct.ac.jp

D.M. Barry
Clarkson's Departments of Chemical & Biomolecular Engineering and Electrical & Computer Engineering, Clarkson University, Potsdam, NY, USA

Center for Advanced Materials Processing (CAMP), Clarkson University, Potsdam, NY, USA
e-mail: dmbarry@clarkson.edu

H. Kanematsu, D.M. Barry (eds.), *Corrosion Control and Surface Finishing*,
DOI 10.1007/978-4-431-55957-3_27

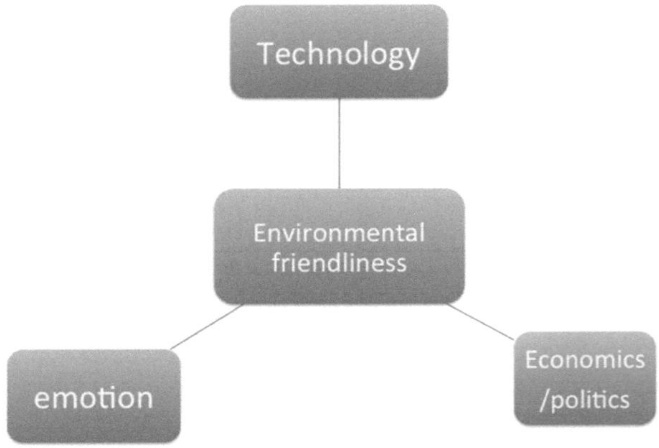

Fig. 1 Three basic aspects to change the concept for surface finishing engineering and technology

Inevitably, the environmental laws and regulations (which seem to be getting stricter) have a serious meaning in regards to the movement of surface finishing engineering and technology in the future. Three main aspects that are changing and affect the surface finishing industry are emotion, technology, and economics/politics, as shown in Fig. 1. As for emotional change, originally we began to live positively with changing environments, meaning we were able to produce something positive.

This should be called civilization. And in that point, we could say that we have been very different from other creatures in nature [1]. By developing civilizations, human beings aimed to get three advantages, as shown in Fig. 2. The emotional aspect to promote environmental friendliness is "relief" in the Fig. 2. On the other hand, two additional merits in a civilized society should be security and convenience. Both merits could be achieved by technology and/or economics with politics. As for "convenience", the weight for technology might be more important than that for the economy. When we come to think about the total balance of these merits, we realize that for decades the emotional aspect corresponding to "relief" has been weak. This means that we placed too much importance on "convenience." Therefore, we need to reestablish the total balance for the merits of human civilization. In that way, we will have to add environmental friendliness to the convenience or functionality component, so that relief and safety can be balanced with the convenience aspect. Viewed in this light, we can understand the necessity of environmental friendliness for the future movement of surface finishing technology and engineering.

As already described, we need to achieve all of the merits in a good balanced manner, so that we can continue to develop our civilization. From the viewpoint, the environmental friendliness will be an indispensable concept for all engineering disciplines. Particularly, the surface of a material is the interface of materials, components, and all other engineered and artificial substances with environments

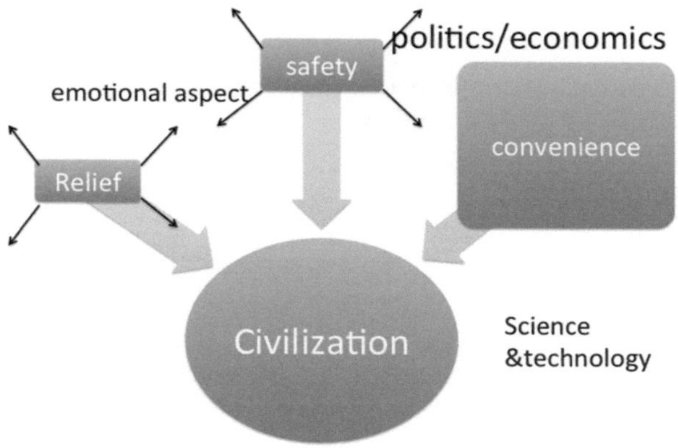

Fig. 2 The sound development of civilization and environmental friendliness

including us (human beings). Therefore, it's not too much to say that the surface will play a key role for the development of our civilization in the future.

On the other side, we need to have a different perspective from such an emotional one, when we think about the change of the concept (environmental friendliness) as scientists and engineers or business people. This is a positive new perspective with which we consider environmental friendliness a growth strategy [2]. For example, we can mention the movement of automobile industries. Nowadays, hybrid cars are very prosperous and some representative automobile giants are tackling with fuel cell or battery operated cars. For the excellent green technology (environmentally friendly technology) could enhance brand images (names) significantly, even though the development and production might cost double, triple, or more. Finally, the increase of brand images (names) would lead to the increase of total profits. And simultaneously, the struggle to overcome technical hurdles would produce technical advantages and technological leadership in the market. In the 1970s, the US government set a very high environmental regulation called the Muskie Act (Clean Air Act) for automobiles. Almost everyone in the industry felt that it was impossible to overcome or meet this requirement. However, a small and new company in Japan devised and developed a new engine and it became the first company to pass the regulation and to get the formal permission for the US market. The name of the company is the Honda Company and it is still prosperous with a high brand image. We think the same principle could be applied to surface finishing engineering and technology. Environmental friendliness could enhance a brand image as well as provide technical advantages. And finally, it would produce a new "premium price". Therefore, we have to consider the concept of environmental friendliness as a good chance for further growth.

This book, *Corrosion Control and Surface Finishing: Environmentally Friendly Approaches,* introduced many aspects of environmental friendliness for surface finishing and corrosion control. However, the discipline (if we are allowed to call it a discipline) is still in a cradle and needs to grow and develop. In addition to aspects of environmental friendliness, the chapter authors provided lots of useful information about the fundamentals of corrosion and surface finishing for corrosion control, conventional and novel environmental evaluation processes for surface finishing, and environmental regulations for metallic materials. Consider this book as a useful guide and great resource for designing future innovative surface finishing/corrosion prevention technologies, etc. with a focus on environmental friendliness.

References

1. Numata M, Sashima M (1992) Environmental problems and environmental education. Kokudo Co. Ltd, Tokyo
2. Watanabe K (2002) Growth strategy with environmental technology. Navigator 7:1–4

MIX
Papier aus verantwortungsvollen Quellen
Paper from responsible sources
FSC® C105338

If you have any concerns about our products,
you can contact us on
ProductSafety@springernature.com

In case Publisher is established outside the EU,
the EU authorized representative is:
Springer Nature Customer Service Center GmbH
Europaplatz 3, 69115 Heidelberg, Germany

Printed by Libri Plureos GmbH
in Hamburg, Germany